Foundation GCSE Mathematics
Revision and Practice

David Rayner

OXFORD

UNIVERSITY PRESS

OXFORD
UNIVERSITY PRESS

Great Clarendon Street, Oxford OX2 6DP

Oxford University Press is a department of the University of Oxford.
It furthers the University's objective of excellence in research, scholarship,
and education by publishing worldwide in

Oxford New York

Auckland Cape Town Dar es Salaam Hong Kong Karachi
Kuala Lumpur Madrid Melbourne Mexico City Nairobi
New Delhi Shanghai Taipei Toronto

With offices in

Argentina Austria Brazil Chile Czech Republic France Greece
Guatemala Hungary Italy Japan South Korea Poland Portugal
Singapore Switzerland Thailand Turkey Ukraine Vietnam

Oxford is a registered trade mark of Oxford University Press
in the UK and in certain other countries

British Library Cataloguing in Publication Data

Data available

ISBN 978 019 915116 5

3 5 7 9 10 8 6 4 2

Printed and bound in Great Britain by Bell and Bain Ltd; Glasgow

Acknowledgements

The image on the cover is reproduced courtesy of Wire Design/Digital Vision.

Preface

This Foundation book is for candidates working through Key Stage 4 towards a GCSE in Mathematics, and has been adapted for the new two-tier specifications for first examination in 2008. It can be used both in the classroom and by students working on their own. There are explanations, worked examples and numerous exercises which, it is hoped, will help students to build up confidence. The questions are graded in difficulty throughout the exercises.

The book can be used either as a course book over the last one or two years before the GCSE examinations or as a revision text in the final year. The contents list shows where all the topics appear and an index at the back of the book provides further reference.

At the end of the book there are several revision exercises which provide mixed questions across the curriculum. There are also multiple choice questions for variety.

The section on Using and Applying Mathematics contains a selection of starters for coursework projects which have been tried and tested. They can be used to provide practice in the strategies involved in attempting 'open-ended' problems.

The author is indebted to the many students and colleagues who have assisted him in this work. He is particularly grateful to Christine Godfrey for her help and many suggestions.

Thanks are also due to the following examination boards for kindly allowing the use of questions from their past mathematics papers and sample questions:

Edexcel Foundation	(Edexcel)
Assessment and Qualifications Alliance	(AQA)
Oxford Cambridge and RSA Examinations	(OCR)
Northern Ireland Council for the Curriculum Examinations and Assessment	(CCEA)
Welsh Joint Education Committee	(WJEC)
Cambridge International Examinations	(CIE)

D. Rayner 2006

Contents

1   Number 1

In this unit you will:
- revise place value of digits
- revise factors and multiples
- learn about highest common factor and least common multiple
- learn about prime numbers and prime factors
- learn about square numbers, cube numbers, square roots and cube roots
- learn index notation
- learn how to use your calculator effectively.

1.1   Properties of numbers

1.1.1   Place value of digits

- The number 576 means $\boxed{500}$ + $\boxed{70}$ + $\boxed{6}$

- The number 2408 means $\boxed{2000}$ + $\boxed{400}$ + $\boxed{8}$

- The number 13 416 is written as 'thirteen thousand, four hundred and sixteen'.

Exercise 1 C

1   You can write the number 427 as $\boxed{400}$ + $\boxed{20}$ + $\boxed{7}$.
   Copy and complete these numbers.

   a   613 = $\boxed{600}$ + $\boxed{10}$ + $\boxed{}$

   b   954 = $\boxed{}$ + $\boxed{50}$ + $\boxed{}$

   c $\boxed{}$ = $\boxed{500}$ + $\boxed{30}$ + $\boxed{9}$

   d   2416 = $\boxed{}$ + $\boxed{}$ + $\boxed{}$ + $\boxed{}$

2   In the number 13 485, write the value of
   a   the figure 8
   b   the figure 3.

3 When you write a cheque you write the amount in words and in figures.

Write these numbers in words.
a 523 **b** 6410 **c** 25 000

4 Write these numbers in figures
  a two hundred and seventeen
  b four thousand, two hundred and fifty
  c five million
  d six thousand and twenty.

5 The populations of three towns are
  Penton 15 614   Quarkby 21 604   Roydon 11 999

  Write the towns in order of their populations with the smallest first.

6 Write these numbers in order of size, smallest first.
  a 314   29   290   85
  b 5010   2000   564   645   1666
  c 60 000   7510   ten thousand   8888

7 **a** Use these cards to make four different 3-digit numbers less than 600.
  b Write the largest 3-digit number you can make with these cards.

2   8   5

8 **a** Use these cards to make the smallest possible 4-digit number.
  b Use these cards to make the largest possible 4-digit number.
  c Use three of the cards to make the largest possible 3-digit number less than 600.

1   3   8   4

9 What four coins can you use to make 44p?

10 What number is half of fifty thousand?

11 What number is ten times as big as 411?

12 Write in figures the number five hundred and ten thousand, two hundred and twelve.

13 Add together in your head 9, 21 and 40 and write your answer.

14 **a** Use these cards to make the largest possible 4-digit number.
 b Use all four cards to make the smallest possible 4-digit number (you cannot start with a zero).

15 Here are five number cards.

 a Use all the cards to make the largest possible **odd** number.
 b Use all the cards to make the smallest possible **even** number.

16 Write the number that is ten more than
 a 247 **b** 3211 **c** 694

17 Write the number that is one thousand more than
 a 392 **b** 25 611 **c** 256 900

18 **a** Prini puts a 2-digit whole number into her calculator. She multiplies the number by 10.

 Write **one** other digit which you know must now be on the calculator.

 b Prini starts again with the same 2-digit number and this time she multiplies it by 1000.

 Write all five digits on the calculator this time.

19 Write the numbers in order, from the smallest to the largest.
 a 2142 2290 2058 2136
 b 5329 5029 5299 5330
 c 25 117 25 200 25 171 25 000 25 500

*20 Here are six number cards.

3   4   6   7   8   9

Copy and complete these sums. Use all six cards each time to make the largest possible answer.

a $\square\square + \square\square + \square\square$

b $\square\square\square + \square\square\square$

*21 Find a number n so that $5n + 7 = 507$.

*22 Find a number x so that $6x + 8 = 68$.

> Remember:
> $5n$ means $5 \times n$
> $6x$ means $6 \times x$

1.1.2 Factors and multiples

- The **factors** of 8 are the numbers that divide exactly into 8.

  So the factors of 8 are 1, 2, 4 and 8.
  The factors of 15 are 1, 3, 5 and 15.

| 1×8 | 2×4 |
| 1×15 | 3×5 |

- The **multiples** of 4 are the numbers in the 4 times table.
  The multiples of 4 are 4, 8, 12, 16, ... and so on.
  The multiples of 7 are 7, 14, 21, 28, ... and so on.

Exercise 2 ⓒ

The factors of 6 are the numbers that divide into 6 exactly.
The factors of 6 are 1, 2, 3 and 6.

| 1×6 | 2×3 |

1 Write the factors of
   a 9   **b** 10   **c** 12

2 Write the factors of
   a 14   **b** 20   **c** 30

EXAMPLE

The factors of 6 are 1, 2, 3, 6.
The factors of 8 are 1, 2, 4, 8.
The numbers 1 and 2 are in both lists. The numbers 1 and 2 are **common factors** of 6 and 8.

3 The table shows the factors of 10 and 15.

Number	Factors
10	1, 2, 5, 10
15	1, 3, 5, 15

Write the common factors of 10 and 15 (the numbers that are factors of 10 and 15).

4 a Copy and complete this table.

Number	Factors
12	
20	

   b Write the common factors of 12 and 20.

5 a The first four multiples of 5 are   5   10   15   20.
   b Write the first four multiples of
      i 3   **ii** 4   **iii** 10

In questions **6** to **9** find the 'odd one out' (the number which is not a multiple of the number given).

6 Multiples of 6:   6, 12, 16, 24.

7 Multiples of 11:   22, 44, 76, 88.

8 Multiples of 4:   8, 12, 22, 28.

9 Multiples of 7:   7, 14, 21, 34.

*10 Copy and complete this sentence.
   'An even number is a multiple of _____.'

*11 n is a odd number.
   a Is $2n$ an odd number or an even number?
   b Is $2n + 1$ an odd number, an even number or could it be either?

1.1.3 L.C.M. and H.C.F.

● The first few **multiples** of 4 are 4, 8, 12, 16, ⑳, 24, 28 . . .
   The first few multiples of 5 are 5, 10, 15, ⑳, 25, 30, 35 . . .

● The Least Common Multiple (L.C.M.) of 4 and 5 is 20.
   It is the lowest number which is in both lists.

- The **factors** of 12 are 1, 2, 3, ④, 6, 12
  The factors of 20 are 1, 2, ④, 5, 10, 20

- The Highest Common Factor (H.C.F.) of 12 and 20 is 4.
  It is the highest number which is in both lists.

Exercise 3Ⓔ

1 a Write the first six multiples of 2.
  b Write the first six multiples of 5.
  c Write the L.C.M. of 2 and 5.

> The L.C.M. is the lowest number that is in both lists.

2 a Write the first four multiples of 4.
  b Write the first four multiples of 12.
  c Write the L.C.M. of 4 and 12.

3 Find the L.C.M. of
  a 6 and 9   **b** 8 and 12   **c** 14 and 35
  d 4 and 6   **e** 5 and 10   **f** 7 and 9

4 The table shows the factors and common factors of 24 and 36.

Number	Factors	Common factors
24	1, 2, 3, 4, 6, 8, 12, 24	} 1, 2, 3, 4, 6, 12
36	1, 2, 3, 4, 6, 9, 12, 18, 36	

> The H.C.F. is the highest number in the list of common factors.

Write the H.C.F. of 24 and 36.

5 The table shows the factors and common factors of 18 and 24.

Number	Factors	Common factors
18	1, 2, 3, 6, 9, 18	} 1, 2, 3, 6
24	1, 2, 3, 4, 6, 8, 12, 24	

Write the H.C.F. of 18 and 24.

6 Find the H.C.F. of
  a 12 and 18   **b** 22 and 55   **c** 45 and 72
  d 18 and 30   **e** 60 and 72   **f** 40 and 50

7 a Find the H.C.F. of 12 and 30.
  b Find the L.C.M. of 8 and 20.
  c Write two numbers whose H.C.F. is 11.
  d Write two numbers whose L.C.M. is 10.

> Don't confuse your L.C.M.s with your H.C.F.s!

*8 Given that $30 = 2 \times 3 \times 5$ and $165 = 3 \times 5 \times 11$, find the highest common factor of 30 and 165 (that is, the highest number that goes into 30 and 165).

*9 If $315 = 3 \times 3 \times 5 \times 7$ and $273 = 3 \times 7 \times 13$, find the highest common factor of 315 and 273.

1.1.4   Square numbers and cube numbers

●   **Square numbers**

$4^2 = 4 \times 4 = 16$

You say '4 squared equals 16'.

$100^2 = 100 \times 100 = 10\,000$

●   **Square root**

The square root of 25 is 5.

You ask 'What do I square to get 25?'

You write $\sqrt{25} = 5$.

Here are some other square roots: $\sqrt{100} = 10$, $\sqrt{1} = 1$, $\sqrt{2} = 1 \cdot 414$ approximately.

Notice that only **square** numbers have square roots that are whole numbers.

●   **Cubes**

The number 64 can be written as

$64 = 4 \times 4 \times 4 = 4^3$

You say '4 cubed is 64' or sometimes '4 to the power 3 is 64'.

Exercise 4 C

1 The first three square numbers are 1, 4, 9.

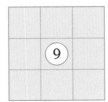

$1 \times 1 = \textcircled{1}$ $2 \times 2 = \textcircled{4}$ $3 \times 3 = \textcircled{9}$

Draw diagrams with labels to show the next two square numbers.

2 The square number 9 is 3×3 and can be written as 3^2 ('3 squared'). Work out

a 5^2 b 10^2 c 6^2

d $3^2 + 2^2$ e $1^2 + 7^2$ f $10^2 - 2^2$

3 What number do you multiply by itself to give these numbers?

a 16   b 81   c 1

4 The symbol for square root is $\sqrt{}$ (for example $\sqrt{36} = 6$). Work out

a $\sqrt{9}$ b $\sqrt{25}$ c $\sqrt{100}$ d $\sqrt{49}$

5 Copy and complete

a $6 + \sqrt{4} = \square$ b $3 + \sqrt{16} = \square$ c $\sqrt{\square} = 6$

6 Answer 'true' or 'false'.

a $2 \times 2 \times 2 = 8$, so 8 is a cube number.

b 27 is a cube number.

c $1 \times 1 = 2$

d $2^3 + 1^3 = 9$

7 Work out

a $10^2 + 2^2$ b $1^3 + 3^3$ c $4^2 + \sqrt{4}$

d $10^3 - 10^2$ e $\sqrt{6^2}$ f $10^4 + 1$

8 In each part copy the four numbers and circle the one that is **not** a square number.

a 4, 9, 48, 64   b 16, 25, 36, 55   c 1, 8, 16, 100

d 4, 10, 49, 81   e 25, 64, 120, 144   f 16, 36, 108, 121

*9 For each pair of numbers here, there is just one **square number** that lies between them. In each case, write that square number.

a 6   12   b 18   33   c 27   47

d 10   22   e 2   8   f 88   108

g 40   50   h 75   95

*10 Shirin has five number cards.

a Write two of the cards to show how Shirin can make a square number.

b Write two of the cards to show how Shirin can make a cube number.

c Write three of the cards to show how Shirin can make a larger cube number.

| 1 | 2 | 4 | 5 | 7 |

*11 Find the smallest value of n for which $1^2 + 2^2 + 3^2 + 4^2 + 5^2 + \ldots + n^2 > 800$

> $>$ means 'is greater than'

1.1.5 Prime numbers

- A prime number is divisible only by itself and by one.

- The first six prime numbers are 2, 3, 5, 7, 11 and 13.
- Notice that the number 1 is **not** a prime number.

Prime factor decomposition

- The factors of a number that are also prime numbers are called prime factors. You find the prime factors of any number by prime factor decomposition.

It is not as hard as it sounds!

EXAMPLE

Find the factors of 90
a with a factor tree   **b** using repeated divisions.

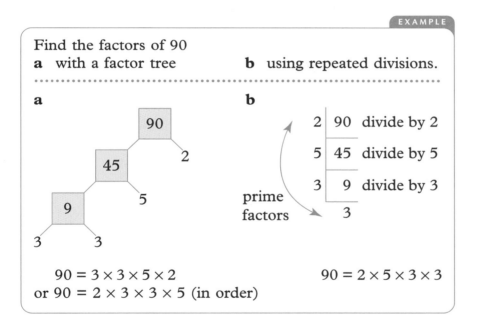

$$90 = 3 \times 3 \times 5 \times 2$$
or $90 = 2 \times 3 \times 3 \times 5$ (in order)

$$90 = 2 \times 5 \times 3 \times 3$$

Exercise 5(E)

1 Which of these numbers are prime numbers?

| 2 | 7 | 9 | 15 | 17 |

2 Write the first ten prime numbers.

3 Write all the even prime numbers up to 100.

4 Find all the prime numbers between 30 and 40.

5 For each of these pairs of numbers, there is just one **prime number** that lies between them. In each case, write that prime number.

    a 8   12   **b** 14 18   **c** 25 30
    d 20 28   **e** 32 40   **f** 44 52
    g 54 60   **h** 38 42

6 In each part copy the four numbers and circle the one that is **not** a prime number.

    a 7, 9, 13, 17   **b** 2, 13, 21, 23
    c 11, 13, 19, 27   **d** 15, 19, 29, 31
    e 31, 37, 39, 41   **f** 23, 43, 47, 49

EXAMPLE

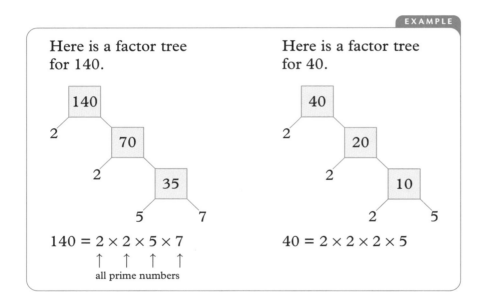

Here is a factor tree for 140.

140 = 2 × 2 × 5 × 7
   ↑ ↑ ↑ ↑
  all prime numbers

Here is a factor tree for 40.

40 = 2 × 2 × 2 × 5

7 You can find prime factors using a factor tree.
Draw a factor tree for each of these numbers.

    a 36   **b** 60   **c** 216   **d** 200   **e** 1500

8 Here is the number 600 written as the product of its prime factors.

600 = 2 × 2 × 2 × 3 × 5 × 5

Use this information to write 1200 as a product of its prime factors.

9 Write each number as a product of its prime factors.
(For example: 30 = 2 × 3 × 5)

    a 28   **b** 32   **c** 34   **d** 81   **e** 294

10 Copy the grid and use a pencil for your answers (so that you can rub out mistakes).
Write every number from 1 to 9, one in each box, so that all the numbers match the conditions for both the row and the column.

	Even number	Square number	Prime number
Factor of 14	2		
Multiple of 3			
Between 3 and 9			

> Remember: Write just **one** number in each box.

11 Copy the grid and write the numbers from 1 to 9, one in each box.

	Factor of 10	More than 3	Factor of 18
Square number			
Even number			
Prime number			

*12 This one is more difficult. Copy the grid and write the numbers from 1 to 16, one in each box. There are several correct solutions. Ask another student to check yours.

	Prime number	Odd number	Factor of 16	Even number
Numbers less than 7				
Factor of 36				
Numbers less than 12				
Numbers between 11–17				

*13 The pair of numbers 11, 13 is a 'prime pair', since 11 and 13 are both prime numbers with only one other number in between.
Similarly, the numbers 17, 19 form another 'prime pair'.

There are four prime pairs between 20 and 100. See if you can find them. Remember, both numbers of the pair **must be prime**.

*14 a Copy these statements and fill in the missing numbers.

$$1 = 1 = 1^2$$
$$1 + 3 = 4 = 2^2$$
$$1 + 3 + 5 = \boxed{} = \boxed{}^2$$
$$1 + 3 + 5 + 7 = \boxed{} = \boxed{}^2$$

b Write the next five lines of the sequence.

*15 Investigation

The numbers 1, 8, 27 are the first three **cube** numbers.

$$1 \times 1 \times 1 = 1^3 = 1 \quad \text{(say '1 cubed')}$$
$$2 \times 2 \times 2 = 2^3 = 8 \quad \text{(say '2 cubed')}$$
$$3 \times 3 \times 3 = 3^3 = 27 \quad \text{(say '3 cubed')}$$

You can add the odd numbers in groups to give an interesting sequence.

$$1 = 1 = 1^3$$
$$3 + 5 = 8 = 2^3$$
$$7 + 9 + 11 = 27 = 3^3$$

Write the next three rows of the sequence to see if the sum of each row always gives a cube number.

*16 Investigation

The letters p and q represent prime numbers.
Is $p + q$ always an even number? Explain your answer.

*17 Investigation

Apart from 1, 3 and 5 all odd numbers less than 100 can be written in the form $p + 2^n$ where p is a prime number and n is greater than or equal to 2.

For example $43 = 11 + 2^5$
$$27 = 23 + 2^2$$

Write the odd numbers from 7 to 21 in the form $p + 2^n$.
Can you write all the odd numbers to 99 in this form?

Exercise 6 Ⓒ

Find the number!

1 a 2-digit number
a prime number
the sum of its digits is 13

2 a 2-digit number
a multiple of both 3 and 4
the sum of its digits is 15

3 a prime number
a factor of 78
a 2-digit number

4 a 3-digit number
a square number
the product of its digits is 2

5 a multiple of 9
a 3-digit number below 200
the product of its digits is 16

6 a 2-digit number
a prime number
the product of its digits is 24

7 a square number
a 3-digit number
the product of its digits is 20

8 a 2-digit number
a factor of 184
a prime number

9 a multiple of 11
a multiple of 7
the product of its digits is 6

10 a 3-digit number
the sum of its digits is 9
a multiple of 37

1.2 Arithmetic without a calculator

1.2.1 Adding and subtracting

You can use different methods to add and subtract without a calculator.

EXAMPLE

```
a      3 6 4
    +    5 3
    ─────────
       4 1 7
       1
```

```
b      2 1 4
       3 0 7
    +    5 2
    ─────────
       5 7 3
         1
```

```
c        4 2 7
    + 5 1 8 6
    ──────────
      5 6 1 3
        1   1
```

Remember to keep the numbers in columns when you are adding and subtracting.

```
d      8 4
    −  6 1
    ───────
       2 3
```

```
e    ⁶⁷̶1 4
    −   2 6
    ─────────
        4 8
```

```
f    2 7⁷8̶¹4
    −    6 3 5
    ──────────
      2 1 4 9
```

Exercise 7 C

Do these additions.

1 27 + 31

2 45 + 22

3 234 + 17

4 316 + 204

5 50 + 911

6 291 + 46

7 299 + 197

8 306 + 205

9 45 + 275

10 903 + 89

11 415 + 207 + 25

12 41 + 607 + 423

13 206 + 114 + 8

14 9 + 19 + 912

15 157 + 16 + 24

16 16 + 2341 + 27

17 3047 + 265

18 274 + 5061

19 2941 + 4067

20 8046 + 147

21 401 + 609 + 21

22 506 + 2615

23 2947 + 4 + 590

24 209 + 607 + 11

25 6672 + 11 + 207

26 994 + 27

27 604 + 12 407

28 9150 + 12 694

29 53 246 + 62 141

30 19 274 + 27 + 584

Exercise 8 C

Do these subtractions.

1 97 − 63	**2** 69 − 41	**3** 83 − 60	**4** 87 − 5
5 192 − 81	**6** 214 − 10	**7** 86 − 29	**8** 52 − 37
9 74 − 18	**10** 91 − 68	**11** 265 − 128	**12** 642 − 181
13 562 − 181	**14** 816 − 274	**15** 509 − 208	**16** 604 − 491
17 808 − 275	**18** 250 − 127	**19** 640 − 118	**20** 265 − 184
21 484 − 219	**22** 6064 − 418	**23** 5126 − 307	**24** 6417 − 29
25 8050 − 218	**26** 406 − 22	**27** 649 − 250	**28** 6009 − 205
29 1717 − 356	**30** 843 − 295		

Find the answer.

31 100 − 99 + 98 − 97 + 96 − ... + 4 − 3 + 2 − 1

Exercise 9 C

Magic squares

In a magic square you get the same number when you add across each row (↔), add down each column (↕) and add diagonally (↗, ↘). Copy and complete these magic squares.

1

3		
2		
7	0	

2

9	7	5
		10

3

8	1	6
4		

4

7		
	6	4
		5

5

12		14
	11	
8		

6

17		
12		16
13		

7

		7	14
	13	2	
	3	16	5
15			4

8

	6		4
10		16	
		2	
1	12	7	14

9

	10		
14	15		8
	4	18	
16	13		6

10

18		12	7
	6		
11	16	5	14
4			

*11

11		7	20	3
	12	25		16
	5	13		9
	18		14	
		19	2	15

*12

16		10	17	4
3	15		9	21
20		14		8
	19			
24			5	12

1.2.2 Multiplying and dividing

Questions on multiplying and dividing are much easier, if you know your multiplication facts:

- '3 nines are 27',
- '4 fours are 16'.

Copy and complete this multiplication table to help you.

×	4	7	3	5	9	11	8	6	2	10
4										
7										
3					27					
5										
9										
11									22	
8			24							
6										
2										
10										

a
$$\begin{array}{r} 3\ 2 \\ \times\ 4 \\ \hline 1\ 2\ 8 \end{array}$$

b
$$\begin{array}{r} 6\ 4 \\ \times\ 4 \\ \hline 2\ 5\ 6 \\ \scriptstyle 1 \end{array}$$

c
$$\begin{array}{r} 3\ 7\ 4 \\ \times\ 6 \\ \hline 2\ 2\ 4\ 4 \\ \scriptstyle 4\ 2 \end{array}$$

d
$$5{\overline{\smash{)}\,7^2 5}}\ \ \substack{1\ 5}$$

e
$$7{\overline{\smash{)}\,3\ 7^2 9^1 4}}\ \ \substack{5\ 4\ 2}$$

f
$$5{\overline{\smash{)}\,6^1 9^4 4}}\ \ \substack{1\ 3\ 8} \quad \text{r}\ 4 \quad \text{or} \quad 138\tfrac{4}{5}$$

Exercise 10 C

Do these multiplications.

1 21×3	**2** 32×3	**3** 42×6	**4** 35×4
5 213×3	**6** 46×5	**7** 205×6	**8** 28×6
9 211×7	**10** 302×7	**11** 213×5	**12** 641×3
13 21×8	**14** 314×6	**15** 131×9	**16** 214×8
17 820×6	**18** 921×4	**19** 2141×6	**20** 3025×5
21 324×8	**22** 643×7	**23** 295×9	**24** 641×10
25 846×10	**26** 275×8	**27** 631×7	**28** 885×9
29 497×8	**30** 2153×6		

Exercise 11 C

Do these divisions.

1 $69 \div 3$	**2** $286 \div 2$	**3** $844 \div 4$	**4** $345 \div 3$
5 $712 \div 4$	**6** $1160 \div 5$	**7** $1581 \div 3$	**8** $2112 \div 4$
9 $415 \div 5$	**10** $994 \div 2$	**11** $1092 \div 4$	**12** $18\,072 \div 3$
13 $3020 \div 5$	**14** $1626 \div 6$	**15** $1660 \div 4$	**16** $1915 \div 5$
17 $4944 \div 6$	**18** $5616 \div 6$	**19** $2247 \div 7$	**20** $10\,710 \div 5$
21 $18\,972 \div 2$	**22** $9256 \div 4$	**23** $1928 \div 8$	**24** $14\,010 \div 2$
25 $5859 \div 7$	**26** $55\,305 \div 9$	**27** $21\,104 \div 8$	**28** $3735 \div 9$

There are 'remainders' in these questions.

29 $76 \div 5$	**30** $87 \div 4$	**31** $57 \div 2$	**32** $373 \div 6$
33 $247 \div 6$	**34** $124 \div 5$	**35** $281 \div 5$	**36** $1173 \div 9$
37 $2143 \div 4$	**38** $6418 \div 5$	**39** $6027 \div 4$	**40** $4135 \div 6$

Speed tests

You can do these questions either:
1 with your book open or
2 with your book closed and your teacher reading out the questions.

In either case write **the answer** only. Be as quick as possible.

Test 1	Test 2	Test 3	Test 4
1 $30 - 8$	**1** $6 + 16$	**1** 8×5	**1** 5×7
2 9×5	**2** $32 - 5$	**2** $17 + 23$	**2** $36 - 18$
3 $40 \div 5$	**3** 9×6	**3** $60 \div 6$	**3** $103 - 20$
4 $24 + 34$	**4** $90 \div 2$	**4** $101 - 20$	**4** $56 \div 7$
5 11×7	**5** $98 + 45$	**5** 49×2	**5** 8×4
6 $60 - 12$	**6** $16 - 7$	**6** $52 + 38$	**6** $53 + 36$
7 9×4	**7** $45 \div 9$	**7** $66 \div 11$	**7** $51 - 22$
8 $27 \div 3$	**8** 13×100	**8** $105 - 70$	**8** $36 \div 3$
9 $55 + 55$	**9** $99 + 99$	**9** 13×4	**9** 20×5
10 $60 - 18$	**10** $67 - 17$	**10** $220 - 30$	**10** $99 + 55$
11 8×6	**11** $570 \div 10$	**11** $100 \div 20$	**11** $200 - 145$
12 $49 \div 7$	**12** 7×3	**12** $2 \times 2 \times 2$	**12** $88 \div 8$
13 $99 + 17$	**13** $55 - 6$	**13** $91 + 19$	**13** 50×100
14 $80 - 59$	**14** $19 + 18$	**14** $200 - 5$	**14** $199 + 26$
15 9×100	**15** $60 \div 5$	**15** 16×2	**15** $80 - 17$

Exercise 12 **©** (Mixed problems)

1 Nicki made a tower using 27 identical discs, each of thickness 5 cm. How high was the tower?

2 What four coins have a total of 37p?

3 Thomas shares £189 equally between seven people. How much does each person receive?

4 At a banquet for 456 people eight guests sat at each table. How many tables were there?

5 A motorist bought 9 litres of petrol at 93p per litre.
   a How much did it cost?
   b What change did she receive from £10?

6 The population of a town decreased from 8716 to 7823. How many people left the town?

7 A man bought five felt tip pens at 28p each and six pads at 84p each. How much did he spend altogether?

8 A tin has a mass of 240 g when empty.
When it is half-full of rice the total mass
is 570 g.
What is its mass when it is full?

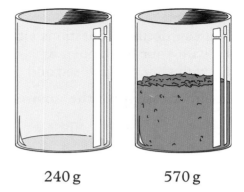

240 g   570 g

9 A well-organised flock of sheep queue up for their daily
ration of 8 kg of grass.
The farmer at the head of the queue has only 2064 kg of
grass to give out.
How many of his flock of 300 sheep will be disappointed?

*10 In a school with 280 students there are 10 more girls
than boys.
  a How many girls are there?
  b How many boys are there?
  c Check that your answers add up to 280.

*11 Find a pair of positive integers a and b for which

$$8a + 65b = 1865$$

> An **integer** is a
> whole number.

Exercise 13 Ⓒ

Cross squares

Each empty square contains either a number or a mathematical
operator ($+$, $-$, $\times$, $\div$). Copy each square and fill in the missing
details.

1

11		4	→	15
×		÷		
		2	→	3
↓		↓		
66			→	132

2

9			17	→	26
×			−		
5	×			→	
↓			↓		
		÷	9	→	5

3

14	+		→	31
×				
4		23	→	92
↓		↓		
	−	40	→	

4

15			→	5
+		×		
		5	→	110
↓		↓		
	−	15	→	22

5

	×	10	→	90
+		÷		
			→	$5\frac{1}{2}$
↓		↓		
20	×		→	100

6

		×		→	52
−			×		
		×	4	→	
↓			↓		
8			8	→	1

7

5			→	60
×		÷		
		24	→	44
↓		↓		
	×	$\frac{1}{2}$	→	50

8

	×	6	→	42
÷		÷		
14	−		→	
↓		↓		
		2	→	1

9

	×	2	→	38
−		÷		
			→	48
↓		↓		
7	−		→	$6\frac{1}{2}$

1.2.3 Inverse operations

The word inverse means 'opposite'.

- The inverse of adding is subtracting $5 + 19 = 24$, $5 = 24 - 19$
- The inverse of subtracting is adding $31 - 6 = 25$, $31 = 25 + 6$
- The inverse of multiplying is dividing $7 \times 6 = 42$, $7 = 42 \div 6$
- The inverse of dividing is multiplying $30 \div 3 = 10$, $30 = 10 \times 3$

Find the missing digits.

a $\boxed{}4 \div 6 = 14$

b $2\boxed{}8 \times 5 = 1340$

c
```
    3 □ 7
  + 2 5 □
  ───────
  □ 3 9
```

a Work out 14×6 because multiplying is the inverse of dividing. Since $14 \times 6 = 84$, the missing digit is 8.

b Work out $1340 \div 5$ because dividing is the inverse of multiplying. Since $1340 \div 5 = 268$, the missing digit is 6.

c Start from the right: $7 + 2 = 9$
   Middle column: $8 + 5 = 13$
   Check
```
    3 8 7
  + 2 5 2
  ───────
    6 3 9
      1
```

Exercise 14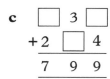

Copy these sums and find the missing digits.

1 **a**
```
    2 8 5
  +□ 1 4
  ───────
    7 □□
```

b
```
    6 3 □
  +□ 5 2
  ───────
    8 □ 9
```

c
```
    □ 3 5
  + 3 4 □
  ───────
    9 □ 9
```

2 **a**
```
    3 5 6
  +5 □ 6
  ───────
    □ 8 □
```

b
```
    2 □ 4
  + 5 3 7
  ───────
    □ 6 1
```

c
```
    3 8 8
  + □ 2 □
  ───────
    8 □ 3
```

3 **a**
```
      4 □
  ×     3
  ───────
    1 4 4
```

b
```
      3 □
  ×     7
  ───────
    2 3 1
```

c
```
    □ □ 1
  ×       5
  ─────────
  1 6 0 5
```

4 **a** $\square\square\square \div 3 = 50$ **b** $\square\square \times 4 = 60$

    c $9 \times \square = 81$ **d** $\square\square\square\square \div 6 = 192$

5 **a**
```
    4 □ 5
  + 2 8 □
  ───────
    □ 3 0
```

b
```
    4 □ 7
  + □ 7 □
  ───────
    6 0 4
```

c
```
    □ 3 □
  +2 □ 4
  ───────
    7 9 9
```

6 **a** $\square\square \times 7 = 245$ **b** $\square\square \times 10 = 580$

    c $32 \div \square = 8$ **d** $\square\square\square \div 5 = 190$

7 **a** $\square\square + 29 = 101$ **b** $\square\square\square - 17 = 91$

    c
```
    □ 8 9
  − 3 □ 6
  ───────
    5 4 □
```

d
```
    3 3 5
  − 2 1 □
  ───────
    □ □ 7
```

8 There is more than one correct answer for each of these
questions. Copy and complete them and ask another
student to check your solution.

    a $\boxed{2}\,\boxed{3} + \square\square - \square\square = 23$

    b $\boxed{8}\,\boxed{5} - \square\square + \square\square = 86$

    c $\boxed{2}\,\boxed{5} \times \square \ \div \square = 25$

    d $\boxed{4}\,\boxed{0} \times \square\square \div \square = 80$

9 Each of these calculations has the same number missing from all three boxes. Copy them and find the missing number in each calculation.

    a ☐ × ☐ − ☐ = 12

    b ☐ ÷ ☐ + ☐ = 9

    c ☐ × ☐ + ☐ = 72

10 Copy the diagrams and work out the missing numbers in these calculations.

    a 5 → ×6 → +9 → ?   **b** ? → +2 → ×5 → 40

    c 2 → +? → ×4 → 36   **d** 7 → ×? → −11 → 10

11 Copy these and in the circle write +, −, × or ÷ to make the calculation correct.

    a $7 \times 4 \bigcirc 3 = 25$ **b** $8 \times 5 \bigcirc 2 = 20$

    c $7 \bigcirc 3 - 9 = 12$ **d** $12 \bigcirc 2 + 4 = 10$

    e $75 \div 5 \bigcirc 5 = 20$

12 Copy these and write the correct signs in the circles.

    a $5 \times 4 \times 3 \bigcirc 3 = 63$

    b $5 + 4 \bigcirc 3 \bigcirc 2 = 4$

    c $5 \times 2 \times 3 \bigcirc 1 = 31$

1.3 Decimals

1.3.1 Decimals and fractions

- Decimal numbers are used as a way of writing fractions.

The decimal number 0·3 is $\frac{3}{10}$.

The decimal number 0·09 is $\frac{9}{100}$.

The decimal number 0·31 is $\frac{31}{100}$ or $\frac{3}{10} + \frac{1}{100}$.

The decimal number 4·27 is 4 units $+ \frac{2}{10} + \frac{7}{100}$.

You can show decimal numbers in a place value table.

	T	U	.	$\frac{1}{10}$	$\frac{1}{100}$
40 =	4	0	.	0	
7 =		7	.	0	
$\frac{7}{10}$ =		0	.	7	
$\frac{5}{100}$ =		0	.	0	5

Exercise 15 C

1 What part of each shape is shaded?
Write your answer as a fraction and as decimal fraction.

a   b   c   d

e   f   g   h

2 Write the value of the underlined figure in each number.
  a 3·<u>5</u>   **b** <u>2</u>6·4   **c** 1<u>7</u>·41   **d** 18·<u>9</u>
  e 1·4<u>1</u>   **f** 0·<u>7</u>4   **g** <u>3</u>4·11   **h** 3·3<u>8</u>

> In 6·<u>2</u>4, 2 = $\frac{2}{10}$

3 Write the next two terms in each sequence.
  a 0·2   0·3   0·4   0·5   . . .
  b 0·2   0·4   0·6   . . .
  c 0·5   1·0   1·5   2·0   . . .
  d 0·1   0·3   0·5   0·7   . . .

4 Write the numbers shown by the arrows as decimals.

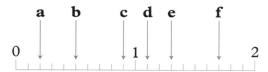

> This is a number line from 0 to 2.

5 Write each fraction as a decimal.

  a $\frac{2}{10}$ **b** $\frac{13}{100}$ **c** $\frac{2}{100}$ **d** $\frac{15}{100}$ **e** $\frac{155}{1000}$ **f** $\frac{227}{1000}$

1.3.2   Writing decimals in order of size

Method

1 Write the decimals in a column lining up the decimal points.
2 Fill any empty spaces with zeros.
3 Write the numbers in order of size.

EXAMPLE

Write 1·6, 0·51, 3 and 1·15 in order.

· ·

1 Write in a column.	**2** Put in zeros.	**3** Write in order.
1·6	1·60	0·51
0·51	0·51	1·15
3	3·00	1·60
1·15	1·15	3·00

Exercise 16 Ⓒ

Write the numbers in order, starting with the smallest.

1 3·7, 4, 1·5, 12
2 31, 3·1, 1·3, 13
3 11, 0·2, 5·2, 6
4 0·4, 0·11, 1, 1·7
5 22, 2·2, 2, 20
6 0·21, 0·31, 0·12
7 0·04, 0·4, 0·35
8 0·67, 0·672, 0·7
9 0·05, 0·045, 0·07
10 0·1, 0·09, 0·089
11 0·75, 0·57, 0·705
12 0·41, 0·041, 0·14

13 0·809, 0·81, 0·8
14 0·006, 0·6, 0·059
15 0·15, 0·143, 0·2
16 0·04, 0·14, 0·2, 0·53
17 1·2, 0·12, 0·21, 1·12
18 2·3, 2·03, 0·75, 0·08
19 0·62, 0·26, 0·602, 0·3
20 0·5, 1·3, 1·03, 1·003
21 0·79, 0·792, 0·709, 0·97
22 1·23, 0·321, 0·312, 1·04
23 0·008, 0·09, 0·091, 0·007
24 2·05, 2·5, 2, 2·046

25 Here are numbers with letters. Put the numbers in order
and write the letters to make a word.

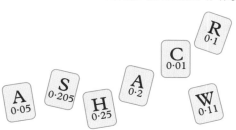

26 Increase these numbers by $\frac{1}{10}$.

   a 32·41   **b** 0·753   **c** 1·06

27   Write these amounts in pounds.

    a 350 pence   **b** 15 pence   **c** 3 pence

    d 10 pence   **e** 1260 pence   **f** 8 pence

28   Write each statement and say whether it is true or false.

    a £5·4 = £5 + 40p   **b** £0·6 = 6p

    c 5p = £0·05   **d** 50p is more than £0·42

Exercise 17 Ⓒ

In questions **1** to **6** write the line which is correct.

1 **a** 0·06 is equal to 0·6   **2** **a** 0·04 is equal to 0·040

    b 0·06 is greater than 0·6   **b** 0·04 is greater than 0·040

    c 0·06 is less than 0·6   **c** 0·04 is less than 0·040

3 **a** 0·14 is equal to 0·41   **4** **a** 0·12 is equal to 0·1

    b 0·14 is greater than 0·41   **b** 0·12 is greater than 0·1

    c 0·14 is less than 0·41   **c** 0·12 is less than 0·1

5 **a** 0·61 is equal to 0·6   **6** **a** 0·6 is equal to 0·60

    b 0·61 is greater than 0·6   **b** 0·6 is greater than 0·60

    c 0·61 is less than 0·6   **c** 0·6 is less than 0·60

In questions **7–34**

$>$ means 'is greater than' (for example $9 > 5$)

$<$ means 'is less than' (for example $7 < 10$)

For each question write 'true' or 'false'.

  7 0·8 = 0·08   **8** 0·7 < 0·71   **9** 0·61 > 0·16   **10** 0·08 > 0·008

11 0·5 = 0·500   **12** 0·4 < 0·35   **13** 0·613 < 0·631   **14** 0·06 > 0·055

15 8 = 8·00   **16** 7 = 0·7   **17** 0·63 > 0·36   **18** 8·2 < 8·022

19 6·04 < 6·40   **20** 0·75 = 0·075   **21** 5 = 0·5   **22** 0·001 > 0·0001

23 0·078 < 0·08   **24** 9 = 9·0   **25** 0·9 > 0·085   **26** 6·2 < 6·02

27 0·05 < 0·005   **28** 0·718 < 0·871   **29** 0·09 > 0·1   **30** 11 = 0·11

31 0·88 > 0·088   **32** 0·65 > 0·605   **33** 2·42 = 2·420   **34** 0·31 = 0·3100

In questions **35** to **38** copy the line and then put each number from the box on the number line.

35

36
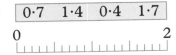

37

38
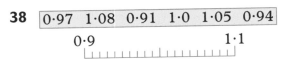

1.3.3 Scale readings

Exercise 18

Write the value shown by the arrow.

1 4 ↓ 5

2 0 ↓ 1

3 9 ↓ 10

4 0 ↓ 2

5 0 ↓ 20

6 10 ↓ 11

7 17 ↓ 19

8 120 ↓ 140

9 0 ↓ 1000

10 0·2 ↓ 0·3

11 1·9 ↓ 2

12 3 ↓ 6

13 0 ↓ 0·1

14 1·7 ↓ 1·9

15 3·1 ↓ 3·2

16 80 ↓ 120

17 0 ↓ 400

18 0 ↓ 200

1.3.4 Adding and subtracting decimals

Remember to line up the decimal points when adding or subtracting.

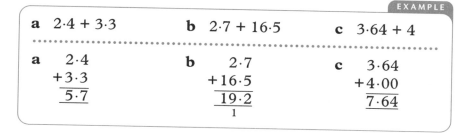

EXAMPLE

a 2·4 + 3·3   **b** 2·7 + 16·5   **c** 3·64 + 4

a   2·4
        +3·3
        ‾‾‾‾‾
         5·7

b   2·7
        +16·5
        ‾‾‾‾‾
         19·2
           1

c   3·64
        +4·00
        ‾‾‾‾‾
         7·64

EXAMPLE

a 3·34 − 1·84   **b** 0·4 − 0·17   **c** 5·03 − 3·47

a $^2\!3.^1 34$
       − 1· 84
       ‾‾‾‾‾‾
        1· 50

b $0.^3 4^1 0$
       −0· 1 7
       ‾‾‾‾‾‾
        0· 2 3

c $^4 5.^9 0^1 3$
       −3· 4 7
       ‾‾‾‾‾‾
        1· 5 6

Exercise 19 C

Do these additions.

1	4·5 + 2·3	**2**	7·2 + 1·6	**3**	8·7 + 3·0	**4**	6·7 + 8·2	**5**	3·7 + 2·9

6	8·5 + 4·31	**7**	3·9 + 4·87	**8**	8·35 + 1·84	**9**	11·7 + 2·84	**10**	4·62 1·14 + 3·31

Now do these.

11 2·84 + 7·3   **12** 18·6 + 2·34   **13** 25·96 + 0·75
14 212·7 + 4·25   **15** 3·6 + 6   **16** 7 + 16·1
17 8 + 3·4 + 0·85   **18** 12 + 5·32 + 0·08   **19** 0·004 + 0·058
20 7·77 + 77·7   **21** 1·9 + 19·1 + 7   **22** 15 + 6·02 + 6
23 0·24 + 0·2 + 2   **24** 245 + 27·9 + 3   **25** 67·1 + 29 + 0·7

26 0·07 + 0·008 + 12   **27** 4·76 + 1 + 0·07   **28** 17 + 0·61 + 5
29 513 + 47·2 + 0·157 **30** 2·6 + 26·6 + 26   **31** 47·4 + 11
32 0·055 + 5 + 15   **33** 3·24 + 32   **34** 9·09 + 999
35 2·63 + 19 + 0·4   **36** 251 + 0·1 + 6·3   **37** 19·7 + 0·8 + 15
38 27 + 2·07 + 0·59   **39** 16·4 + 27 + 0·15   **40** 374 + 200·6 + 9

Exercise 20 C

Do these subtractions.

1	8·6 − 1·4	**2**	7·7 − 2·3	**3**	8·8 − 1·7	**4**	7·4 − 1·9	**5**	8·4 − 1·6

6	7·2 − 4·5	**7**	6·3 − 1·8	**8**	5·4 − 2·7	**9**	17·25 − 8·14	**10**	8·27 − 1·19

Now do these.

11 4·81 − 3·7   **12** 6·92 − 2·56   **13** 8·27 − 5·86   **14** 19·7 − 8·9
15 3·6 − 2·24   **16** 8·4 − 2·17   **17** 8·24 − 5·78   **18** 19·6 − 7·36
19 15·4 − 7   **20** 23·96 − 8   **21** 8 − 5·2   **22** 9 − 6·8
23 13 − 2·7   **24** 25 − 3·2   **25** 0·325 − 0·188 **26** 0·484 − 0·43
27 7 − 0·35   **28** 6 − 1·28   **29** 2·38 − 1·81   **30** 11 − 7·4

By adding and subtracting, find the value of

*31 $4\cdot24 + 7\cdot0 - 4\cdot01$ ***32** $8\cdot36 + 1\cdot28 - 3\cdot11$

*33 $0\cdot35 + 0\cdot63 - 0\cdot55$ ***34** $5\cdot27 + 0\cdot761 - 1\cdot23$

*35 $4\cdot76 + 0\cdot09 - 3\cdot55$ ***36** $2\cdot6 + 9 - 3\cdot64$

*37 $3\cdot5 + 8 - 2\cdot16$ ***38** $0\cdot31 + 8 - 5\cdot88$

*39 $14\cdot9 + 19 - 17\cdot25$ ***40** $11 + 0\cdot73 - 4\cdot2$

41 Four rods are joined end to end. Their individual lengths are 18·3 cm, 75·2 cm, 11 cm and 0·7 cm. What is their combined length?

42 Sue buys three computer games costing £11·45, £23·99 and £5·60. How much change does she get from £50?

1.3.5 Multiplying and dividing by 10, 100, 1000

• To **multiply**, move the decimal point to the **right**.	$3\cdot24 \times 10 = 32\cdot4$ $10\cdot61 \times 10 = 106\cdot1$ $4\cdot134 \times 100 = 413\cdot4$ $8\cdot2 \times 100 = 820$
• To **divide**, move the decimal point to the **left**.	$15\cdot2 \div 10 = 1\cdot52$ $624\cdot9 \div 100 = 6\cdot249$ $509 \div 1000 = 0\cdot509$

Exercise 21 ⓒ

Do these multiplications.

1 $0\cdot634 \times 10$ **2** $0\cdot838 \times 10$ **3** $0\cdot815 \times 100$ **4** $0\cdot074 \times 100$

5 $7\cdot245 \times 1000$ **6** $0\cdot032 \times 1000$ **7** $0\cdot63 \times 10$ **8** $1\cdot42 \times 100$

9 $0\cdot041 \times 100$ **10** $0\cdot3 \times 100$ **11** $0\cdot71 \times 1000$ **12** $3\cdot95 \times 10$

Do these divisions.

13 $6\cdot24 \div 10$ **14** $8\cdot97 \div 10$ **15** $17\cdot5 \div 100$ **16** $23\cdot6 \div 100$

17 $127 \div 1000$ **18** $705 \div 1000$ **19** $13 \div 10$ **20** $0\cdot8 \div 10$

21 $0\cdot7 \div 100$ **22** $218 \div 10$ **23** $35 \div 1000$ **24** $8\cdot6 \div 1000$

Now do these.

25 $0\cdot95 \times 100$ **26** $11\cdot11 \times 10$ **27** $3\cdot2 \times 10$ **28** $0\cdot07 \times 1000$

29 $57\cdot6 \div 10$ **30** $999 \div 100$ **31** 66×10 **32** $100 \div 100$

33 $42 \div 1000$ **34** $0\cdot62 \times 10\,000$ **35** $0\cdot9 \div 100$ **36** $555 \div 10\,000$

37 Here are some number cards.

$$\boxed{0} \quad \boxed{1} \quad \boxed{2} \quad \boxed{3} \quad \boxed{4} \quad \boxed{5} \quad \boxed{\cdot} \quad \boxed{0}$$

a Jason picks the cards $\boxed{1}$ $\boxed{3}$ and $\boxed{4}$ to make the number 314. What extra card could he take to make a number ten times as big as 314?

*b Mel chose three cards to make the number 5·2.

    i What cards could she take to make a number ten times as big as 5·2?

    ii What cards could she take to make a number 100 times as big as 5·2?

    iii What cards could she take to make a number which is $\frac{1}{100}$ of 5·2?

1.3.6 Multiplying decimals by whole numbers

EXAMPLE

a $3{\cdot}2 \times 4$	**b** $1{\cdot}8 \times 2$	**c** $5{\cdot}4 \times 6$	**d** $0{\cdot}71 \times 5$

a $\begin{array}{r} 3{\cdot}2 \\ \times\,4 \\ \hline 12{\cdot}8 \end{array}$
    b $\begin{array}{r} 1{\cdot}8 \\ \times\,2 \\ \hline 3{\cdot}6 \\ \hline \scriptstyle 1 \end{array}$
    c $\begin{array}{r} 5{\cdot}4 \\ \times\,6 \\ \hline 32{\cdot}4 \\ \hline \scriptstyle 2 \end{array}$
    d $\begin{array}{r} 0{\cdot}75 \\ \times\,5 \\ \hline 3{\cdot}55 \end{array}$

Exercise 22 C

Do these multiplications.

1 $1{\cdot}3 \times 3$	**2** $2{\cdot}4 \times 2$	**3** $5{\cdot}1 \times 4$	**4** $6{\cdot}1 \times 5$
5 $7{\cdot}2 \times 4$	**6** $1{\cdot}3 \times 6$	**7** $0{\cdot}7 \times 7$	**8** $3{\cdot}3 \times 3$
9 $0{\cdot}84 \times 4$	**10** $1{\cdot}63 \times 5$	**11** $3{\cdot}1 \times 7$	**12** $0{\cdot}14 \times 6$
13 $11{\cdot}7 \times 7$	**14** $2{\cdot}14 \times 8$	**15** $1{\cdot}73 \times 5$	**16** $2{\cdot}72 \times 3$
17 $50{\cdot}1 \times 4$	**18** $70{\cdot}7 \times 7$	**19** $11{\cdot}9 \times 2$	**20** $2{\cdot}09 \times 4$

21 A woman buys five books, each costing £3·25. What is the total cost?

22 Mo buys 100 stamps each costing £0·31. What is the total cost?

23 At a restaurant seven people each pay £7·15 for food and £2·30 for drinks. What is the total bill for the seven people?

24 Calculate the area of this picture.

Remember: Area = length × width

1.3.7 Multiplying decimals by decimals

The answer has the same number of decimal places as the total number of decimal places in the question.

EXAMPLE

a 0·2 × 0·8   **b** 0·4 × 0·07   **c** 5 × 0·06

a 0·2 × 0·8
(2 × 8 = 16)
So 0·2 × 0·8 = 0·16

b 0·4 × 0·07
(4 × 7 = 28)
So 0·4 × 0·07 = 0·028

c 5 × 0·06
(5 × 6 = 30)
So 5 × 0·06 = 0·3

Exercise 23Ⓔ

Do these multiplications without a calculator.

1 0·2 × 0·3	**2** 0·5 × 0·3	**3** 0·4 × 0·3	**4** 0·2 × 0·03
5 0·6 × 3	**6** 0·7 × 5	**7** 0·9 × 2	**8** 8 × 0·1
9 0·4 × 0·9	**10** 0·02 × 0·7	**11** 2·1 × 0·6	**12** 4·7 × 0·5
13 21·3 × 0·4	**14** 5·2 × 0·6	**15** 4·2 × 0·03	**16** 212 × 0·6
17 0·85 × 0·2	**18** 3·27 × 0·1	**19** 12·6 × 0·01	**20** 0·02 × 17
21 0·05 × 1·1	**22** 52 × 0·01	**23** 65 × 0·02	**24** 0·5 × 0·002

1.3.8 Dividing decimals by whole numbers

EXAMPLE

a 5·6 ÷ 4   **b** 0·7 ÷ 5   **c** 52·5 ÷ 6

a
$$\begin{array}{r} 1\cdot 4 \\ 4\overline{)5\cdot {}^16} \end{array}$$

b
$$\begin{array}{r} 0\cdot 1\ 4 \\ 5\overline{)0\cdot 7{}^20} \end{array}$$

c
$$\begin{array}{r} 8\cdot\ 7\ 5 \\ 6\overline{)5\ 2\cdot {}^45{}^30} \end{array}$$

Exercise 24Ⓔ

Do these divisions without a calculator.

1 8·4 ÷ 4	**2** 9·2 ÷ 4	**3** 7·5 ÷ 3	**4** 7·5 ÷ 5
5 91·4 ÷ 2	**6** 20·7 ÷ 6	**7** 7·6 ÷ 2	**8** 13·5 ÷ 5
9 17·2 ÷ 8	**10** 10·8 ÷ 9	**11** 9·2 ÷ 5	**12** 7·8 ÷ 6
13 16·8 ÷ 7	**14** 29·4 ÷ 7	**15** 23·4 ÷ 9	**16** 18·6 ÷ 3
17 34·0 ÷ 5	**18** 51·2 ÷ 8	**19** 27·6 ÷ 6	**20** 25·2 ÷ 7
21 0·9 ÷ 5	**22** 0·7 ÷ 5	**23** 0·7 ÷ 2	**24** 0·6 ÷ 4

1.3.9 Dividing decimals by decimals

EXAMPLE

a 9·36 ÷ 0·4

a Multiply both numbers by 10 so that you can divide by a **whole number**. (Move the decimal points to the right.)
So work out 93·6 ÷ 4

$$\begin{array}{r} 2\ 3\cdot 4 \\ 4\overline{)9^13\cdot{}^16} \end{array}$$

b 0·0378 ÷ 0·07

b Multiply both numbers by 100 so that you can divide by a whole number. (Move the decimal points to the right.)
So work out 3·78 ÷ 7

$$\begin{array}{r} 0\cdot 5\ 4 \\ 7\overline{)3\cdot{}^37^28} \end{array}$$

Exercise 25Ⓔ

Do these divisions without a calculator.

1 0·84 ÷ 0·4	**2** 0·93 ÷ 0·3	**3** 0·872 ÷ 0·2	**4** 0·8 ÷ 0·2
5 2·8 ÷ 0·7	**6** 1·25 ÷ 0·5	**7** 8 ÷ 0·5	**8** 40 ÷ 0·2
9 7 ÷ 0·1	**10** 0·368 ÷ 0·04	**11** 0·915 ÷ 0·03	**12** 0·248 ÷ 0·04
13 0·625 ÷ 0·05	**14** 8·54 ÷ 0·07	**15** 1·272 ÷ 0·006	**16** 4·48 ÷ 0·08
17 0·12 ÷ 0·002	**18** 7·5 ÷ 0·005	**19** 0·09 ÷ 0·3	**20** 0·77 ÷ 1·1
21 0·055 ÷ 0·11	**22** 21·28 ÷ 7	**23** 22·48 ÷ 4	**24** 3·12 ÷ 4
25 0·7 ÷ 5	**26** 3 ÷ 0·8	**27** 0·3 ÷ 4	**28** 1·2 ÷ 8
29 0·732 ÷ 0·6	**30** 0·1638 ÷ 0·001	**31** 1·05 ÷ 0·6	**32** 7·52 ÷ 0·4

33 A cake weighing 4·8 kg is cut into several pieces each weighing 0·6 kg. How many pieces are there?

34 A phone call costs £0·04. How many calls can I make if I have £3·52?

*35 A sheet of paper is 0·01 cm thick. How many sheets are there in a pile of paper 5·8 cm thick?

Exercise 26(E)

1 $3\cdot7 + 0\cdot62$	**2** $8\cdot45 - 2\cdot7$	**3** $11\cdot3 - 2\cdot14$	**4** $2\cdot52 \times 0\cdot4$
5 $3\cdot74 \div 5$	**6** $17 + 3\cdot24$	**7** $12 - 1\cdot8$	**8** $23\cdot6 \div 8$
9 $82\cdot1 \times 0\cdot06$	**10** $0\cdot034 \times 1000$	**11** $62\cdot1 \div 100$	**12** $11\cdot4 - 3\cdot16$
13 $0\cdot153 \times 0\cdot8$	**14** $2\cdot16 + 9\cdot99$	**15** $18\cdot606 \div 7$	**16** $6\cdot042 \times 11$
17 $34\cdot1 \times 1000$	**18** $0\cdot41 \div 100$	**19** $52\cdot6 \times 0\cdot04$	**20** $0\cdot365 - 0\cdot08$
21 $2\cdot329\,56 \div 9$	**22** $654 \times 0\cdot005$	**23** $0\cdot7 + 0\cdot77 + 0\cdot777$	**24** $54 \div 100$
25 $27 \times 0\cdot001$	**26** $6\cdot007 \times 1\cdot1$	**27** $8\cdot2 - 1\cdot64$	**28** $47\cdot04 \div 6$

Exercise 27(E)

Cross numbers

Make three copies of the cross number grid. Label them **A**, **B** and **C** and then fill in the answers using the clues.

A

Across
- **1** 13×7
- **2** $0\cdot214 \times 10\,000$
- **4** $265 - 248$
- **5** $2 \times 2 \times 2 \times 2 \times 2 \times 2$
- **7** $90 - (9 \times 9)$
- **8** 14×5
- **9** $2226 \div 7$
- **11** $216 \div (18 \div 3)$
- **12** $800 - 363$
- **14** $93 - (6 \times 2)$
- **15** $0\cdot23 \times 100$
- **16** $8 \times 8 - 1$

Down
- **1** $101 - 7$
- **2** $2\cdot7 \div 0\cdot1$
- **3** $44\cdot1 + 0\cdot9$
- **4** $(2 \times 9) - (8 \div 2)$
- **6** 9^2
- **8** $6523 + 917$
- **9** $418 \div 11$
- **10** $216 + (81 \times 100)$
- **13** $2 \times 2 \times 2 \times 3 \times 3$

B

Across
- **1** $2\cdot4 \times 40$
- **2** $1600 - 27$
- **4** $913 - 857$
- **5** $2 + (9 \times 9)$
- **7** $0\cdot4 \div 0\cdot05$
- **8** $27 \times 5 - 69$
- **9** $4158 \div 7$
- **11** $2^6 + 6$
- **12** $5\cdot22 \div 0\cdot03$
- **14** $201 - 112$
- **15** 7 million $\div 100\,000$
- **16** $\frac{1}{4}$ of 372

Down
- **1** $558 \div 6$
- **2** $6\cdot4 \div 0\cdot4$
- **3** $0\cdot071 \times 1000$
- **4** $11\cdot61 + 4\cdot2 + 37\cdot19$
- **6** $(7 - 3\cdot1) \times 10$
- **8** $8 \times 8 \times 100 - 82$
- **9** $0\cdot08 \times 700$
- **10** $40 \times 30 \times 4 - 1$
- **13** $\frac{1}{5}$ of 235

C

Across
- **1** $2\cdot6 \times 10$
- **2** $6\cdot314 \times 1000$
- **4** $600 - 563$
- **5** $0\cdot25 \times 100$
- **7** $3 \div 0\cdot5$
- **8** $0\cdot08 \times 1000$
- **9** $3\cdot15 \div 0\cdot01$
- **11** $1\cdot1 \times 70$
- **12** $499 + 103$
- **14** $1 \div 0\cdot1$
- **15** $0\cdot01 \times 5700$
- **16** $1000 - 936$

Down
- **1** $0\cdot2 \times 100$
- **2** $6\cdot7 \div 0\cdot1$
- **3** $1800 \div 100$
- **4** $21 \div 0\cdot6$
- **6** $420 \times 0\cdot05$
- **8** $0\cdot8463 \times 10\,000$
- **9** $0\cdot032 \times 1000$
- **10** $5\cdot706 \div 0\cdot001$
- **13** 5^2

1.3.10 Using one calculation to find another

Once you know the answer to one calculation, you can use it to solve others.

EXAMPLE

You are told that $42 \times 15 = 630$. Use this to work out 4.2×15.

....

Since 4·2 is $42 \div 10$ you can see that $4.2 \times 15 = 63$.

Similarly you can use the fact that $6.4 \times 27 = 172.8$ to find these answers.

a $6.4 \times 2.7 = 17.28$ $(2.7 = 27 \div 10)$
b $0.64 \times 27 = 17.28$ $(0.64 = 6.4 \div 10)$
c $172.8 \div 27 = 6.4$ ($\div$ is the inverse of $\times$)

Exercise 28Ⓔ

1 You are told that $32 \times 1.9 = 60.8$; find
   a 32×19 b 320×19 c 3.2×1.9

2 You are told that $37.6 \times 54 = 2030.4$; find
   a 3.76×54 b 37.6×5.4 c 376×54

3 You are told that $82.3 \times 2.3 = 189.29$; find
   a 823×2.3 b 8.23×23 c 82.3×0.23

4 You are told that $59.3 \times 61 = 3617.3$; find
   a 5.93×6.1 b 0.593×6.1 c 59.3×0.0061

5 You are told that $36.2 \times 134 = 4850.8$; find
   a 3620×134 b 36.2×1.34 c $\dfrac{4850.8}{134}$

6 You are told that $81.6 \times 215 = 17\,544$; find
   a 8.16×2150 b 81.6×2.15 c $\dfrac{17\,544}{81.6}$

1.4 Negative numbers

1.4.1 Using negative numbers

- If the weather is very cold and the temperature is 3 degrees below zero, it is written $-3\,°$.

- If a golfer is 5 under par for his round, the scoreboard will show -5.

An easy way to begin calculations with negative numbers is to think about changes in temperature.

a Suppose the temperature is $-2°$ and it rises by $7°$.
The new temperature is $5°$.
You can write $-2 + 7 = 5$.

b Suppose the temperature is $-3°$ and it falls by $6°$.
The new temperature is $-9°$.
You can write $-3 - 6 = -9$.

Exercise 29 Ⓒ

In questions **1** to **12** move up or down the thermometer to find the new temperature.

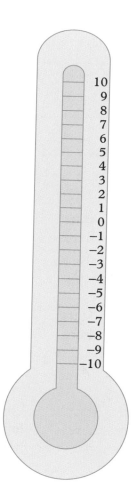

 1 The temperature is $+8°$ and it falls by $3°$.

 2 The temperature is $+4°$ and it falls by $5°$.

 3 The temperature is $+2°$ and it falls by $6°$.

 4 The temperature is $-1°$ and it falls by $6°$.

 5 The temperature is $-5°$ and it rises by $1°$.

 6 The temperature is $-8°$ and it rises by $4°$.

 7 The temperature is $-3°$ and it rises by $7°$.

 8 The temperature is $+4°$ and it rises by $8°$.

 9 The temperature is $+9°$ and it falls by $14°$.

10 The temperature is $-13°$ and it rises by $13°$.

11 The temperature is $-6°$ and it falls by $5°$.

12 The temperature is $-25°$ and it rises by $10°$.

13 Write these temperatures from the coldest to the hottest.
   a $-2°C, 5°C, -7°C$
   b $-1°C, 2°C, 0°C$
   c $-8°C, 3°C, -11°C$
   d $-4°C, -1°C, -2°C, -7°C$
   e $4°C, -5°C, -2°C, -4°C$

14 Write these numbers in order of size, starting with the lowest.
   a $6, -3, -5, 2, 0$
   b $-8, 11, -6, 3, -1$
   c $-10, -15, 5, -20, -2$
   d $23, -10, -5, -15, 18$

15 Copy these sequences and write the next three numbers in each.

a 10, 8, 6, 4, —, —, —

b 12, 9, 6, 3, —, —, —

c 3, 2, 1, 0, −1, —, —, —

d 4, 2, 0, −2, —, —, —

e 12, 6, 0, —, —, —

f −3, −2, −1, —, —, —

g −8, −6, −4, —, —, —

h 10, 6, 2, —, —, —

> These can all be solved by adding or subtracting numbers.

1.4.2 Adding and subtracting with negative numbers

For adding and subtracting you can use a number line.

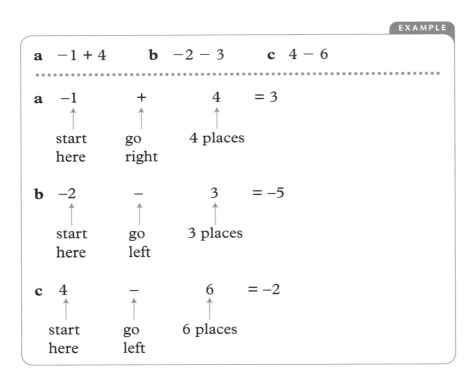

EXAMPLE

a −1 + 4   b −2 − 3   c 4 − 6

a −1   +   4   = 3
   start   go   4 places
   here   right

b −2   −   3   = −5
   start   go   3 places
   here   left

c 4   −   6   = −2
   start   go   6 places
   here   left

Exercise 30 C

Do these.

1 −6 + 2	**2** −7 − 5	**3** −3 − 8	**4** −5 + 2
5 −6 + 1	**6** 8 − 4	**7** 4 − 9	**8** 11 − 19
9 4 + 15	**10** −7 − 10	**11** 16 − 20	**12** −7 + 2
13 −6 − 5	**14** 10 − 4	**15** −4 + 0	**16** −6 + 12
17 −7 + 7	**18** 2 − 20	**19** 8 − 11	**20** −6 − 5
21 −8 − 4	**22** −3 + 7	**23** −6 + 10	**24** −5 + 5

| **25** −11 + 3 | **26** 7 − 10 | **27** −5 + 8 | **28** −12 + 0 |
| **29** −1 + 19 | **30** 20 − 25 | **31** −6 − 60 | **32** −2 + 100 |

● When you have two (+) or (−) signs together use this rule

++ = +	+− = −
−− = +	−+ = −

a 3 − (−6) = 3 + 6
        = 9
b −4 + (−5) = −4 − 5
          = −9
c −5 − (+7) = −5 − 7
          = −12

d 8 + (−8) = 8 − 8
          = 0
e 0 − (−11) = 0 + 11
          = 11
f −10 − (−3) = −10 + 3
          = −7

Exercise 31 Ⓒ

Do these calculations.

1 −3 + (−5)	**2** −5 − (+2)	**3** 4 − (+3)	**4** −3 − (−4)
5 6 − (−3)	**6** 16 + (−5)	**7** −4 + (−4)	**8** 20 − (−22)
9 −6 − (−10)	**10** 95 + (−80)	**11** −3 − (+4)	**12** −5 − (+4)
13 6 + (−7)	**14** −4 + (−3)	**15** −7 − (−7)	**16** 3 − (−8)

17 −8 + (−6)	**18** 7 − (+7)	**19** 12 − (−5)	**20** 9 − (+6)
21 −3 − (−2)	**22** 8 + (−11)	**23** 10 − (−2)	**24** −7 + (−2)
25 9 − (+6)	**26** 7 + (−7)	**27** 0 − (−8)	**28** −6 − (−8)

29 Copy and complete these addition squares.

a

+	−5	1	6	−2
3	−2	4		
−2				
6				
−10				

b

+			−4	
5			1	
			0		5
		7		6	17
			−4	

1.4.3 Multiplying and dividing with negative numbers

- When two numbers with the same sign are multiplied together, the answer is positive.

  $+7 \times (+3) = +21$
  $-6 \times (-4) = +24$

- When two numbers with different signs are multiplied together, the answer is negative.

  $-8 \times (+4) = -32$
  $+7 \times (-5) = -35$

- When dividing numbers, the rules are the same as in multiplication.

  $-70 \div (-2) = +35$
  $+12 \div (-3) = -4$
  $-20 \div (+4) = -5$

Exercise 32 C

1 $-3 \times (+2)$	**2** $-4 \times (+1)$	**3** $+5 \times (-3)$	**4** $-3 \times (-3)$
5 $-4 \times (2)$	**6** $-5 \times (3)$	**7** $6 \times (-4)$	**8** $3 \times (2)$
9 $-3 \times (-4)$	**10** $6 \times (-3)$	**11** $-7 \times (3)$	**12** $-5 \times (-5)$
13 $6 \times (-10)$	**14** $-3 \times (-7)$	**15** $8 \times (6)$	**16** $-8 \times (2)$
17 $-7 \times (6)$	**18** $-5 \times (-4)$	**19** $-6 \times (7)$	**20** $11 \times (-6)$
21 $8 \div (-2)$	**22** $-9 \div (3)$	**23** $-6 \div (-2)$	**24** $10 \div (-2)$
25 $-12 \div (-3)$	**26** $-16 \div (4)$	**27** $4 \div (-1)$	**28** $8 \div (-8)$
29 $16 \div (-8)$	**30** $-20 \div (-5)$	**31** $-16 \div (1)$	**32** $18 \div (-9)$
33 $36 \div (-9)$	**34** $-45 \div (-9)$	**35** $-70 \div (7)$	**36** $-11 \div (-1)$
37 $-16 \div (-1)$	**38** $100 \div (-10)$	**39** $-2 \div (-2)$	**40** $50 \div (-10)$
41 $-8 \times (-8)$	**42** $-9 \times (3)$	**43** $10 \times (-60)$	**44** $-8 \times (-5)$
45 $-12 \div (-6)$	**46** $-18 \times (-2)$	**47** $-8 \div (4)$	**48** $-80 \div (10)$

49 Copy and complete these multiplication squares.

a

$\times$	4	-3	0	-2
-5				
2				
10				
-1				

b

$\times$			-1	
3			-3	
		-15		18
	-14		-7	-42
		10		

Questions on negative numbers are more difficult when different sorts are mixed together. The questions in these three short tests are mixed.

Test 1

1 $-8 - 8$	**2** $-8 \times (-8)$	**3** -5×3	**4** $-5 + 3$
5 $8 - (-7)$	**6** $20 - 2$	**7** $-18 \div (-6)$	**8** $4 + (-10)$
9 $-2 + 13$	**10** $+8 \times (-6)$	**11** $-9 + (+2)$	**12** $-2 - (-11)$
13 $-6 \times (-1)$	**14** $2 - 20$	**15** $-14 - (-4)$	**16** $-40 \div (-5)$
17 $5 - 11$	**18** -3×10	**19** $9 + (-5)$	**20** $7 \div (-7)$

Test 2

1 $-2 \times (+8)$	**2** $-2 + 8$	**3** $-7 - 6$	**4** $-7 \times (-6)$
5 $+36 \div (-9)$	**6** $-8 - (-4)$	**7** $-14 + 2$	**8** $5 \times (-4)$
9 $11 + (-5)$	**10** $11 - 11$	**11** $-9 \times (-4)$	**12** $-6 + (-4)$
13 $3 - 10$	**14** $-20 \div (-2)$	**15** $16 + (-10)$	**16** $-4 - (+14)$
17 $-45 \div 5$	**18** $18 - 3$	**19** $-1 \times (-1)$	**20** $-3 - (-3)$

Test 3

1 $-10 \times (-10)$	**2** $-10 - 10$	**3** $-8 \times (+1)$	**4** $-8 + 1$
5 $5 + (-9)$	**6** $15 - 5$	**7** $-72 \div (-8)$	**8** $-12 - (-2)$
9 $-1 + 8$	**10** $-5 \times (-7)$	**11** $-10 + (-10)$	**12** $-6 \times (+4)$
13 $6 - 16$	**14** $-42 \div (+6)$	**15** $-13 + (-6)$	**16** $-8 - (-7)$
17 $5 \times (-1)$	**18** $2 - 15$	**19** $21 + (-21)$	**20** $-16 \div (-2)$

1.5 Order of operations

1.5.1 The BIDMAS rule

Mathematicians all over the world regularly exchange their ideas and the results of their theories, even though much of the time they are unable to speak the same language! They can communicate mathematically because it has been agreed that everyone follows certain rules.

- Think about this question:
  'What is five add seven multiplied by three?'

  If you add first, you get: $5 + 7 \times 3$
  $$= 12 \times 3$$
  $$= 36$$

  If you multiply first, you get: $5 + 7 \times 3$
  $$= 5 + 21$$
  $$= 26$$

If everyone came up with different answers to the same mathematical question, life would be rather stressful as people would have to argue constantly over who was correct.

This table shows the order in which to do mathematical operations to ensure you all get the same answers.

B rackets	( )	do first	'B'
I ndices	3^2	do next	'I'
D ivision **M** ultiplication	÷ ×	do this pair next	'D' 'M'
A ddition **S** ubtraction	+ −	do this pair last	'A' 'S'

Indices mean powers and roots, see Section 1.6.

Remember the word '**B I D M A S**'.

EXAMPLE

a $40 \div 5 \times 2$
$= 8 \times 2$
$= 16$

For ÷ and × do in the order they appear

b $9 + 8 - 7$
$= 17 - 7$
$= 10$

For + and − do in the order they appear

c $5 + 2 \times 3$
$= 5 + 6$
$= 11$

× before +

d $10 - 8 \div 2$
$= 10 - 4$
$= 6$

÷ before −

Exercise 33 Ⓒ

Do these. Show every step in your working.

1 $5 + 3 \times 2$ **2** $4 - 1 \times 3$ **3** $7 - 4 \times 3$
4 $2 + 2 \times 5$ **5** $9 + 2 \times 6$ **6** $13 - 11 \times 1$
7 $7 \times 2 + 3$ **8** $9 \times 4 - 12$ **9** $2 \times 8 - 7$
10 $4 \times 7 + 2$ **11** $13 \times 2 + 4$ **12** $8 \times 5 - 15$

13 $6 + 10 \div 5$ **14** $7 - 16 \div 8$ **15** $8 - 14 \div 7$
16 $5 + 18 \div 6$ **17** $2 \times 18 \div 6$ **18** $6 - 12 \div 4$
19 $20 \div 4 + 2$ **20** $15 \div 3 - 7$ **21** $24 \div 6 - 8$

22 $30 \div 6 + 9$ **23** $8 \div 2 + 9$ **24** $28 \div 7 - 4$
25 $13 + 3 \times 13$ **26** $9 + 26 \div 13$ **27** $10 \times 8 - 70$
28 $96 \div 4 - 4$ **29** $36 \div 9 + 1$ **30** $1 \times 2 + 3$

a $8 + 3 \times 4 - 6$

$= 8 + (3 \times 4) - 6$
$= 8 + 12 - 6$
$= 14$

× and ÷ before
+ and −

b $3 \times 2 - 8 \div 4$

$= (3 \times 2) - (8 \div 4)$
$= 6 - 2$
$= 4$

c $\dfrac{8 + 6}{2} = \dfrac{14}{2}$

$= 7$

A horizontal line
acts as a bracket.

Notice that the brackets make the working easier.

Exercise 34 Ⓒ

Evaluate these. Show every step in your working.

1 $2 + 3 \times 4 + 1$ **2** $4 + 8 \times 2 - 10$ **3** $7 + 2 \times 2 - 6$
4 $25 - 7 \times 3 + 5$ **5** $17 - 3 \times 5 + 9$ **6** $11 - 9 \times 1 - 1$
7 $1 + 6 \div 2 + 3$ **8** $6 + 28 \div 7 - 2$ **9** $8 + 15 \div 3 - 5$
10 $5 - 36 \div 9 + 3$ **11** $6 - 24 \div 4 + 0$ **12** $8 - 30 \div 6 - 2$
13 $3 \times 4 + 1 \times 6$ **14** $4 \times 4 + 14 \div 7$ **15** $2 \times 5 + 8 \div 4$

16 $21 \div 3 + 5 \times 4$ **17** $10 \div 2 + 1 \times 3$ **18** $15 \div 5 + 18 \div 6$
19 $5 \times 5 - 6 \times 4$ **20** $2 \times 12 - 4 \div 2$ **21** $7 \times 2 - 10 \div 2$
22 $35 \div 7 - 5 \times 1$ **23** $36 \div 3 - 1 \times 7$ **24** $42 \div 6 - 56 \div 8$
25 $72 \div 9 + 132 \div 11$ **26** $19 + 35 \div 5 - 16$ **27** $50 - 6 \times 7 + 8$
28 $30 - 9 \times 2 + 40$ ***29** $4 \times 11 - 28 \div 7$ ***30** $13 \times 11 - 4 \times 8$

In questions **31** to **50** remember to do the operation in the brackets first.

31 $3 + (6 \times 8)$ **32** $(3 \times 8) + 6$ **33** $(8 \div 4) + 9$
34 $3 \times (9 \div 3)$ **35** $(5 \times 9) - 17$ **36** $10 + (12 \times 8)$
37 $(16 - 7) \times 6$ **38** $48 \div (14 - 2)$ **39** $64 \div (4 \times 4)$
40 $81 + (9 \times 8)$ **41** $67 - (24 \div 3)$ **42** $(12 \times 8) + 69$
43 $(6 \times 6) + (7 \times 7)$ **44** $(12 \div 3) \times (18 \div 6)$ **45** $(5 \times 12) - (3 \times 9)$
46 $(20 - 12) \times (17 - 9)$ **47** $100 - (99 \div 3)$ **48** $1001 + (57 \times 3)$

49 $(3 \times 4 \times 5) - (72 \div 9)$ **50** $(2 \times 5 \times 3) \div (11 - 5)$ **51** $\dfrac{15 - 7}{2}$

52 $\dfrac{160}{7 + 3}$ **53** $\dfrac{19 + 13}{6 - 2}$ **54** $\dfrac{5 \times 7 - 9}{13}$

1.6 Powers and roots

1.6.1 Indices

- Indices are a neat way of writing products.
  $2 \times 2 \times 2 \times 2 \times = 2^4$ (2 to the power 4)
  $5 \times 5 \times 5 = 5^3$ (5 to the power 3)
  $3 \times 3 \times 3 \times 3 \times 3 \times 10 \times 10 = 3^5 \times 10^2$

- Numbers like 3^2, 5^2, 11^2 are **square numbers**.
  Numbers like 2^3, 6^3, 11^3 are **cube numbers**.

> You should learn all the square numbers 1, 4, 9, 16, $\cdots$ up to 225.

- To work out $3 \cdot 2^2$ on a calculator, press $\boxed{3\cdot 2}$ $\boxed{x^2}$ $\boxed{=}$

  To work out 3^4 on a calculator, press

  $\boxed{3}$ $\boxed{x^y}$ $\boxed{4}$ $\boxed{=}$ or $\boxed{3}$ $\boxed{\wedge}$ $\boxed{4}$ $\boxed{=}$

Exercise 35 Ⓒ

Write these using indices.

  1 $3 \times 3 \times 3 \times 3$ **2** 5×5
  3 $6 \times 6 \times 6$ **4** $10 \times 10 \times 10 \times 10 \times 10$
  5 $1 \times 1 \times 1 \times 1 \times 1 \times 1 \times 1$ **6** $8 \times 8 \times 8 \times 8$
  7 $7 \times 7 \times 7 \times 7 \times 7 \times 7$ **8** $2 \times 2 \times 2 \times 5 \times 5$
  9 $3 \times 3 \times 7 \times 7 \times 7 \times 7$ **10** $3 \times 3 \times 10 \times 10 \times 10$
11 $5 \times 5 \times 5 \times 5 \times 11 \times 11$ **12** $2 \times 3 \times 2 \times 3 \times 3$
13 $5 \times 3 \times 3 \times 5 \times 5$ **14** $2 \times 2 \times 3 \times 3 \times 3 \times 11 \times 11$

15 Work these out without using a calculator.
    a 4^2 **b** 6^2 **c** 10^2 **d** 3^3 **e** 10^3

16 Use the $\boxed{x^2}$ button on a calculator to work out these.
    a 9^2 **b** 21^2 **c** $1 \cdot 2^2$ **d** $0 \cdot 2^2$ **e** $3 \cdot 1^2$
    f 100^2 **g** 25^2 **h** $8 \cdot 7^2$ **i** $0 \cdot 9^2$ **j** $81 \cdot 4^2$

17 Find the areas of these squares.

    a    **b**   **c**

13 cm   2·5 cm   11·4 cm

*18 Write these in index form.

    a $a \times a \times a$ **b** $n \times n \times n \times n$ **c** $s \times s \times s \times s \times s$

    d $p \times p \times q \times q \times q$ **e** $b \times b \times b \times b \times b \times b \times b$

*19 Use a calculator to work out these.

    a 6^3 **b** 2^8 **c** 3^5 **d** 10^5 **e** 4^3

    f $0{\cdot}1^3$ **g** $1{\cdot}7^4$ **h** $3^4 \times 7$ **i** $5^3 \times 10$

*20 A scientist has a dish containing 10^9 germs.
One day later there are 10 times as many germs.
How many germs are in the dish now?

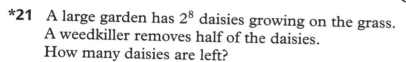

*21 A large garden has 2^8 daisies growing on the grass.
A weedkiller removes half of the daisies.
How many daisies are left?

*22 A maths teacher won the National Lottery and, as a leaving
present, he decided to set a final test to a class of 25 students.
The person coming 25th won 2p, the 24th won 4p, the 23rd
8p, the 22nd 16p and so on, doubling the amount each time.

    a Write 2, 4, 8, 16 as powers of 2.

    b How much, in pounds, would be given to the person
who came first in the test?

*23 Sean says 'If you work out the product of any four
consecutive numbers and then add one, the answer will
be a square number.'

    For example: $1 \times 2 \times 3 \times 4 = 24$

                      $24 + 1 = 25$, which is a square number.

    Is Sean right? Test his theory on four (or more) sets of four
consecutive numbers.

1.6.2 Square roots and cube roots

A square has an area of $529\,\text{cm}^2$.
How long is a side of the square?

....

In other words, what number **multiplied by itself** makes 529?
The answer is the **square root** of 529.

On a calculator press $\boxed{\sqrt{}}$ $\boxed{529}$ $\boxed{=}$

(On older calculators you may need to press $\boxed{529}$ $\boxed{\sqrt{}}$.)

The side of the square is 23 cm.

$529\,\text{cm}^2$	?

?

EXAMPLE

A cube has a volume of $512\,\text{cm}^3$.
How long is a side of the cube?

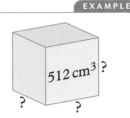

$512\,\text{cm}^3$?

....

The answer is the **cube root** of 512.

On a calculator press $\boxed{\sqrt[3]{}}$ $\boxed{512}$ $\boxed{=}$

The side of the cube is 8 cm. (Check $8 \times 8 \times 8 = 512$)

Exercise 36Ⓔ

1 Work out, without a calculator.
   a $\sqrt{16}$ **b** $\sqrt{36}$ **c** $\sqrt{1}$ **d** $\sqrt{100}$

2 Find the sides of the squares.

a
Area
$= 81\,\text{cm}^2$ x
x

b
Area
$= 49\,\text{cm}^2$ y
y

c
Area
$= 144\,\text{cm}^2$ z
z

3 Use a calculator to find these square roots, correct to 1 dp.
   a $\sqrt{10}$ **b** $\sqrt{29}$ **c** $\sqrt{107}$ **d** $\sqrt{19.7}$
   e $\sqrt{2406}$ **f** $\sqrt{58.6}$ **g** $\sqrt{0.15}$ **h** $\sqrt{0.727}$

4 A square photo has an area of $150\,\text{cm}^2$. Find the length of each side of the photo, correct to the nearest mm.

5 A square field has an area of 20 hectares. How long is each side of the field, correct to the nearest m?
(1 hectare $= 10\,000\,\text{m}^2$)

6 The area of square A is equal to the sum of the areas of squares B and C. Find the length x, correct to 1 dp.

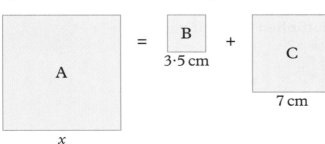

A

$=$ B
   3·5 cm

$+$ C
    7 cm

x

7 Find these cube roots.

   a $\sqrt[3]{64}$ **b** $\sqrt[3]{125}$ **c** $\sqrt[3]{1000}$

8 A cube has a volume of 200 cm³.

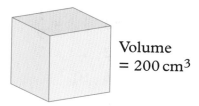

Volume
= 200 cm³

  Find the length of the sides of the cube, correct to 1 dp.

*9 **A challenge!**

The reciprocal of 2 is $\frac{1}{2}$. The reciprocal of 7 is $\frac{1}{7}$. The reciprocal of x is $\frac{1}{x}$.

Find the square root of the reciprocal of the square root of the reciprocal of ten thousand.

1.6.3   Negative indices and the zero index

Look at this sequence.

$$2^4 \quad 2^3 \quad 2^2 \quad 2^1 \quad\quad 2^0 \quad\quad 2^{-1}$$

$$16 \longrightarrow 8 \longrightarrow 4 \longrightarrow 2 \longrightarrow \boxed{1} \longrightarrow \boxed{\tfrac{1}{2}}$$

From left to right the index goes down by one each time as the number is divided by two each time.

You can see that $2^0 = 1$

and that $\qquad\qquad 2^{-1} = \dfrac{1}{2}$

●   The **reciprocal** of 3 is $\frac{1}{3}$. The reciprocal of 10 is $\frac{1}{10}$.

    The reciprocal of n is $\frac{1}{n}$ (which can be written n^{-1}).

●   In general $x^0 = 1$ for any value of x that is not zero.

$$x^{-1} = \frac{1}{x}$$

$$\text{So } 2^{-3} = \frac{1}{2^3} \qquad\qquad 3^{-2} = \frac{1}{3^2}$$

Exercise 37Ⓔ

In questions **1** to **12**, work out the value of each number.

1 3^{-1}	**2** 4^{-1}	**3** 10^{-1}	**4** 1^{-4}	**5** 3^{-2}	**6** 4^{-2}
7 10^{-2}	**8** 8^0	**9** 7^{-2}	**10** $(-6)^0$	**11** 9^{-2}	**12** 1^{-7}

In questions **13** to **32** answer 'true' or 'false'.

13 $2^3 = 8$ **14** $3^2 = 6$ **15** $5^3 = 125$ **16** $2^{-1} = \dfrac{1}{2}$

17 $10^{-2} = \dfrac{1}{20}$ **18** $3^{-3} = \dfrac{1}{9}$ **19** $2^2 > 2^3$ **20** $2^3 < 3^2$

21 $2^{-2} > 2^{-3}$ **22** $3^{-2} < 3^3$ **23** $1^9 = 9$ **24** $(-3)^2 = -9$

25 $5^{-2} = \dfrac{1}{10}$ **26** $10^{-3} = \dfrac{1}{1000}$ **27** $10^{-2} > 10^{-3}$ **28** $5^{-1} = 0.2$

29 $10^{-1} = 0.1$ **30** $2^{-2} = 0.25$ **31** $5^0 = 1$ **32** $16^0 = 0$

1.6.4 Multiplying and dividing

- To multiply powers of the same number **add** the indices.

EXAMPLE

$3^2 \times 3^4 = (3 \times 3) \times (3 \times 3 \times 3 \times 3) = 3^6$
$2^3 \times 2^2 = (2 \times 2 \times 2) \times (2 \times 2) = 2^5$
$7^3 \times 7^5 = 7^8$ (add the indices)

- To divide powers of the same number **subtract** the indices.

EXAMPLE

$2^4 \div 2^2 = \dfrac{2 \times 2 \times 2 \times 2}{2 \times 2} = 2^2$

$\left. \begin{array}{l} 5^6 \div 5^2 = 5^4 \\ 7^8 \div 7^3 = 7^5 \end{array} \right\}$ (subtract the indices)

Exercise 38(E)

Write these in simpler form.

1 $5^2 \times 5^4$ **2** $6^3 \times 6^2$ **3** $10^4 \times 10^5$ **4** $7^5 \times 7^3$

5 $3^6 \times 3^4$ **6** $8^3 \times 8^3$ **7** $2^3 \times 2^{10}$ **8** $3^6 \times 3^{-2}$

9 $5^4 \times 5^{-1}$ **10** $7^7 \times 7^{-3}$ **11** $5^{-3} \times 5^5$ **12** $3^{-2} \times 3^{-2}$

13 $6^{-3} \times 6^8$ **14** $5^{-2} \times 5^{-8}$ **15** $7^{-3} \times 7^9$ **16** $7^4 \div 7^2$

17 $6^7 \div 6^2$ **18** $8^5 \div 8^4$ **19** $5^{10} \div 5^2$ **20** $10^7 \div 10^5$

21 $9^6 \div 9^8$ **22** $3^8 \div 3^{10}$ **23** $2^6 \div 2^2$ **24** $3^3 \div 3^5$

25 $7^2 \div 7^8$ **26** $3^{-2} \div 3^2$ **27** $5^{-3} \div 5^2$ **28** $8^{-1} \div 8^4$

29 $5^{-4} \div 5^1$ **30** $6^2 \div 6^{-2}$ **31** $3^4 \div 3^4$ **32** $5^2 \div 5^2$

33 $\dfrac{3^4 \times 3^5}{3^2}$ **34** $\dfrac{2^8 \times 2^4}{2^5}$ **35** $\dfrac{7^3 \times 7^3}{7^4}$ **36** $\dfrac{5^9 \times 5^{10}}{5^{20}}$

1.6.5   Further rules of indices

- To raise a power of a number to a further power, **multiply** the indices.

EXAMPLE

a $(x^2)^3 = x^6$ b $(a^4)^2 = a^8$

c $3x^2 \times 4x^5 = 12x^7$
$(3 \times 4) \quad (2+5)$

d $4a^7 \times 5a^2 = 20a^9$
$(4 \times 5) \quad (7+2)$

e $12x^5 \div 3x^2 = 4x^3$
$(12 \div 3) \quad (5-2)$

f $(3a^2)^3 = 3^3 \times a^6 = 27a^6$

Exercise 39Ⓔ

Write these in simpler form.

1 $(3^3)^2$ **2** $(5^4)^3$ **3** $(7^2)^5$ **4** $(8^2)^{10}$
5 $(x^2)^3$ **6** $(a^5)^3$ **7** $(n^7)^2$ **8** $(y^3)^3$
9 $(2^{-1})^2$ **10** $(3^{-2})^2$ **11** $(7^{-1})^{-2}$ **12** $(x^3)^{-1}$
13 $2a^2 \times 3a^3$ **14** $4n^3 \times 5n^1$ **15** $7x^4 \times 2x$ **16** $8y^5 \times 3y^2$
17 $5n^3 \times n^4$ **18** $6y^2 \times 2$ **19** $3p^3 \times 3p^2$ **20** $2p \times 5p^5$
21 $(2x^2)^3$ **22** $(3a^2)^3$ **23** $(4y^3)^2$ **24** $(5x^4)^2$

Solve the equations to find the value of x.

25 $x^2 = 9$ **26** $x^5 = 1$ **27** $x^3 = 27$ **28** $x^5 = 0$
29 $2^x = 8$ **30** $3^x = 3$ **31** $5^x = 25$ **32** $10^x = 1000$

33 $2^x = \dfrac{1}{2}$ **34** $4^x = \dfrac{1}{4}$ **35** $7^x = 1$ **36** $3x^3 = 24$

37 $10x^3 = 640$ **38** $2x^3 = 0$ **39** $10^x = 0 \cdot 1$ **40** $5^x = 1$

1.7   Using a calculator

1.7.1   Money and time on a calculator

- To work out £27·30 ÷ 7, key in

| 2 | 7 | · | 3 | 0 | ÷ | 7 | = |

The answer is 3·9. Remember this means £3·90.

> EXAMPLE
>
> A machine takes 15 minutes to make one toy. How long will it take to make 1627 toys?
>
> ....
>
> 15 minutes is one quarter of an hour and $\frac{1}{4} = 0.25$ as a decimal.
>
> Key in 0·25 $\times$ 1627 $=$
>
> The answer is 406·75.
> It will take 406·75 hours or 406 hours 45 minutes.

- 6 minutes is $\frac{6}{60}$ of an hour. $\frac{6}{60} = \frac{1}{10} = 0.1$ hours.

  In the same way 27 minutes $= \frac{27}{60}$ of an hour.

  $\frac{27}{60} = 0.45$ hours.

Exercise 40 C

1 Do these calculations and give your answer in **pounds**.
   a £1·22 × 5   **b** £153·60 ÷ 24   **c** £12·35 − £7·65
   d 20p × 580   **e** 6p × 2155   **f** £10 ÷ 250

2 Write these time intervals in hours, as decimals.
   a 2 h 30 min   **b** 4 h 15 min   **c** 3 h 45 min
   d 6 min   **e** 12 min   **f** 15 min

6 min $= \frac{6}{60}$ hour
$= 0.1$ hour
15 min $= \frac{15}{60}$ hour
$= 0.25$ hour

3 Do these calculations and give your answer in **hours**.
   a 2 h 45 min × 9   **b** 3 h 15 min × 7   **c** 14 h 30 min ÷ 5
   d 15 min × 11   **e** 30 min × 5   **f** 5 h 15 min ÷ 3

1.7.2   Order of operations

Where there is a mixture of operations to do, remember **BIDMAS**.

- Work out brackets first   (B)
- Work out indices next   (I)
- Work out ÷, ×   (DM)
- Then +, −   (AS)
  It is a good idea to check your answers by estimation or by performing the inverse operation.

Exercise 41 C

Use a calculator and give the answers correct to one decimal place.

1 $2 \cdot 5 \times 1 \cdot 67$

2 $19 \cdot 6 - 3 \cdot 7311$

3 $0 \cdot 792^2$

4 $0 \cdot 13 + 8 \cdot 9 - 3 \cdot 714$

5 $2 \cdot 4^2 - 1 \cdot 712$

6 $5 \cdot 3 \times 1 \cdot 7 + 3 \cdot 7$

7 $0 \cdot 71 \times 0 \cdot 92 - 0 \cdot 15$

8 $9 \cdot 6 \div 1 \cdot 72$

9 $8 \cdot 17 - 1 \cdot 56 + 7 \cdot 4$

10 $\sqrt{4 \cdot 52}$

11 $\sqrt{198}$

12 $\sqrt{\dfrac{2 \cdot 63}{1 \cdot 9}}$

In questions **13** to **30** remember BIDMAS.

13 $2 \cdot 5 + 3 \cdot 1 \times 2 \cdot 4$

14 $7 \cdot 81 + 0 \cdot 7 \times 1 \cdot 82$

15 $8 \cdot 73 + 9 \div 11$

16 $11 \cdot 7 \div 9 - 0 \cdot 74$

17 $7 \div 0 \cdot 32 + 1 \cdot 15$

18 $2 \cdot 6 + 5 \cdot 2 \times 1 \cdot 7$

19 $2 \cdot 9 + \dfrac{8 \cdot 3}{1 \cdot 83}$

20 $1 \cdot 7^2 + 2 \cdot 62$

21 $5 \cdot 2 + \dfrac{11 \cdot 7}{1 \cdot 85}$

22 $9 \cdot 64 + 26 \div 12 \cdot 7$

23 $1 \cdot 27 + 3 \cdot 1^2$

24 $4 \cdot 2^2 \div 9 \cdot 4$

25 $0 \cdot 151 + 1 \cdot 4 \times 9 \cdot 2$

26 $1 \cdot 7^3$

27 $8 \cdot 2 + 3 \cdot 2 \times 3 \cdot 3$

28 $3 \cdot 2 + \dfrac{1 \cdot 41}{6 \cdot 72}$

29 $\dfrac{1 \cdot 9 + 3 \cdot 71}{2 \cdot 3}$

30 $\dfrac{8 \cdot 7 - 5 \cdot 371}{1 \cdot 14}$

1.7.3 Using brackets

Most calculators have brackets buttons like these $\boxed{(}\;\boxed{)}$.

When you use the bracket buttons the calculator does the calculation inside the brackets $\boxed{(}\;\boxed{)}$ first.
Try it.

Don't forget to press the $\boxed{=}$ button at the end to give the final answer.

EXAMPLE

a $8 \cdot 72 - (1 \cdot 4 \times 1 \cdot 7)$

b $\dfrac{8 \cdot 51}{(1 \cdot 94 - 0 \cdot 711)}$

$\boxed{8 \cdot 72}\;\boxed{-}\;\boxed{(}\;\boxed{1 \cdot 4}\;\boxed{\times}$

$\boxed{1 \cdot 7}\;\boxed{)}\;\boxed{=}$

$8 \cdot 72 - (1 \cdot 4 \times 1 \cdot 7)$
$= 6 \cdot 3$ to 1 dp

$\boxed{8 \cdot 51}\;\boxed{\div}\;\boxed{(}\;\boxed{1 \cdot 94}\;\boxed{-}$

$\boxed{0 \cdot 711}\;\boxed{)}\;\boxed{=}$

$\dfrac{8 \cdot 51}{(1 \cdot 94 - 0 \cdot 711)} = 6 \cdot 9$ to 1 dp

1.7.4 Using the $\boxed{\text{ANS}}$ button

You can use the $\boxed{\text{ANS}}$ button as a short-term memory. It holds the answer from the previous calculation.

EXAMPLE

Work out $\dfrac{5}{1\cdot2 - 0\cdot761}$, using the $\boxed{\text{ANS}}$ button.

....

Find the bottom line first

$\boxed{1\cdot2}$ $\boxed{-}$ $\boxed{0\cdot761}$ $\boxed{\text{EXE}}$ $\boxed{5}$ $\boxed{\div}$ $\boxed{\text{ANS}}$ $\boxed{\text{EXE}}$

The calculator reads $11\cdot389\,521\,64$.
Notice that the $\boxed{\text{EXE}}$ button works the same as the $\boxed{=}$ button.

Exercise 42 C

Do these and give the answers correct to 1 decimal place.
Use the brackets buttons or the $\boxed{\text{ANS}}$ button.

1 $18\cdot41 - (7\cdot2 \times 1\cdot3)$

2 $11\cdot01 + (2\cdot6 \div 7)$

3 $(1\cdot27 + 5\cdot6) \div 1\cdot4$

4 $9\cdot6 + (11\cdot2 \div 4)$

5 $(8\cdot6 \div 3) - 1\cdot4$

6 $11\cdot7 - (2\cdot6 \times 2\cdot7)$

7 $7\cdot41 - \left(\dfrac{7\cdot3}{1\cdot4}\right)$

8 $\left(\dfrac{8\cdot91}{1\cdot7}\right) - 2\cdot63$

9 $\dfrac{1\cdot41}{(1\cdot7 + 0\cdot21)}$

10 $(1\cdot56 + 1\cdot9) \div 2\cdot45$

11 $3\cdot2 \times (1\cdot9 - 0\cdot74)$

12 $8\cdot9 \div (1\cdot3 - 0\cdot711)$

13 $(8\cdot72 \div 1\cdot4) \times 1\cdot49$

14 $(2\cdot67 + 1\cdot2 + 5) \times 1\cdot13$

15 $23 - (9\cdot2 \times 1\cdot85)$

16 $\dfrac{(8\cdot41 + 1\cdot73)}{1\cdot47}$

17 $\dfrac{7\cdot23}{(8\cdot2 \times 0\cdot91)}$

18 $\dfrac{(11\cdot4 - 7\cdot87)}{17}$

In questions **19** to **40** use the $\boxed{x^2}$ button when you need it.

19 $2\cdot6^2 - 1\cdot4$

20 $8\cdot3^2 \times 1\cdot17$

21 $7\cdot2^2 \div 6\cdot67$

22 $(1\cdot4 + 2\cdot67)^2$

23 $(8\cdot41 - 5\cdot7)^2$

24 $(2\cdot7 \times 1\cdot31)^2$

25 $8\cdot2^2 - (1\cdot4 + 1\cdot73)$

26 $\dfrac{2\cdot6^2}{(1\cdot3 + 2\cdot99)}$

27 $4\cdot1^2 - \left(\dfrac{8\cdot7}{3\cdot2}\right)$

28 $\dfrac{(2\cdot7 + 6\cdot04)}{(1\cdot4 + 2\cdot11)}$

29 $\dfrac{(8 \cdot 71 - 1 \cdot 6)}{(2 \cdot 4 + 9 \cdot 73)}$

30 $\left(\dfrac{2 \cdot 3}{1 \cdot 4}\right)^2$

31 $9 \cdot 72^2 - (2 \cdot 9 \times 2 \cdot 7)$

32 $(3 \cdot 3 + 1 \cdot 3^2) \times 9$

33 $(2 \cdot 7^2 - 2 \cdot 1) \div 5$

34 $\left(\dfrac{2 \cdot 84}{7}\right) + \left(\dfrac{7}{11 \cdot 2}\right)$

35 $\dfrac{(2 \cdot 7 \times 8 \cdot 1)}{(12 - 8 \cdot 51)}$

36 $\left(\dfrac{2 \cdot 3}{1 \cdot 5}\right) - \left(\dfrac{6 \cdot 3}{8 \cdot 9}\right)$

*37 $(1 \cdot 31 + 2 \cdot 705) - 1 \cdot 3^2$

*38 $(2 \cdot 71 - 0 \cdot 951) \times 5 \cdot 62$

*39 $\dfrac{(8 \cdot 5 \times 1 \cdot 952)}{(7 \cdot 2 - 5 \cdot 96)}$

*40 $\left(\dfrac{80 \cdot 7}{30 \cdot 3}\right) - \left(\dfrac{11 \cdot 7}{10 \cdot 2}\right)$

Exercise 43 **C**

Puzzle

If you work out $25 \times 503 \times 4 + 37$ on a calculator you should get the number 50 337. If you turn the calculator upside down (and use a little imagination) you can see the word 'LEEDS'.

Find the words given by the clues below.

1 $83 \times 85 + 50$ (Lots of this in the garden)

2 $211 \times 251 + 790$ (Tropical or Scilly)

3 $19 \times 20 \times 14 - 2 \cdot 66$ (Not an upstanding man)

4 $(84 + 17) \times 5$ (Dotty message)

5 $0 \cdot 014\,43 \times 7 \times 4$ (Three times as funny)

6 $79 \times 9 - 0 \cdot 9447$ (Greasy letters)

7 $50 \cdot 19 - (5 \times 0 \cdot 0039)$ (Not much space inside)

8 $2 \div 0 \cdot 5 - 3 \cdot 295$ (Rather lonely)

9 $6 \cdot 2 \times 0 \cdot 987 \times 1\,000\,000 - 860^2 + 118$ (Flying ace)

10 $7420 \times 7422 + 118^2 - 30$ (Big Chief)

11 $(13 \times 3 \times 25 \times 8 \times 5) + 7$ (Dwelling for masons)

12 $71^2 - 11^2 - 5$ (Sad gasp)

13 $904^2 + 89\,621\,818$ (Prickly customer)

14 $(559 \times 6) + (21 \times 55)$ (What a surprise!)

15 $566 \times 711 - 23\,617$ (Bolt it down)

16 $\dfrac{9999 + 319}{8 \cdot 47 + 2 \cdot 53}$ (Sit up and plead)

17 $\dfrac{2601 \times 6}{4^2 + 1^2}$; $(401 - 78) \times 5^2$ (two words) (Not a great man)

18 $0 \cdot 4^2 - 0 \cdot 1^2$ (Little Sidney)

19 $\dfrac{(27 \times 2000 - 2)}{(0 \cdot 63 \div 0 \cdot 09)}$ (Not quite a mountain)

20 $(5^2 - 1^2)^4 - 14\,239$ (Just a name)

21 $48^4 + 102^2 - 4^2$ (Pursuits)

22 $615^2 + (7 \times 242)$ (Almost a goggle)

23 $14^4 - 627 + 29$ (Good book, by God!)

24 $0 \cdot 034 \times 11 - 0 \cdot 002\,92$; $9^4 - (8 \times 71)$ (two words) (Nice for breakfast)

25 $(426 \times 474) + (318 \times 487) + 22\,018$ (Close to a bubble)

26 $\dfrac{36^3}{4} - 1530$ (Foreign-sounding girl's name)

27 $(594 \times 571) - (154 \times 132) - 38$ (Female Bobby)

28 $(7^2 \times 100) + (7 \times 2)$ (Lofty)

29 $240^2 + 134$; $241^2 - 7^3$ (two words) (Devil of a chime)

30 $1384 \cdot 5 \times 40 - 1 \cdot 991$ (Say this after sneezing)

31 $(2 \times 2 \times 2 \times 2 \times 3)^4 + 1929$ (Unhappy ending)

32 $141\,918 + 83^3$ (Hot stuff in France)

1.8 Solving numerical problems 1

1.8.1 Mixed questions

Exercise 44 **C**

1 Write the reading on each scale.

2 There are 5 people in a team.
How many teams can you
make from 118 people?

3 When James Wilkinson was born his dad planted a tree.
James died in 1920, aged 75. How old was the tree in 1975?

4 Washing-up liquid is sold in 200 ml containers. Each
container costs 57p. How much will it cost to buy 10 litres
of the liquid?

5 A train is supposed to leave London at 11:24 and arrive in
Brighton at 12:40. The train was delayed and arrived
$2\frac{1}{4}$ hours late. At what time did the train arrive?

6 Big Ben stopped for repairs at 17:15 on Tuesday and
restarted at 08:20 on Wednesday.
For how long had it been stopped?

7 How much would I pay for nine litres of paint if two litres
cost £2·30?

8 Here is a 'magic square'. The numbers in each row,
column and diagonal add up to to the same number.

Copy and complete these magic squares.

8	1	6
3	5	7
4	9	2

a

6	13	
	9	
		12

b

11			10
2	13	16	
		4	
7	12		6

*9 A piece of wire 48 cm long is bent to form a rectangle in which the length is twice the width.
Find the area of the rectangle.

10 A rectangular floor 3 m by 4 m is covered with square tiles, each of side length 50 cm.
How many tiles are there?

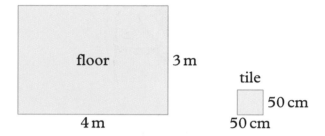

floor   3 m

tile

50 cm

4 m   50 cm

Exercise 45 ⓒ

1 Copy each calculation and find the missing digits.

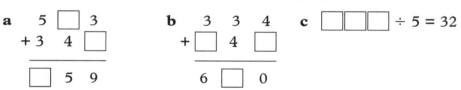

   a   5 ▢ 3   **b**   3 3 4   **c**   ▢▢▢ ÷ 5 = 32

       + 3   4   ▢   + ▢   4   ▢

       ▢   5   9   6   ▢   0

2 The fourteenth number in the sequence 1, 2, 4, 8, 16 ... is 8192. What is
  a the fifteenth number
  b the thirteenth number?

3 The manager of a toy shop bought 10 000 model cars for £4950.
The models were sold at 95p each.
What was the total profit made?

4 Work out these, without using a calculator.
  a $0·6 + 2·72$ **b** $3·21 - 1·6$
  c $2·8 - 1·34$ **d** $8 - 3·6$
  e $100 \times 0·062$ **f** $27·4 \div 10$

5 Six people can travel in one car and there are altogether 106 people to transport. How many cars are needed?

6 Copy and complete these 'magic squares'.

a

−1	−2	
	0	
		1

b

		−1
	2	
5	0	

c

0		−4
	−1	
2		

7 Write these numbers in order of size, smallest first.
 0·2   0·5   0·05   0·201   0·21

8 The scale shows temperatures in °C.
Write the temperatures for the arrows marked A, B, C and D.

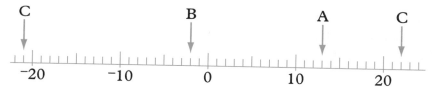

9 A baby falls asleep at 20:55 on Monday night and wakes up at 06:10 on Tuesday morning. For how long was the baby asleep?

10 Here are five number cards.

What is the largest **odd** number you can make with these five cards?

Exercise 46 C

1 Copy and complete this bill.

$6\frac{1}{2}$ kg of potatoes at 40p per kg = £☐

2 kg of beef at ☐ per kg = £7·20

☐ jars of coffee at 95p per jar = £6·65

Total = £☐

2 Write in index form (for example $2 \times 2 \times 2 = 2^3$).
  a $3 \times 3 \times 3 \times 3$ **b** $1 \times 1 \times 1 \times 1 \times 1 \times 1 \times 1$ **c** $7 \times 7 \times 7 \times 7 \times 7$
  d half of 2^5 **e** 10 times 10^5 **f** 1% of 1 million

3 Do these without using a calculator.
    a $(-3) \times (-3)$ **b** $-7 + 2$ **c** $12 \div (-2)$
    d $8 - 18$ **e** $5 - (-2)$ **f** $5 \times (-2)$

4 How many 50 ml bottles can be filled from a jar containing one litre of liquid?

5 **a** Which four coins make a total of 77p?
    b Which five coins make a total of 86p?
    c Which five coins make a total of £1·57?

*6 Two numbers m and z are such that z is greater than 10 and m is less than 8. Arrange the numbers 9, z and m in order of size, starting with the smallest.

*7 One day a third of the class is absent and 16 children are present.
How many children are in the class when no one is away?

*8 Copy the table and then write the numbers 1 to 9, one in each box, so that all the numbers satisfy the conditions for both the row and the column.

	Prime number	Multiple of 3	Factor of 16
Number greater than 5			
Odd number			
Even number			

*9 A man is 35 cm taller than his daughter, who is 5 cm shorter than her mother. The man was born in 1949 and is 1·80 m tall. How tall is the wife?

10 In a simple code A = 1, B = 2, C = 3, ... Z = 26. Decode these messages.

> Start by writing the alphabet with all the code numbers.

    a 23,   8,   1,   20
       20,   9, 13,   5
        4, 15
      23,   5
       6,   9, 14,   9, 19,   8.

    b 19, 4^2, (3×7), 18, $(90 - 71)$
       1^3, (9×2), $(2^2 + 1^2)$
      18, $\left(\frac{1}{5} \text{ of } 105\right)$, 2, $\left(1 \div \frac{1}{2}\right)$, 3^2, 19, 2^3.

    c 23, $(100 \div 20)$
        1, $(2 \times 3 \times 3)$, $(2^2 + 1^2)$
        21, $(100 - 86)$, $(100 \div 25)$, 5, $(2^4 + 2)$
        1, (5×4), $\left(10 \div \dfrac{1}{2}\right)$, 1, $(27 \div 9)$, $(99 \div 9)$.

Exercise 47 Ⓒ

1 A special new cheese is on offer at £3·48 per kilogram. Mrs Mann buys half a kilogram. How much change does she receive if she pays with a £5 note?

2 A cup and a saucer together cost £2·80. The cup costs 60p more than the saucer. How much does the cup cost?

3 A garden 9 m by 12 m is to be treated with fertiliser. One cup of fertiliser covers an area of 2 m² and one bag of fertiliser is sufficient for 18 cups.
  a Find the area of the garden.
  b Find the number of bags of fertiliser needed.

4 Copy and complete this pattern.

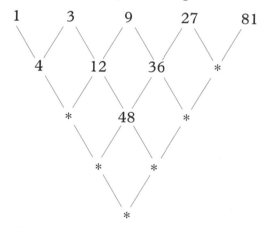

5 Six lamp posts lie at equal distances from each other along a straight road. If the distance between each pair of lamp posts is 20 m, how far is it from the first lamp post to the sixth?

6 Here is a list of numbers.

    2   5   8   22   25   27   44   49

Write the numbers that are
  a factors of 15   **b** multiples of 9   **c** prime numbers
  d square numbers   **e** cube numbers.

7 A coach journey started at
20:30 on Tuesday and finished
at 07:00 on the next day.

How long was the journey
in hours and minutes?

8 Work out, without using a calculator.
  a $0.6 - 0.06$ **b** 0.04×1000
  c $0.4 \div 100$ **d** $7.2 - 5$
  e 10% of £90   **f** 25% of £160.

9 Here are three number cards. [2] [7] [8]

You can use the three cards to make the number 728.
  a Use the three cards to make a number which is **more**
    than 728.
  b Use the three cards to make a number which is **less**
    than 728.
  c Use the three cards to make an **odd** number.

10 Find two numbers which
  a multiply to give 12 and add up to 7
  b multiply to give 42 and add up to 13
  c multiply to give 32 and add up to 12
  d multiply to give 48 and add up to 26.

Exercise 48 ©

1 In a simple code $A = 1, B = 2, C = 3$ and so on. When the word
'BAT' is written in code its total score is $(2 + 1 + 20) = 23$.
  a Find the score for the word 'ZOOM'.
  b Find the score for the word 'ALPHABET'.
  c Find a word with a score of 40.

2 How many cubes, each of edge 1 cm, are required to fill a
box with internal dimensions 5 cm by 8 cm by 3 cm?

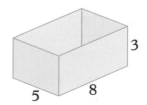

3 A swimming pool 20 m by 12 m contains water to a uniform
depth of $1\frac{1}{2}$ m. 1 m³ of water weighs 1000 kg. What is the
weight of the water in the pool?

4   The houses in a street are numbered from 1 to 60.
How many times does the number '2' appear?

5   Draw a large copy of this square.

1	2	3	4

Your task is to fill up all 16 squares using four 1s, four 2s,
four 3s and four 4s. Each number may appear only once in
any row (↔) or column (↕). The first row has been drawn
already. Ask another student to check your solution.

6   Between the times 11:57 and 12:27 the milometer of a car
changes from 23 793 miles to 23 825 miles.
At what average speed is the car travelling?

7   Copy these shapes and find which ones you can draw without
going over any line twice and without taking the pencil from
the paper? Write 'yes' or 'no' for each shape.

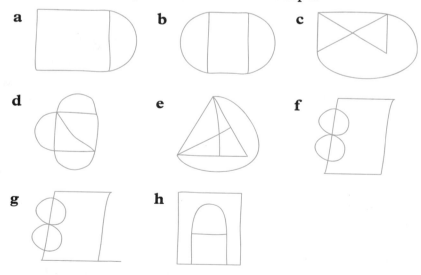

a   **b**   **c**

d   **e**   **f**

g   **h**

*8   A generous, but not very bright, teacher decides to award 1p
to the person coming 10th in a test, 2p to the person coming
9th, 4p to the person coming 8th and so on, doubling the
amount each time. How much does the teacher award to the
person who came top?

1.8.2 Timetables

Exercise 49 Ⓒ

1 For how many minutes do each of these programmes last
   a 'The Money Programme'
   b 'Fawlty Towers'
   c 'Face the Press'
   d '100 Great Sporting Moments'?

2 How much of a video tape would you use if you recorded 'The Jewel in the Crown' and 'The Writing on the Wall'?

3 At what time does 'Comrades' start on the 24-hour clock?

4 There were four films on the two channels. What was the title of the shortest film?

5 A video tape is 3 hours long. How much of the tape is left after taping the two films in the afternoon on Channel 4?

6 How much time is devoted to sport on BBC 2? (Include 'Under Sail'.)

7 For how many hours and minutes does Channel 4 broadcast programmes?

8 What is the starting time on the 24-hour clock of the programme in which 'Basil' appears?

9 How many programmes were repeats?

10 For how long are 'Pages from Ceefax' broadcast?

11 What is the starting time on the 24-hour clock of the programme in which the 'Redskins' appear?

12 How much of a two-hour video tape is left after taping 'Windmill' and 'The Natural World'?

*13 For how many hours and minutes does BBC 2 broadcast programmes?

BBC 2

9.0	**PAGES FROM CEEFAX.**
10.20	**OPEN UNIVERSITY.**
11.25	**PAGES FROM CEEFAX.**
11.50	**CHAMPION THE WONDER HORSE*:** Lost River (rpt). A drought brings danger.
12.15	**WINDMILL:** Archive film on animals.
1.10	**STATES OF MIND:** Jonathan Miller talks to Professor Richard Gregory (rpt).
2.0	**RUGBY SPECIAL:** Highlights of a County Championship match and a Welsh Cup match.
2.30	**TENNIS:** Cup Final.
4.15	**UNDER SAIL:** New series.
4.35	**RACHMANINOV MASTERCLASS.**
5.20	**THINKING ALOUD:** James Bond joins a discussion on espionage.
6.0	**NEWS REVIEW,** with Moira Stewart.
6.30	**THE MONEY PROGRAMME:** Guns for Sale. A look at Britain's defence industry.
7.15	**THE NATURAL WORLD:** City of Coral. A voyage beneath the Caribbean.
8.5	**COMRADES:** Educating Rita. The first of 12 films about life in Russia profiles a young trainee teacher.
8.50	**100 GREAT SPORTING MOMENTS:** Kelly Holmes: Golds in the Athens Olympics.
9.10	**FAWLTY TOWERS:** Basil and Sybil fall out over alterations to the hotel (rpt).
9.40	**FILM:** A Dangerous Summer (see Film Guide).
11.5	**TENNIS:** Cup Final.
11.55	**MUSIC AT NIGHT. 12.10 CLOSE.**

CHANNEL 4

1.5	**IRISH ANGLE — HANDS:** Basket Maker.
1.30	**FACE THE PRESS:** Graham Kelly, Secretary of the Football League, questioned by Ian Wooldridge of the Daily Mail and Brian Glanville of the Sunday Times.
2.0	**POB'S PROGRAMME,** with Patricia Hodge.
2.30	**FILM*:** Journey Together (see Film Guide).
4.15	**FILM*:** The London Blackout Murders, with John Abbot (see Film Guide).
5.15	**NEWS; WEATHER,** followed by **THE BUSINESS PROGRAMME.**
6.0	**AMERICAN FOOTBALL:** Dallas Cowboys and Washington Redskins.
7.15	**THE HEART OF THE DRAGON:** Understanding (rpt).
8.15	**THE JEWEL IN THE CROWN (T):** The Towers of Silence (rpt).
9.15	**THE WRITING ON THE WALL:** Who Governs? The political events of 1974 recalled by Robert Kee.
10.25	**FILM*:** Seven Days to Noon (see Film Guide). **12.10 CLOSE.**

Test yourself

1 a Write 28 as the product of its prime factors.
 b Find the least common multiple (L.C.M.) of 28 and 42.

(AQA, 2004)

2 Look at this list of numbers.
 3, 4, 6, 8, 10, 16, 27, 35
 a A number and its square root both appear in the list.
 What is the number and what is its square root?
 b A number and its cube root both appear in the list.
 What is the number and what is its cube root?

(CCEA)

3 Work out these.
 a 426 + 37 + 384
 b 800 − 472
 c 132 × 8
 d 0·2 × 0·4

(AQA, 2004)

4 Write these decimals in order of size, smallest first.
 0·7   0·75   0·705   0·075

(MEI, Spec)

5 The table shows the average minimum temperatures, in °C, in January in some cities.

Archangel	Athens	Darwin	Moscow	Ulan Bator
−20°	9°	27°	−15°	−32°

 a What is the lowest temperature listed?
 b What was the difference between the temperatures in Darwin and in Moscow?
 c Hong Kong is 35 °C warmer than Archangel.
 What is the temperature in Hong Kong?

(OCR, 2003)

6 N is a negative number.
 Is N^2 positive or negative? Explain your answer.

(OCR, 2004)

7 Find the highest common factor of 54 and 72.

(Edexcel, 2003)

8 Ken had **one thousand and twenty pounds**.
Lisa had **eight pounds and six pence**.
Write down, in figures, how much money Ken and Lisa
each had.

(Edexcel, 2004)

9 Here is a list of numbers.
   10   11   12   13   14   15   16   17   18   19

   a From this list, write
   i an even number
   ii two numbers that add to give 34
   iii two numbers that subtract to give 7
   iv a square number.
   b i From the list, write a multiple of 5.
   ii Explain how you know that this is a multiple of 5.
   c Which number in the list is a factor of 32?

(OCR, 2003)

10 a   56   180   15   0   30   125   14
   From the list, write
   i a factor of 28
   ii the square root of 3136
   iii a common factor of 60 and 45
   iv the cube of 5.

   b Steve has a box containing 100 tiles.
   Each tile is a one centimetre square.
   He takes out some tiles.
   He arranges these tiles as a square.
   Jo arranges the rest of the tiles to make a **bigger** square.
   How long is each side of Jo's square?

(OCR, 2004)

11 Magazines are stored in piles of 100.
Each magazine is 0·4 cm thick.
Calculate the height of one pile of magazines.

(AQA, 2003)

12 a Copy the square. Fill in the empty boxes so that
   each row, each column and each diagonal adds up to 0.
   b Multiply together the three numbers in the top row
   and write down your answer.

(CIE)

−1	4	−3
	0	

13   Here is part of a railway timetable.

Manchester	07:53	09:17	10:35	11:17	13:30	14:36	16:26
Stockport	08:01	09:26	10:43	11:25	13:38	14:46	16:39
Macclesfield	08:23	09:38	10:58	11:38	13:52	14:58	17:03
Congleton	08:31	–	–	11:49	–	15:07	17:10
Kidsgrove	08:37	–	–	–	–	–	17:16
Stoke-on-Trent	08:49	10:00	11:23	12:03	14:12	15:19	17:33

A train leaves Manchester at 10:35.
a   At what time should this train arrive in Stoke-on-Trent?
Doris has to go to a meeting in Stoke-on-Trent.
She will catch the train in Stockport.
She needs to arrive in Stoke-on-Trent before 2 p.m. for her meeting.
b   Write the time of the latest train she can catch in Stockport.
c   Work out how many minutes it should take the 14:36 train from Manchester to get to Stoke-on-Trent.
The 14:36 train from Manchester to Stoke-on-Trent takes less time than the 16:26 train from Manchester to Stoke-on-Trent.
d   How many minutes less?

(Edexcel, 2004)

14   Every day, a quarter of a million babies are born in the world.
a   Write a quarter of a million using figures.
b   Work out the number of babies born in 28 days.
Give your answer in millions.

(Edexcel, 2003)

15   Using the information that

$97 \times 123 = 11\,931$

write the value of
a 9.7×12.3
b $0.97 \times 123\,000$
c $11.931 \div 9.7$

(Edexcel, 2003)

In the remaining questions you may use a calculator.

16 Jam is sold in two sizes.
A large pot of jam costs 88p and weighs 822 g.
A small pot of jam costs 47p and weighs 454 g.
Which pot of jam is better value for money?
You **must** show all your working.

(AQA)

17 a Calculate
  i $12 \cdot 1^2$
  ii $\sqrt{46}$ (Give your answer to 1 decimal place.)
  iii $\dfrac{5}{0 \cdot 5 \times 0 \cdot 4}$

  b Complete the sentence below.
One million is the same as ten to the power of ___.

(OCR, 2003)

18 Use your calculator to work out
$$(2 \cdot 3 + 1 \cdot 8)^2 \times 1 \cdot 07$$

Write all the figures on your calculator display.

(Edexcel, 2003)

19 a A shop sells DVDs for £13·50 each.
How many DVDs can you buy for £100 and how much change should you get?
  b Stickers cost 12p for a packet.
Aaron buys some packets of stickers and pays with a £1 coin.
He gets 28p change.
How many packets of stickers did he buy?

(OCR, 2005)

20 Put brackets in this expression so that its value is 45·024
$$1 \cdot 6 + 3 \cdot 8 \times 2 \cdot 4 \times 4 \cdot 2$$

(Edexcel, 2003)

2   Algebra 1

In this unit you will:
- use letters for numbers in expressions
- simplify expressions
- use brackets in algebra calculations
- learn the main operations using algebra
- learn how to solve equations
- substitute a value into a formula or expression
- revise sequences and identify and use sequence rules
- find the general term of a sequence.

2.1   Basic algebra

2.1.1   Using letters for numbers in expressions

- Suppose there is an unknown number of people on a bus.
  Call this number x.
  If two more people get on the bus there will be $x + 2$ people
  on the bus.

> - $x + 2$ is an **expression**. An expression has no equals sign
>   whereas an equation does have an equals sign.

- Suppose a piece of cheese weighs w kilograms.
  If you cut off 2 kilograms you have $w - 2$ kilograms left.

- If I start with a number n and then double it, I will have $2n$
  ($2n$ means $2 \times n$).
  If I then add 8, I will have $2n + 8$.

> - When you multiply, write the number before the letter. So write
> $2n$, not $n2$.

Exercise 1 C

Find the expression I am left with.

1 I start with n and then double it. $n \longrightarrow \boxed{\times 2} \longrightarrow ?$

2 I start with y and then add 4.

3 I start with x and then take away 7.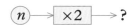

> You can draw
> a function machine
> to help you.

4 I start with x and then add 100.

5 I start with y and then multiply by 5.

6 I start with s and multiply by 100.

7 I start with t and then treble it.

8 I start with z and then add 11.

9 I start with p and then take away 9.

10 I start with n and then add x.

11 I start with n and then multiply by 4.

12 I start with x, double it and then add 3. $x \longrightarrow \boxed{\times 2} \longrightarrow \boxed{+3} \longrightarrow ?$

13 I start with n, double it and then take away 12.

14 I start with m, treble it and then add 2.

15 I start with y and multiply by 20.

16 I start with x, treble it and then add 3.

17 I start with y, double it and then take away 7.

18 I start with k, treble it and then add 10.

EXAMPLE

Find the expression I am left with if
a I start with n, add 2 and then multiply the result by 4.
b I start with n, subtract 3 and then divide the result by 7.
c I start with n, add 6 and then square the result.
....

a $n \rightarrow n + 2 \rightarrow 4(n + 2)$
b $n \rightarrow n - 3 \rightarrow \dfrac{n - 3}{7}$
c $n \rightarrow n + 6 \rightarrow (n + 6)^2$

Exercise 2 C

Find the expression I am left with.

1 I start with x, add 4 and then multiply the result by 3.

2 I start with x, add 3 and then multiply the result by 5.

3 I start with y, add 11 and then multiply the result by 6.

4 I start with x, add 3 and then divide the result by 4.

Use brackets.

Write this as $\dfrac{x + 3}{4}$.

5 I start with x, subtract 7 and then divide the result by 3.

6 I start with y, subtract 8 and then divide the result by 5.

7 I start with $4a$, add 3, multiply the result by 2 and then divide the final result by 4.

8 I start with m, subtract 6, multiply the result by 3 and the divide the final result by 4.

9 I start with x, square it and then subtract 6.

10 I start with x, square it, add 3 and then divide the result by 4.

11 I start with n, add 2 and then square the result.

> Use brackets.

12 I start with w, subtract x and then square the result.

13 I start with x, square it, subtract 7 and then divide the result by 3.

14 I start with x, subtract 9, square the result and then add 10.

15 I start with y, add 7, square the result and then divide by x.

16 I start with a, subtract x, cube the result and then divide by y.

17 A piece of wood is l cm long. If I cut off a piece 3 cm long, how much wood remains?

18 A piece of string is 15 cm long. How much remains after I cut off a piece of length x cm?

19 A delivery van weighs l kg. At a depot it picks up goods weighing 200 kg and later delivers goods weighing m kg. How much does it weigh after making the delivery?

20 A box usually contains n chocolates. The shopkeeper puts an extra 2 chocolates into each box. A girl buys 4 boxes. How many chocolates does she have?

21 A brick weighs w kg. How much do six bricks weigh?

22 A sack weighs l kg. How much do x sacks weigh?

23 A man shares a sum of n pence equally between six children. How much does each child receive?

24 A sum of £p is shared equally between you and four others. How much does each person receive?

25 A cake weighing 12 kg is cut into n equal pieces. How much does each piece weigh?

2.1.2 Simplifying expressions

You can **simplify** the expression $3n + 2n$ to $5n$. This is because $3n + 2n$ means '3 ns plus 2 ns', which is equivalent to 5 ns.

You can write the expression $8x + x$ as $8x + 1x$ and then simplify it to $9x$.

> ● You can only simplify like terms. These are parts of an expression that have the same letter.

The expression $3x + 8x$ has two **like** terms. You can simplify it to $11x$.
The expression $5x + 2y$ has two **unlike** terms and you cannot simplify it.

Here are some more examples

EXAMPLE

a $8n + 2n - n = 9n$ **b** $2x + 3y + 4x = 6x + 3y$
c $4x + 3 + 2x - 1 = 6x + 2$ **d** $n + m - n + 3m = 4m$

Exercise 3 C

Collect like terms together.

1 $2a + 3a$ **2** $6a + 5a$
3 $2a + 3a + 4a$ **4** $5n + n$
5 $7n - 2n$ **6** $10n - n$
7 $3x + 2x - x$ **8** $4x + 10x + 2x$
9 $3x - x + x$ **10** $4a + b + 2a + 3b$
11 $6a + 4b + 3a + 2b$ **12** $3x + y + 2x + 4y$
13 $2x + 5y + 7x + 2y$ **14** $4m + 2n + m + n$
15 $6m + 3n + 10m - n$ **16** $3x + 4 + 2x + 7$
17 $11x + 12 + 2x - 4$ **18** $7x + 8 + x + 2$
19 $12x + 3x + 4y + x$ **20** $2x + 3 + 3x + 5$
21 $4x + 8 + 5x - 3$ **22** $5x - 3 + 2x + 7$
23 $6x + 1 + x + 3$ **24** $4x - 3 + 2x + 10 + x$
25 $5x + 8 + x + 4 + 2x$ **26** $7x - 9 + 2x + 3 + 3x$
27 $5x + 7 - 3x - 2$ **28** $4x - 6 - 2x + 1$
29 $10x + 5 - 9x - 10 + x$ **30** $4a + 6b + 3 + 9a - 3b - 4$
31 $8m - 3n + 1 + 6n + 2m + 7$ **32** $6p - 4 + 5q - 3p - 4 - 7q$
33 $12s - 3t + 2 - 10s - 4t + 12$ **34** $a - 2b - 7 + a + 2b + 8$
35 $3x + 2y + 5z - 2x - y + 2z$ **36** $6x - 5y + 3z - x + y + z$
37 $2k - 3m + n + 3k - m - n$ **38** $12a - 3 + 2b - 6 - 8a + 3b$
39 $3a + x + e - 2a - 5x - 6e$ **40** $m + 7n - 5 - 4n + 8 - m$

In questions **41** and **46** find the perimeter of each shape.
Give the answers in the simplest form.

> Perimeter = sum of all the sides.

41
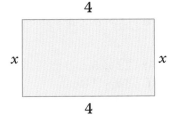

42
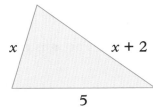

43
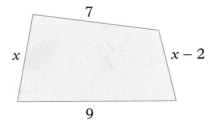

44
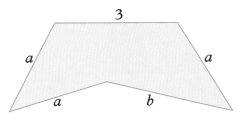

*45
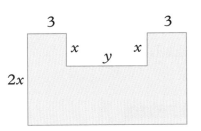

*46
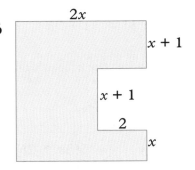

2.1.3 Multiplying with letters and numbers

$3n$ means   3 lots of n which is $3 \times n$.
$6a$ means   6 lots of a which is $6 \times a$.

When you multiply a letter by a number you write them next to each other.

You write $5 \times n$ as $5n$ You write $a \times b$ as ab
You write $4 \times m \times n$ as $4mn$ You write $3 \times a \times n$ as $3an$

- You can simplify expressions by multiplying them by each other.

$2a \times 5b = 10ab$ $(2 \times 5 = 10,\ a \times b = ab)$
$4m \times 3n = 12mn$ $(4 \times 3 = 12, m \times n = mn)$
$2n \times 3n = 6n^2$ $(2 \times 3 = 6,\ n \times n = n^2)$

Exercise 4 C

1 Write in a simpler form. Part **a** is done for you.
    a $a \times b = ab$ **b** $n \times m$ **c** $x \times y$ **d** $h \times t$
    e $a \times d \times n$ **f** $3 \times a \times b$ **g** $4 \times n \times m$ **h** $3 \times a \times b \times c$

2 Write in a simpler form. Part **a** is done for you.
    a $2a \times 3b = 6ab$ **b** $3a \times 5b$ **c** $2c \times 6d$
    d $5n \times 2m$ **e** $2p \times 7q$ **f** $3a \times 10n$
    g $10s \times 2t$ **h** $12a \times 3b$ **i** $4u \times 8v$

3 Write in a simpler form. Part **a** is done for you.
    a $3y \times y = 3 \times y \times y = 3y^2$ **b** $y \times y$ **c** $2x \times 3x$
    d $4t \times 6t$ **e** $6a \times a$ **f** $5y \times 2y$
    g $x \times 2x$ **h** $y \times 3y$ **i** $10x \times 10x$

2.1.4 Using brackets

The expression $3(a + b)$ means $3 \times a + 3 \times b$.
        So $3(a + b) = 3a + 3b$.

●   This is called expanding the brackets or removing the brackets.

Here are some more examples.

> EXAMPLE
>
> $2(x + 2y) = 2x + 4y$ $5(3a + 2b) = 15a + 10b$
> $5(x + 2) = 5x + 10$ $4(2x - 1) = 8x - 4$
> $n(n + 2) = n^2 + 2n$ $2n(3n + 1) = 6n^2 + 2n$

Exercise 5 C

Remove the brackets.
   1 $2(x + 3)$ **2** $3(x + 5)$ **3** $4(x + 6)$
   4 $2(2x + 1)$ **5** $5(2x + 3)$ **6** $4(3x - 1)$
   7 $6(2x - 2)$ **8** $3(5x - 2)$ **9** $5(3x - 4)$
 10 $7(2x - 3)$ **11** $2(2x + 3)$ **12** $3(2x + 1)$
 13 $5(x + 4)$ **14** $6(2x + 2)$ **15** $2(4x - 1)$

 16 $2(a + 3b)$ **17** $3(2a + 5b)$ **18** $5(2m + 3n)$
 19 $7(2a - 3b)$ **20** $11(a + 2b)$ **21** $8(3a + 2b)$
 22 $x(x + 5)$ **23** $x(x - 2)$ **24** $x(x - 3)$
 25 $x(2x + 1)$ **26** $x(3x - 2)$ **27** $x(3x + 5)$
 28 $2x(x - 1)$ **29** $2x(x + 2)$ **30** $3x(2x + 3)$

Remove the brackets and simplify.

31 $3(x + 2) + 4(x + 1)$
32 $5(x - 2) + 3(x + 4)$
33 $2(a - 3) + 3(a + 1)$
34 $5(a + 1) + 6(a + 2)$
35 $7(a - 2) + (a + 4)$
36 $3(t - 2) + 5(2 + t)$
37 $3(x + 2) + 2(x + 1)$
38 $4(x + 3) + 3(x + 2)$
39 $5(x - 2) + 3(x - 2)$
40 $4(a - 2) + 2(2a + 1)$
41 $x(2x + 1) + 3(x + 2)$
42 $x(2x - 3) + 5(x + 1)$
43 $a(3a + 2) + 2(2a - 2)$
44 $y(5y + 1) + 3(y - 1)$
45 $x(2x + 1) + x(3x + 1)$
46 $a(2a + 3) + a(a + 1)$

2.1.5 Subtracting terms in brackets

$$-(a + b) \text{ means } -1 \times (a + b) = -a - b$$
$$-(2a - b) \text{ means } -1 \times (2a - b) = -2a + b$$
$$3a + 2b - (2a + b) = 3a + 2b - 2a - b = a + b$$

Exercise 6 Ⓒ

1 Remove the brackets. Part **a** is done for you.

  a $-(m + n) = -m - n$
  b $-(a + b)$
  c $-(2a + b)$
  d $-(m - n)$
  e $-(a - b)$
  f $-(3a - b)$
  g $-(a + b - c)$
  h $-(2a + b - 2c)$
  i $-(3x - y - 2)$

Remove the brackets and simplify.

2 $2a + 5b - (a + b)$
3 $5a + 2b - (a + b)$
4 $6a + 8b - (2a + b)$
5 $3a + 7b - (2a + 3b)$
6 $2(a + 2b) - (a + b)$
7 $3(2a + b) - (a + 2b)$
8 $3(m + n) - (2m + n)$
9 $5(m + 2n) - (m - n)$
10 $7(2m + n) - 3(m + n)$
11 $6(m + 3n) - 10n$
12 $3x - 2(x + y)$
13 $10x - 3(2x + y)$
14 $5(a + 3b) + 4(a + 5b)$
15 $2(3x + 4y) - 3(x - y)$

16 Find an expression for the perimeter of each rectangle.

a

3

2a + 1

b

4

5x − 2

c
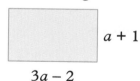
a + 1

3a − 2

17 Find an expression for
  a the perimeter of this picture
  b the area of this picture.

3x + y

2

*18   The three rods, A, B and C, in the diagram have lengths
of x, $(x + 1)$ and $(x - 2)$ cm.

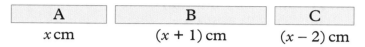

In the diagrams below write the length l in terms of x.
Give your answers in their simplest form.

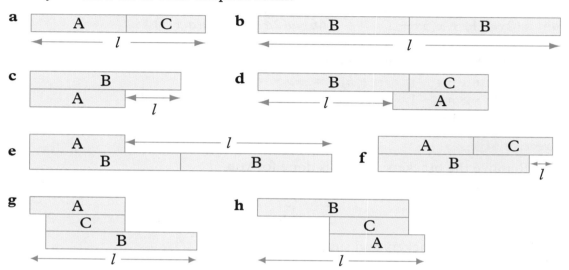

2.1.6   Algebra overview

- You can multiply letters and numbers.
$$4 \times n = 4n$$
$$a \times b = ab$$
$$n \times (a + b) = n (a + b)$$
$$= na + nb$$

- You can add like terms.

$$2a + 3b + 6a + b = 8a + 4b$$
$$a^2 + 2a^2 = 3a^2$$
$$3x + 8 + 2x - 2 = 5x + 6$$

- You can cancel fractions.
$$\frac{\cancel{5} \times 2}{\cancel{5}} = 2 \qquad \frac{6 \times \cancel{n}}{\cancel{n}} = 6$$
$$\frac{\cancel{a} \times a \times a}{\cancel{a}} = a^2 \qquad \frac{8n}{2} = 4n$$
$$\frac{n + n + n}{n} = \frac{3\cancel{n}}{\cancel{n}} = 3$$

Exercise 7 C

In questions **1** to **18** answer 'always true' or 'sometimes false'.

1 $6 \times a = 6a$ **2** $3 \times n = 3 + n$ **3** $n \times n = n^2$

4 $a + b = b + a$ **5** $n \times n \times n = 3n$ **6** $a \times 7 = 7a$

7 $n - m = m - n$ **8** $a + a = 2a$ **9** $n + n + n = n^3$

10 $m + m^2 = m^3$ **11** $3c - c = 3$ **12** $a + 2b = 2b + a$

13 $a(m + n) = am + an$ **14** $n \div 2 = 2 \div n$ **15** $n \times n \times m = n^2 m$

16 $2n^2 = 4n^2$ **17** $(2n)^2 = 4n^2$ **18** $2n + 2n = 4n^2$

19 Here are some algebra cards.

$\boxed{n + n}$ $\boxed{n \times n^2}$ $\boxed{3n \div 3}$ $\boxed{n \div 4}$

$\boxed{n \times n \times n}$ $\boxed{n^2 \div n}$ $\boxed{4 \div n}$ $\boxed{5n - n}$ $\boxed{4n - 2n}$

a Which cards will always be the same as $\boxed{2n}$?

b Which cards will always be the same as $\boxed{n^3}$?

c Which cards will always be the same as $\boxed{n}$?

d Which card will always be the same as $\boxed{\dfrac{4}{n}}$?

e Draw a new card which will always be the same as $\boxed{n^2 + n^2}$

20 In the expression $3n + 7$, two operations are done in the following order:

$n \longrightarrow \boxed{\times 3} \longrightarrow \boxed{+ 7} \longrightarrow 3n + 7$

Draw similar diagrams to show the correct order of operations for these expressions.

a $6n - 1$ **b** $8n + 10$ **c** $\dfrac{n}{2} + 3$

d $3(2n + 5)$ **e** $5(2n - 4)$ **f** $\dfrac{(n + 4)}{7}$

In questions **21** to **35** simplify the expressions.

21 $\dfrac{3n}{n}$ **22** $\dfrac{a}{a}$ **23** $\dfrac{n^2}{n}$

24 $6n - 5n$ **25** $a + b + c + a$ **26** $3n^2 - n^2$

27 $mn + mn$ **28** $\dfrac{n \times n \times n}{n}$ **29** $\dfrac{n + n + n}{n}$

30 $a \times a^2$ **31** $6n \div 6$ **32** $3t + 4 - 3p - 1$

33 $\dfrac{2a}{2a}$ **34** $2n + 2(n + 1)$ **35** $n + 4 + 4 + n$

2.2 Solving equations

2.2.1 Using letters for numbers in equations

Many problems in mathematics are easier to solve if you use letters instead of numbers. This is called using **algebra**.

- Here is an equation. $n + 10 = 75$
  In this equation the letter n represents a definite number so that 'n plus 10 equals 75'.
  So, in this equation the value of n is 65.

- Here is another equation. $2x + 1 = 7$
  In this equation the letter x represents a definite number so that 'two times x plus one equals 7'.
  So, in this equation the value of x must be 3.

Equations are like weighing scales that are balanced. The scales remain balanced if the same weight is added or taken away from both sides.

EXAMPLE

On the left is an unknown weight, x, plus a 2 kg weight.
On the right there is a 2 kg weight and a 3 kg weight.

If the two 2 kg weights are taken away the scales are still balanced. So the weight x is 3 kg.
You can write this as the equation $x + 2 = 2 + 3$
$$x = 3$$

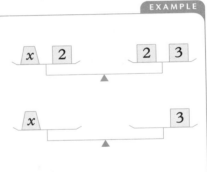

Exercise 8 C

The scales are balanced. Work out the weight of the object, x, in each case. Each small weight □ is 1 kg.

1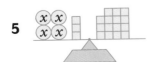

2

3

4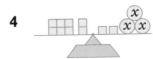

5

6

7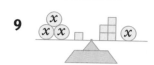

8

9 ![scale problem 9]

10 ![scale problem 10]

2.2.2 Rules for solving equations

You solve equation in the same way as you solve the weighing scale problems.

> The main rule when solving equations is
> - 'Do the same thing to both sides'

You can **add** the same thing to both sides.
You can **subtract** the same thing from both sides.
You can **multiply** both sides by the same thing.
You can **divide** both sides by the same thing.

EXAMPLE

Solve the equations.

a $x + 7 = 12$
   (-7) (-7) Subtract 7 from both sides.
   $x = 5$

b $x - 3 = 11$
   $(+3)$ $(+3)$ Add 3 to both sides.
   $x = 14$

c $2x = 12$
   $(\div 2)$ $(\div 2)$ Divide both sides by 2.
   $x = 6$

d $3x + 1 = 19$
   (-1) (-1) Subtract 1 from both sides.
   $3x = 18$
   $(\div 3)$ $(\div 3)$ Divide both sides by 3.
   $x = 6$

Exercise 9 C

Solve these equations.

1 $x + 7 = 10$
2 $x + 3 = 20$
3 $x - 7 = 7$
4 $x - 5 = 11$
5 $6 + x = 13$
6 $8 + x = 15$
7 $7 = x + 4$
8 $7 = x - 6$
9 $1 = x - 3$
10 $7 + x = 7$
11 $x - 11 = 20$
12 $14 = 6 + x$
13 $2x = 20$
14 $3x = 24$
15 $5x = 40$
16 $3x + 1 = 10$
17 $4x + 2 = 22$
18 $5x + 3 = 18$
19 $2n - 3 = 1$
20 $3n - 4 = 8$
21 $6n - 1 = 5$
22 $4a + 3 = 3$
23 $6a + 7 = 19$
24 $4 + 2a = 6$

> In question 7,
> $7 = x + 4$
> is the same as
> $x + 4 = 7$.

Exercise 10 Ⓒ

Solve these equations.
Some of the answers are fractions.

1 $5x - 3 = 1$		**2** $3x - 7 = 0$		**3** $2x + 5 = 20$	
4 $6x - 9 = 2$		**5** $7x + 6 = 6$		**6** $9x - 4 = 1$	
7 $11x - 10 = 1$		**8** $15y + 2 = 5$		**9** $7y + 8 = 10$	
10 $4y - 11 = -8$		**11** $3z - 8 = -6$		**12** $4p + 25 = 30$	
13 $5t - 6 = 0$		**14** $9m - 13 = 1$		**15** $4 + 3x = 5$	

16 $7 + 2x = 8$		**17** $5 + 20x = 7$		**18** $3 + 8x = 0$	
19 $50y - 7 = 2$		**20** $200y - 51 = 49$		**21** $5u - 13 = -10$	
22 $9x - 7 = -11$		**23** $11t + 1 = 1$		**24** $3 + 8y = 40$	
25 $12 + 7x = 2$		**26** $6 = 3x - 1$		**27** $8 = 4x + 5$	
28 $9 = 2x + 7$		**29** $11 = 5x - 7$		**30** $0 = 3x - 1$	
31 $40 = 11 + 14x$		**32** $-4 = 5x + 1$		**33** $-8 = 6x - 3$	
34 $13 = 4x - 20$		**35** $-103 = 2x + 7$			

2.2.3 Equations with *x* on both sides

EXAMPLE

Solve the equations

a $8x - 3 = 3x + 1$

b $3x + 9 = 18 - 7x$

....

a $8x - 3 = 3x + 1$ add 3 and subtract 3x
$$8x - 3x = 1 + 3$$
$$5x = 4$$
$$x = \frac{4}{5} \quad \text{divide by 5}$$

b $3x + 9 = 18 - 7x$
$$3x + 7x = 18 - 9 \quad \text{add 7x and subtract 9}$$
$$10x = 9$$
$$x = \frac{9}{10} \quad \text{divide by 10}$$

Exercise 11 Ⓔ

Solve these equations.

1 $7x - 3 = 3x + 8$		**2** $5x + 4 = 2x + 9$		**3** $6x - 2 = x + 8$	
4 $8x + 1 = 3x + 2$		**5** $7x - 10 = 3x - 8$		**6** $5x - 12 = 2x - 6$	
7 $4x - 23 = x - 7$		**8** $8x - 8 = 3x - 2$		**9** $11x + 7 = 6x + 7$	
10 $9x + 8 = 10$		**11** $5 + 3x = x + 8$		**12** $4 + 7x = x + 5$	
13 $6x - 8 = 4 - 3x$		**14** $5x + 1 = 7 - 2x$		**15** $6x - 3 = 1 - x$	

16 $3x - 10 = 2x - 3$		**17** $5x + 1 = 6 - 3x$		**18** $11x - 20 = 10x - 15$	
19 $6 + 2x = 8 - 3x$		**20** $7 + x = 9 - 5x$		**21** $3y - 7 = y + 1$	
22 $8y + 9 = 7y + 8$		**23** $7y - 5 = 2y$		**24** $3z - 1 = 5 - 4z$	
25 $8 = 13 - 4x$		**26** $10 = 12 - 2x$		**27** $13 = 20 - 9x$	
28 $8 = 5 - 2x$		**29** $5 + x = 7 - 8x$		**30** $3x + 11 = 2 - 3x$	

> **EXAMPLE**
>
> Solve the equations
> **a** $3(x - 1) = 2(x + 7)$ **b** $5(2x + 1) = 3(x-2) + 20$
>
> ....
>
> **a** $3(x - 1) = 2(x + 7)$ **b** $5(2x + 1) = 3(x-2) + 20$
> $3x - 3 = 2x + 14$ $10x + 5 = 3x - 6 + 20$
> $3x - 2x = 14 + 3$ $10x - 3x = -6 + 20 - 5$
> $x = 17$ $7x = 9$
> $x = 1\frac{2}{7}$

Exercise 12Ⓔ

Solve these equations.

1 $2(x + 1) = x + 5$
3 $5(x - 3) = 3(x + 2)$
5 $5(x - 3) = 2(x - 7)$
7 $10(x - 3) = x$

2 $4(x - 2) = 2(x + 1)$
4 $3(x + 2) = 2(x - 1)$
6 $6(x + 2) = 2(x - 3)$
8 $3(2x - 1) = 4(x + 1)$

9 $4(2x + 1) = 5(x + 3)$
11 $5(x + 1) + 3 = 3(x - 1)$
13 $5(2x + 1) - 5 = 3(x + 1)$
15 $2(x - 10) = 4 - 3x$

10 $3(x - 1) + 7 = 2(x + 1)$
12 $7(x - 2) - 3 = 2(x + 2)$
14 $3(4x - 1) - 3 = x + 1$
16 $3x + 2(x + 1) = 3x + 12$

17 $4x - 2(x + 4) = x + 1$
19 $5x - 2(x - 2) = 6 - 2x$
21 $4(x + 3) + 2(x - 1) = 4$
23 $5(x - 3) + 3(x + 2) = 7x$
25 $4(3x - 1) - 3(3x + 2) = 0$

18 $2x - 3(x + 2) = 2x + 1$
20 $3(x + 1) + 2(x + 2) = 10$
22 $3(x - 2) - 2(x + 1) = 5$
24 $3(2x + 1) - 2(2x + 1) = 10$

2.2.4 Equations with fractions

> **EXAMPLE**
>
> Solve the equations
> **a** $\frac{7}{x} = 8$ **b** $\frac{3x}{4} = 2$
>
> ....
>
> **a** $\frac{7}{x} = 8$ **b** $\frac{3x}{4} = 2$
> $7 = 8x$ multiply by x $3x = 8$ multiply by 4
> $\frac{7}{8} = x$ $x = \frac{8}{3}$
> $x = 2\frac{2}{3}$

Exercise 13Ⓔ

Solve these equations.

1 $\dfrac{3}{x} = 5$　　　　**2** $\dfrac{4}{x} = 7$　　　　**3** $\dfrac{11}{x} = 12$　　　　**4** $\dfrac{6}{x} = 11$

5 $\dfrac{2}{x} = 3$　　　　**6** $\dfrac{5}{y} = 9$　　　　**7** $\dfrac{7}{y} = 9$　　　　**8** $\dfrac{4}{t} = 3$

9 $\dfrac{3}{a} = 6$　　　　**10** $\dfrac{8}{x} = 12$　　　　**11** $\dfrac{3}{p} = 1$　　　　**12** $\dfrac{15}{q} = 10$

13 $\dfrac{x}{4} = 6$　　　　**14** $\dfrac{x}{5} = 3$　　　　**15** $\dfrac{y}{5} = -2$　　　　**16** $\dfrac{a}{7} = 3$

17 $\dfrac{t}{3} = 7$　　　　**18** $\dfrac{m}{4} = \dfrac{2}{3}$　　　　**19** $\dfrac{x}{7} = \dfrac{5}{8}$　　　　**20** $\dfrac{2x}{3} = 1$

21 $\dfrac{4x}{5} = 3$　　　　**22** $\dfrac{x+1}{4} = 3$　　　　**23** $\dfrac{x-3}{2} = 5$　　　　**24** $\dfrac{4+x}{7} = 2$

25 $\dfrac{4}{x+1} = 3$　　　　**26** $3 = \dfrac{9}{x+2}$　　　　**27** $\dfrac{5}{x-2} = 3$　　　　**28** $\dfrac{7}{x} = \dfrac{1}{4}$

29 $\dfrac{3}{x} = \dfrac{2}{3}$　　　　**30** $\dfrac{2x+1}{3} = 1$　　　　**31** $\dfrac{x}{2} = 110$　　　　**32** $\dfrac{500}{y} = -1$

33 $-99 = \dfrac{98}{f}$　　　　**34** $\dfrac{x}{3} + 5 = 7$　　　　**35** $\dfrac{x}{5} - 2 = 4$　　　　**36** $\dfrac{2x}{3} + 4 = 5$

37 $\dfrac{x}{6} - 10 = 4$　　　　**38** $\dfrac{6}{x} + 1 = 2$　　　　**39** $\dfrac{5}{x} - 7 = 0$　　　　**40** $5 + \dfrac{3}{x} = 10$

2.2.5 Setting up equations

> **EXAMPLE**
>
> If you multiply a 'mystery' number by 2 and then add 3 the answer is 14. Find the 'mystery' number.
>
> ....
>
> Let the mystery number be x.
> Then $2x + 3 = 14$
> $\qquad\quad 2x = 11$
> $\qquad\quad\ \ x = 5\dfrac{1}{2}$
> The 'mystery' number is $5\dfrac{1}{2}$.

Exercise 14Ⓔ

Find the 'mystery' number in each question by forming an equation and then solving it.

1 If you multiply the number by 3 and then add 4, the answer is 13.

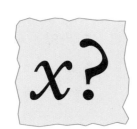

2 If you multiply the number by 4 and then add 5, the answer is 8.

3 If you multiply the number by 2 and then subtract 5, the answer is 4.

4 If you multiply the number by 10 and then add 19, the answer is 16.

5 If you add 3 to the number and then multiply the result by 4, the answer is 10.

6 If you subtract 11 from the number and then treble the result, the answer is 20.

7 If you double the number, add 4 and then multiply the result by 3, the answer is 13.

8 If you treble the number, take away 6 and then multiply the result by 2, the answer is 18.

9 If you double the number and subtract 7 you get the same answer as when you add 5 to the number.

10 If you multiply the number by 5 and subtract 4, you get the same answer as when you add 3 to the number and then double the result.

11 If you multiply the number by 6 and add 1, you get the same answer as when you add 5 to the number and then treble the result.

12 If you add 5 to the number and then multiply the result by 4, you get the same answer as when you add 1 to the number and then multiply the result by 2.

EXAMPLE

The length of a rectangle is twice the width. If the perimeter is 36 cm, find the width.

a Let the width of the rectangle be x cm.
Then the length of the rectangle is $2x$ cm.

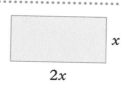

b Set up an equation.
$x + 2x + x + 2x = 36$

c Solve $6x = 36$
                $x = 6$

The width of the rectangle is 6 cm.

Exercise 15Ⓔ

Answer these questions by setting up an equation and then solving it.

1 Find x if the perimeter is 7 cm.

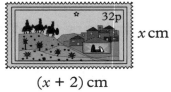

x cm

$(x + 2)$ cm

2 Find x if the perimeter is 10 cm.

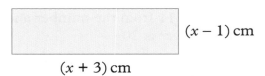

$(x - 1)$ cm

$(x + 3)$ cm

3 The length of a rectangle is 3 times its width. If the perimeter of the rectangle is 11 cm, find its width.

> Let the width be x cm.

4 The length of a rectangle is 4 cm more than its width. If its perimeter is 13 cm, find its width.

5 The width of a rectangle is 5 cm less than its length. If the perimeter of the rectangle is 18 cm, find its length.

6 Find x in these shapes.

a

Area = 18 cm² | x cm

5 cm

b

Area = 15 cm² | $(x + 3)$ cm

5 cm

c

Area = 35 cm²

x

10 cm

7 Find x in these triangles.

a
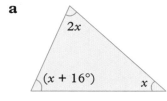
$2x$

$(x + 16°)$ x

b
$(2x - 1)$ x

$(3x - 5°)$

c
$(x + 14°)$

$(2x + 10°)$

76   x

8 The angles of a triangle are $32°$, x and $(4x + 3°)$. Find the value of x.

9 The sum of three consecutive whole numbers is 168. Let the first number be x. Form an equation and hence find the three numbers.

> Consecutive means next to each other. For example: 3, 4, 5

10 The sum of four consecutive whole numbers is 170.
Find the numbers.

11 In this triangle AB = x cm.
BC is 3 cm shorter than AB.
AC is twice as long as BC.
   a Write, in terms of x, the lengths of
   i   BC
   ii AC.
   The perimeter of the triangle is 41 cm.
   b Write an equation in x and solve
   it to find x.

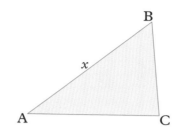

12 This is a rectangle. Work out x and hence find
the perimeter of the rectangle.

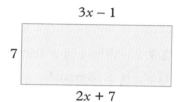

13 Here are four expressions involving an unknown
number n.

   A   B   C   D
   $2n + 1$ $n - 5$ $2n + 3$ $3n + 1$

   a Find the value of n if the expressions A and B
   are equal.
   b Find the value of n if the expressions C and D
   are equal.
   c Which two expressions could never be equal for **any**
   value of n?

14 Find the length of the sides of this equilateral
triangle.

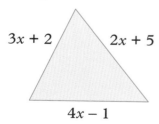

15 Petra has £12 and Suki has nothing. They both receive the
same money for doing a delivery job.
Now Petra has three times as much as Suki.
How much did they get for the job?

16   The area of rectangle A is twice the area of rectangle B. Find x.

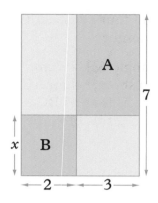

2.3   Using a formula

2.3.1   What is a formula?

Here is a **formula** for calculating a person's wage.

> Wage earned = Hours worked × Rate per hour

This formula will work for **any** value of 'hours worked' and for **any** value of 'rate per hour'.
For example if Anna works for 40 hours at a rate per hour of £8, then,

Anna's wage earned = 40 × £8
= £320

Exercise 16 Ⓒ

1   The formula for calculating a person's wage is

Wage earned = Hours worked × Rate per hour.

a   Calculate Amir's wage if he worked 30 hours and his rate per hour is £7.
b   Calculate Sam's wage if she worked 40 hours and her rate per hour is £9.

2   The formula for the circumference of a circle is

Circumference ≈ diameter × 3

Calculate the approximate circumference of a circle with diameter 20 cm.

≈ means 'is approximately equal to'.

3 The formula for the perimeter of an equilateral triangle is

Perimeter = 3 × side length

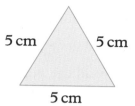

a Find the perimeter of an equilateral triangle whose side length is 5 cm.

b Find the perimeter of an equilateral triangle whose side length is 11 cm.

4 The formula for the area A of a triangle is

$$A = \frac{1}{2} \times \text{base} \times \text{height}$$

Find the area of each of these triangles.

a base = 6 cm, height = 10 cm

b base = 4 cm, height = 8 cm

c base = 10 cm, height = 7 cm

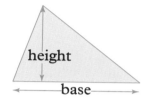

5 You can work out a baby's weight with the formula

weight = weight at birth + 500 grams per month after the birth.

Calculate the weights of these babies

a weight at birth = 2500 grams, age 3 months

b weight at birth = 3100 grams, age 6 months

c weight at birth = 2750 grams, age 10 months.

6 The cost per month of Dave's phone is

Cost = (ten pounds) + (number of texts × 5 pence).

Find the cost for a month when he sent

a 200 texts

b 1000 texts.

2.3.2 Using letters in a formula

You can use letters for the quantities in a formula.
Here is a square of side s.
The formula for the perimeter is $P = 4s$.

The formula for the area is $A = s \times s = s^2$

So for a square with side 9 cm, $P = 4 \times 9 = 36$ cm
and $A = 9 \times 9 = 81$ cm^2.

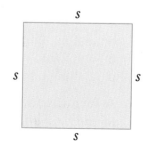

Exercise 17 Ⓒ

1 The formula for the perimeter of this rectangle is

$P = 2a + 2b$

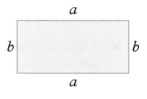

Find the perimeter of a rectangle when
a $a = 4\,cm$ and $b = 3\,cm$
b $a = 7\,cm$ and $b = 1\,cm$.

2 The formula for the area, A, of the rectangle in question **1** is

$$A = ab$$

Find the area of a rectangle where
a $a = 5\,cm$ and $b = 2\,cm$
b $a = 7\,cm$ and $b = 3\,cm$.

3 The formula for the height of a tree is $h = 5t + 3$.
Find the value of h when
a $t = 2$ **b** $t = 10$ **c** $t = 6$.

4 The formula for the volume, V, of a rectangular box with dimensions l, w and h (for length, width and height) is

$$V = lwh$$

Copy and complete the table.

	l	*w*	*h*	*V*
a	4	5	3	
b	7	2	10	
c	8	11	1	

5 Use the formulae and values given to find the value of c.
a $c = mx$; $m = 8$, $x = 7$
b $c = ny + a$; $n = 3$, $y = 4$, $a = 5$
c $c = 2ht$; $h = 6$, $t = 10$
d $c = a^2 + b^2$; $a = 5$, $b = 7$

> Formulae is the plural of formula.

6 A formula for estimating the volume of a cylinder of radius r and height h is

$$V = 3r^2h$$

a Find the value of V when $r = 10$ and $h = 2$.
b Find the value of V when $r = 5$ and $h = 4$.

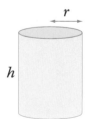

*7 A builder uses a rule to work out how long it will take him to tile a wall.

> I need 50 minutes to prepare and then 8 minutes for each tile.

Write this rule in algebra using T for the total time in minutes he needs to lay n tiles.

EXAMPLE

A formula connecting velocity with acceleration and time is $v = u + at$.
Find the value of v when $u = 3$
$$a = 4$$
$$t = 6.$$

....

$v = u + at$
$v = 3 + (4 \times 6)$
$v = 27$

> **Velocity** is speed in a certain direction.

Exercise 18Ⓔ

1 A formula involving force, mass and acceleration is $F = ma$.
Find the value of F when $m = 12$ and $a = 3$.

2 The height of a growing tree is given by the formula
$h = 2t + 15$. Find the value of h when $t = 7$.

3 The time required to cook a joint of meat is given by the
formula, $T = (\text{mass of joint}) \times 3 + \dfrac{1}{2}$. Find the value of T
when $(\text{mass of joint}) = 2\dfrac{1}{2}$.

4 An important formula in physics states that $I = mv - mu$.
Find the value of I when $m = 6$, $v = 8$, $u = 5$.

5 The distance travelled by an accelerating car is given by the
formula $s = \left(\dfrac{u + v}{2}\right)t$. Find the value of s when $u = 17$,
$v = 25$ and $t = 4$.

6 The formula for the total surface area, A, of a solid cuboid is

$A = 2bc + 2ab + 2ac$

Find the value of A when $a = 2$, $b = 3$, $c = 4$.

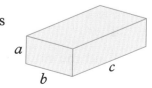

7 The formula for the height of a stone thrown upwards is

$h = ut - 5t^2$

Find the value of h when $u = 70$ and $t = 3$.

*8 The formula for the speed of an accelerating particle is

$v^2 = u^2 + 2as$

Find the value of v when $u = 11$, $a = 5$ and $s = 6$.

*9 The time period, T, of a simple pendulum is given

by the formula $T = 2\pi\sqrt{\left(\dfrac{l}{g}\right)}$, where l is the length

of the pendulum and g is the gravitational acceleration.
Find T when $l = 0.65$, $g = 9.81$ and $\pi = 3.142$.

> Use a calculator.

*10 The sum, S, of the squares of the integers from 1 to n is
given by $S = \dfrac{1}{6}n(n + 1)(2n + 1)$. Find S when $n = 12$.

2.3.3 Substituting into expressions

An algebraic **expression** has letters and numbers.
For example, $2n$, $5n + 1$ and $a^2 + b^2$ are all expressions.

> Note that there is
> no equals sign in an
> expression.

- You can find the value of an expression by replacing each letter
  with a number.

Exercise 19 C

In questions **1** to **10** find the value of each expression. The first
one is done for you.

1 $2x + 1$ if $x = 5$
     $2 \times 5 + 1$ $= 10 + 1 = 11$

2 $3x - 1$ if $x = 10$

3 $5x - 2$ if $x = 7$

4 $4x + 3$ if $x = 0$

5 $10 + a$ if $a = 35$

6 $7 - a$ if $a = 1$

7 $12 - b$ if $b = 3$

8 $16 + b$ if $b = 30$

9 $4 + 3c$ if $c = 100$

10 $20 - 2c$ if $c = 5$

11 Find the value of these expressions when $n = 3$.
   a $n^2 + 2$ **b** $2n^2$ **c** n^3

> Remember:
> $2n^2 = 2(n^2)$

12 Find the value of these expressions when $a = 0.5$.
   a $4a + 1$ **b** $4 - a$ **c** $3(a + 0.5)$

*13 Find the value of these expressions when $x = 2$.

    a $\dfrac{x + 2}{x}$ **b** $\dfrac{x + 4}{x - 1}$ **c** $3x^2$

*14 Find the value of these expressions when $n = 2$.

    a $n^2 + 5$ **b** $3n - 6$ **c** $(n - 1)^2$

    d $\dfrac{5n}{2}$ **e** $\dfrac{n + 10}{12}$ **f** $n^2 + 3n$

EXAMPLE

If $n = 5$ and $x = -2$, work out the value of these expressions.

a $2n + x$

    $= 10 + -2$

    $= 10 - 2$

    $= 8$

b nx

    $= 5 \times (-2)$

    $= -10$

c $3 - x$

    $= 3 - (-2)$

    $= 3 + 2$

    $= 5$

Exercise 20 C

In questions **1** to **10** find the value of the expressions when $m = 4$ and $t = -2$.

1 $t + m$ **2** $2(m + 3)$ **3** $5(m - 4)$ **4** t^2 **5** $2m + t$

6 $3t$ **7** $m^2 + 1$ **8** $2(t + 3)$ **9** $\dfrac{m + 12}{m}$ **10** $(t + 3)^2$

In questions **11** to **34** find the value of the expressions when $a = 5$

$b = 4$

$c = 1$

$d = -2$.

11 $5a - c$ **12** $2b + a$ **13** $a + d$ **14** $3c - b$

15 $4b + c$ **16** $2d - a$ **17** $5b + 10$ **18** $a + b + c$

19 $b - c$ **20** $7 - 2a$ **21** $25 + 5b$ **22** $3a - 4d$

23 $a^2 + b^2$ **24** $ac + b$ **25** $6 - 2c$ **26** $d^2 + 4$

27 $ab + c$ **28** $5d - 2c$ **29** $b^2 + cd$ **30** $5a + b + d$

31 $bd + c^2$ **32** $2(a - c)$ **33** $3(a + d)$ **34** $a(c + b)$

●   Reminders

    $a^2 = a \times a$ $a^3 = a \times a \times a$

    $2a^2 = 2(a^2)$ $(2a)^2 = 2a \times 2a$

  $a(b - c)$ Work out the term in brackets first.

    $\dfrac{a + b}{c}$ The division line works like a bracket, so work out $a + b$ first.

Exercise 21Ⓔ

Work out the values of these expressions when $m = 2$
$$t = -2$$
$$x = -3$$
$$y = 4.$$

1	m^2	**2**	t^2	**3**	x^2	**4**	y^2
5	m^3	**6**	t^3	**7**	x^3	**8**	y^3
9	$2m^2$	**10**	$(2m)^2$	**11**	$2t^2$	**12**	$(2t)^2$
13	$2x^2$	**14**	$(2x)^2$	**15**	$3y^2$	**16**	$4m^2$

17	$5t^2$	**18**	$6x^2$	**19**	$(3y)^2$	**20**	$3m^3$
21	$x^2 + 4$	**22**	$y^2 - 6$	**23**	$t^2 - 3$	**24**	$m^3 + 10$
25	$x^2 + t^2$	**26**	$2x^2 + 1$	**27**	$m^2 + xt$	**28**	my^2
29	$(mt)^2$	**30**	$(xy)^2$	**31**	$(xt)^2$	**32**	yx^2

33	$m - t$	**34**	$t - x$	**35**	$y - m$	**36**	$m - y^2$
37	$t + x$	**38**	$2m + 3x$	**39**	$3t - y$	**40**	$xt + y$
41	$3(m + t)$	**42**	$4(x + y)$	**43**	$5(m + 2y)$	**44**	$2(y - m)$
45	$m(t + x)$	**46**	$y(m + x)$	**47**	$x(y - m)$	**48**	$t(2m + y)$
49	$m^2(y - x)$	**50**	$t^2(x^2 + m)$				

Exercise 22Ⓔ

If $w = -2$, $x = 3$, $y = 0$, $z = 2$, work out

1 $\dfrac{w}{z} + x$ **2** $\dfrac{w + x}{z}$ **3** $y\left(\dfrac{x + z}{w}\right)$ **4** $x^2(z + wy)$

5 $x(x + wz)$ **6** $w^2\sqrt{(z^2 + y^2)}$ **7** $2(w^2 + x^2 + y^2)$ **8** $2x(w - z)$

9 $\dfrac{z}{w} + x$ **10** $\dfrac{z + w}{x}$ **11** $\dfrac{x + w}{z^2}$ **12** $\dfrac{y^2 - w^2}{xz}$

13 $z^2 + 4z + 5$ **14** $\dfrac{1}{w} + \dfrac{1}{z} + \dfrac{1}{x}$ **15** $\dfrac{4}{z} + \dfrac{10}{w}$ **16** $\dfrac{yz - xw}{xz - w}$

2.4 Sequences

2.4.1 What is a sequence?

A sequence is a list of numbers that has a pattern to it. Each number in a sequence is called a **term**. Here are three sequences.

1 3 5 7	19 16 13 10
+2 +2 +2	−3 −3 −3

1   2   4   8
× 2   × 2   × 2

You find the next term by adding 2.

You find the next term by subtracting 3.

You find the next term by multiplying by 2.

Exercise 23 C

1 The numbers in boxes form a sequence. Copy them and
find the next term.

> You could use a
> number line to
> help you.

a | 8 | | 10 | | 12 | | 14 | | |

b | 2 | | 5 | | 8 | | 11 | | |

c | 11 | | 9 | | 7 | | 5 | | |

d | 1 | | 5 | | 9 | | 13 | | |

In questions **2** to **17** write the sequence and find the next term.

2 1, 4, 7, 10, __   3 4, 8, 12, 16, __
4 2, 5, 8, 11, __   5 21, 17, 13, 9, __
6 5, 10, 15, 20, __   7 15, 13, 11, 9, __
8 2, 8, 14, 20, __   9 9, 18, 27, 36, __

10 31, 26, 21, 16, __   11 1, 2, 4, 8, 16, __
12 1, 3, 9, 27, __   13 2, 20, 200, 2000, __
14 80, 40, 20, 10, __   15 200, 100, 50, 25, __
16 88, 99, 110, __   17 39, 36, 33, 30, __

Write each sequence and find the missing number.

18 3   8   13   ☐   23   19 3   ☐   12   24   48
20 ☐   12   8   4   0   21 3   6   ☐   12   15
22 100   50   ☐ $12\frac{1}{2}$ 23 1   ☐   9   27   ☐
24 ☐   8   16   24   ☐   25 ☐   1   10   100   ☐

2.4.2 Sequence rules

● For the sequence 9, 13, 17, 21, 25 . . ., the first term is 9
and the term-to-term rule is 'add 4'.

● For the sequence 3, 6, 12, 24, 48, . . ., the term-to-term
rule is 'double' or 'multiply by 2'.

Exercise 24 C

1 Write the first five terms for each sequence.
a Start with 2 and add 3 each time.
b Start with 30 and subtract 5 each time.
c Start with 1 and multiply by 2 each time.
d Start with 1 and multiply by 10 each time.
e Start with 35 and subtract 7 each time.
f Start with 64 and divide by 2 each time.
g Start with −10 and add 2 each time.

2 The first term of a sequence is 11 and the term-to-term rule is 'add 5'. Write the first five terms of the sequence.

3 The table shows the first term and the rule of several sequences. Write the first five terms of each sequence.

	First term	Rule
a	96	add 2
b	100	subtract 11
c	10	multiply by 2
d	−6	add 3

4 Write the rule for each sequence.
   a 2, 5, 8, 11, 14, . . .
   b 30, 25, 20, 15, 10, . . .
   c 3, 6, 12, 24, . . .
   d 1, 8, 15, 22, 29, . . .
   e 21, 17, 13, 9, 5, . . .

5 Write each sequence and find the missing number.
   a 1, 6, 11, ☐, 21   **b** 80, 40, 20, ☐
   c 3, 30, 300, ☐, 30 000   **d** ☐, 5, 8, 11, 14
   e 2, 4, 8, ☐, 32   **f** 12, 8, 4, ☐, −4
   g 2, 8, ☐, 20, 26   **h** −1, 2, 5, ☐, 11

6 Write the rule for each sequence.
   a 1, 5, 9, 13, 17, . . .   **b** 65, 55, 45, 35, . . .
   c 1, 4, 16, 64, . . .   **d** 84, 42, 21, $10\frac{1}{2}$, . . .
   e 37, 32, 27, 22, . . .   **f** 2, 3, 4, 5, . . .
   g 200, 100, 50, 25, . . .   **h** 5500, 550, 55, 5·5, . . .

2.4.3 Harder sequences

Here are three sequences where the rule is more complicated.

- 5   8   12   17
   +3 +4 +5   . . . so the next term is $17 + 6 = 23$.

- 15 14 16 13 17
   −1 +2 −3 +4   . . . so the next term is $17 − 5 = 12$.

- 1 1 2 3 5 8 . . . the rule is **add the previous two numbers each time**. The next two terms are 13 and 21.

> This is called the Fibonacci sequence.

Exercise 25Ⓔ

1 Write each sequence and find the next term.

    a 1, 3, 6, 10, __   **b** 0, 4, 9, 15, 22, __
    c 5, 7, 10, 14, __   **d** 1, 2, 4, 7, 11, __
    e 50, 49, 47, 44, __   **f** 10, 11, 9, 12, 8, __
    g 8, 4, 2, 1, __   **h** 8, 10, 6, 12, 4, __

2 This is a Fibonacci sequence: 1, 1, 2, 3, 5, 8, . . .
You make each term by adding the previous two
numbers each time. Write the sequence and find the
next four terms.

3 Here is the start of a sequence: 1, 3, 4, . . .
You get each new term by adding the last two terms.
For example $4 = 1 + 3$.
The next term will be 7.
    a Write the next six terms.
    b Use the same rule to write the next four terms of
        a sequence which starts 2, 5, 7, . . .

4 a Write the next two lines of this sequence.
       $3 \times 4 = 3 + 3^2$
       $4 \times 5 = 4 + 4^2$
       $5 \times 6 = 5 + 5^2$
            $=$
            $=$

    b Complete these lines.
       $10 \times 11 =$
       $30 \times 31 =$

5 Copy the pattern and write the next three lines.
    $1 + 9 \times 0 \quad = \quad 1$
    $2 + 9 \times 1 \quad = \quad 11$
    $3 + 9 \times 12 \quad = \quad 111$
    $4 + 9 \times 123 \quad = 1111$
    $5 + 9 \times 1234 =$

6 For the sequence 2, 3, 8, . . . you get each new term by
squaring the last term and then subtracting 1.
Write the next two terms.

7 To get the sequence 3, 3, 5, 4, 4 you count the letters in
'one, two, three, four, five, . . .'.
Write the next three terms.

*8   Here is the sequence of the first six odd and even numbers.

	1st	2nd	3rd	4th	5th	6th
odd	1	3	5	7	9	11
even	2	4	6	8	10	12

Find   **a**   the 8th even number
       b   the 8th odd number
       c   the 13th even number
       d   the 13th odd number.

*9   **a**   Write the next three lines of this pattern.

$$1^3 = 1^2 \qquad\qquad = 1$$
$$1^3 + 2^3 = (1 + 2)^2 \qquad = 9$$
$$1^3 + 2^3 + 3^3 = (1 + 2 + 3)^2 = 36$$

   b   Work out as simply as possible

$$1^3 + 2^3 + 3^3 + 4^3 + 5^3 + 6^3 + 7^3 + 8^3 + 9^3 + 10^3.$$

*10   You can add the odd numbers 1, 3, 5, 7, 9, ... to give an interesting sequence.

$$1 \qquad\qquad = 1 = 1 \times 1 \times 1$$
$$3 + 5 \qquad\qquad = 8 = 2 \times 2 \times 2$$
$$7 + 9 + 11 \qquad = 27 = 3 \times 3 \times 3$$
$$13 + 15 + 17 + 19 = 64 = 4 \times 4 \times 4$$

1, 8, 27, 64 are **cube** numbers.
You write $2^3 = 8$ ('two cubed equals eight')
$4^3 = 64$
Or the other way round:

$\sqrt[3]{8} = 2$ ('the cube root of eight equals two')
$\sqrt[3]{27} = 3$

   a   Continue adding the odd numbers in the same way as before.
         Do you **always** get a cube number?
   b   Write the value of
         i $\sqrt[3]{125}$ **ii** $\sqrt[3]{1000}$ **iii** 11^3

2.4.4 Using algebra to find the nth term of a sequence

● For the sequence 3, 6, 9, 12, 15, ... the rule is 'add 3'.

Term number   1   2   3   4   5

Term   3   6   9   12   (15)   ← This is term number
                        +3   +3   +3   +3   5 or the 5th term.

Here is the **mapping diagram** for the sequence.

Term number		Term
1	⟶	3
2	⟶	6
3	⟶	9
4	⟶	12
⋮		⋮
10	⟶	30

You can also find each term by multiplying the term number by 3.
So the 10th term is 30, the 33rd term is 99.
You can draw this as a function machine.

Term number ⟶ [×3] ⟶ Term

● A general term in the sequence is the nth term, where n stands for any number.
● The nth term of this sequence is $3n$.

You can find the nth term by putting n into the function machine.

● Here is a more difficult sequence: 5, 9, 13, 17, ...

The rule is 'add 4' so, in the mapping diagram, there is a column for 4 times the term number.

Term number (n)		$4n$		Term
1	⟶	4	⟶	5
2	⟶	8	⟶	9
3	⟶	12	⟶	13
4	⟶	16	⟶	17

You can see that each term is 1 more than $4n$.

Term number ⟶ [×4] ⟶ [+1] ⟶ Term

So, the 10th term is $(4 \times 10) + 1 = 41$
the 15th term is $(4 \times 15) + 1 = 61$
the nth term is $(4 \times n) + 1 = 4n + 1$

Exercise 26Ⓔ

1 Write each sequence and match it with the correct formula
for the nth term.

a 2, 4, 6, 8,...
b 10, 20, 30, 40,...
c 3, 6, 9, 12,...
d 11, 22, 33, 44,...
e 100, 200, 300, 400,...
f 6, 12, 18, 24,...
g 22, 44, 66, 88,...
h 30, 60, 90, 120,...

$22n$

$30n$

$100n$

$3n$

$10n$

$6n$

$2n$

$11n$

> You can draw
> a function machine
> to help you.

2 Copy and complete these mapping diagrams.

a

Term number (n)	Term
1 ⟶	5
2 ⟶	10
3 ⟶	15
4 ⟶	20
⋮	⋮
11 ⟶	☐
⋮	⋮
n ⟶	☐

b

Term number (n)	Term
1 ⟶	9
2 ⟶	18
3 ⟶	27
4 ⟶	36
⋮	⋮
20 ⟶	☐
⋮	⋮
n ⟶	☐

c

Term number (n)	Term
1 ⟶	100
2 ⟶	200
3 ⟶	300
⋮	⋮
12 ⟶	☐
⋮	⋮
n ⟶	☐

3 Copy and complete these mapping diagrams. Note that an
extra column has been written.

a

Term number (n)	$2n$	Term
1 ⟶	2 ⟶	5
2 ⟶	4 ⟶	7
3 ⟶	6 ⟶	9
4 ⟶	8 ⟶	11
⋮	⋮	⋮
10 ⟶	☐ ⟶	☐
⋮	⋮	⋮
n ⟶	☐ ⟶	☐

b

Term number (n)	$3n$	Term
1 ⟶	3 ⟶	4
2 ⟶	6 ⟶	7
3 ⟶	9 ⟶	10
4 ⟶	12 ⟶	13
⋮	⋮	⋮
20 ⟶	☐ ⟶	☐
⋮	⋮	⋮
n ⟶	☐ ⟶	☐

4 Here you are given the nth term. Copy and complete the diagrams.

a

Term number (n)	6n	Term
1 $\longrightarrow$	6 $\longrightarrow$	8
2 $\longrightarrow$	12 $\longrightarrow$	14
3 $\longrightarrow$	$\square$ $\longrightarrow$	$\square$
4 $\longrightarrow$	$\square$ $\longrightarrow$	$\square$
$\vdots$	$\vdots$	$\vdots$
n $\longrightarrow$	6n $\longrightarrow$	6n + 2

b

Term number (n)	5n	Term
1 $\longrightarrow$	5 $\longrightarrow$	3
2 $\longrightarrow$	10 $\longrightarrow$	8
3 $\longrightarrow$	$\square$ $\longrightarrow$	$\square$
4 $\longrightarrow$	$\square$ $\longrightarrow$	$\square$
$\vdots$	$\vdots$	$\vdots$
n $\longrightarrow$	5n $\longrightarrow$	5n − 2

5 Here are the first five terms of a sequence:
6, 11, 16, 21, 26, ...
  a Draw a mapping diagram like those in question **4**.
  b Write
    i the 10th term   **ii** the nth term.

> The rule for the sequence is 'add 5', so write a column for '5n' in the diagram.

6 Here are three sequences.
A : 1, 4, 7, 10, 13, ...
B : 6, 10, 14, 18, 22, ...
C : 5, 12, 19, 26, 33, ...
For each sequence   **a** draw a mapping diagram
                       b find the nth term.

7

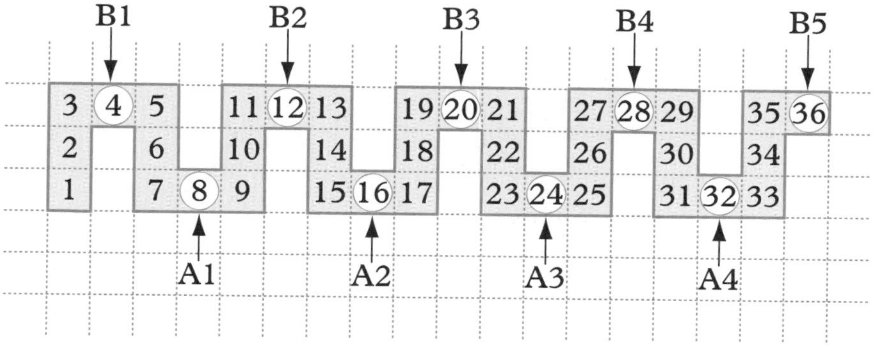

Look at the sequence made from A1, A2, A3, A4, ... and the sequence made from B1, B2, B3, B4, ...
Write
a A10
b B10
c the nth term in the 'A' sequence
d the nth term in the 'B' sequence.

8

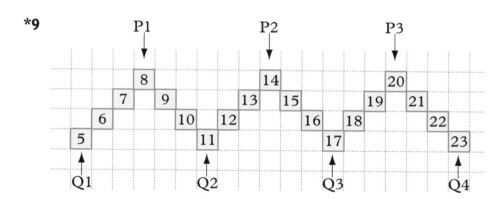

Look at the sequence made from C1, C2, C3, C4, ... and
the sequence made from D1, D2, D3, D4, ...
Write
a C20
b D20
c the nth term in the 'C' sequence
d the nth term in the 'D' sequence.

*9

a Find the nth term in the 'P' sequence
b Find the nth term in the 'Q' sequence.

2.4.5 Finding a formula or rule for a sequence

Here is a sequence of 'houses' made from matches.

The table records the number of houses, h, and the number of matches, m.

h	m
1	5
2	9
3	13
4	17

4 more matches each time

The number in the h column goes up one at a time. Look at the number in the m column.
In this case, the numbers in the m column go up by 4 each time.
This suggests that a column for $4h$ might help.

Now you can see that m is one more that $4h$.

So the formula linking m and h is $\mathbf{m = 4h + 1}$

$\times 4$ $+1$

h	$4h$	m
1	4	5
2	8	9
3	12	13
4	16	17

EXAMPLE

The table shows how r changes with n.
What is the formula that links r with n?

n	r
2	3
3	8
4	13
5	18

n 2   3   4   5

r 3   8   13   18
   $+5$ $+5$ $+5$

Because r goes up by 5 each time, write another column for $5n$.
The table shows that r is always 7 less than $5n$, so the formula linking r with n is $r = 5n - 7$.

n	$5n$	r
2	10	3
3	15	8
4	20	13
5	25	18

If the numbers in the first column do not go up by one each time, this method does not work. In that case you have to think of something clever!

Exercise 27Ⓔ

1 This sequence shows patterns of red tiles, r, and white tiles, w.
   The table shows the numbers of red and white tiles.

 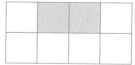 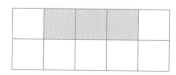

r	w
1	5
2	6
3	7
4	8

What is the formula for w in terms of r? Write it as '$w = \ldots$'

2 This is a different sequence with red tiles, r, and white tiles, w, and the related table.

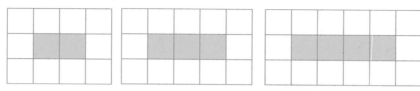

r	w
2	10
3	12
4	14
5	16

What is the formula? Write it as '$w = \ldots$'

3 Here is a sequence of Is.

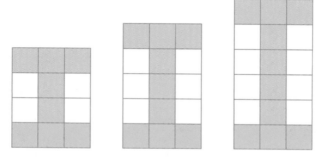

Make your own table for red tiles, r, and white tiles, w. What is the formula for w in terms of r?

4 In this sequence there are matches, m, and triangles, t.

Make a table for t and m. It starts like this.

Continue the table up to $t = 10$ and find a formula for m in terms of t.
Write '$m = \ldots$'.

t	m
1	3
2	5
⋮	⋮

5 Here is a different sequence of matches and triangles.

Make a table and find a formula connecting m and t.

6 In this sequence there are triangles, t, and squares, s, around the outside.

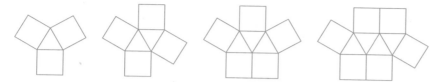

What is the formula connecting t and s?

7 Look at these tables. In each case, find a formula connecting the two letters.

a

n	p
1	3
2	8
3	13
4	18

Write '$p = \ldots$'

b

n	k
2	17
3	24
4	31
5	38

Write '$k = \ldots$'

c

n	w
3	17
4	19
5	21
6	23

Write '$w = \ldots$'

*8 This is one part of a sequence of cubes, c, made from matches, m.

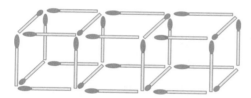

Find a formula connecting m and c.

Test yourself

1 Simplify these expressions.
   a $3x + 2x - x$
   b $5x + 3y - 2x + 4y$
   c $3 \times a \times 4$

   (AQA, 2004)

2 Match each of the algebra expressions on the left with the one that is the same on the right. One has been done for you.

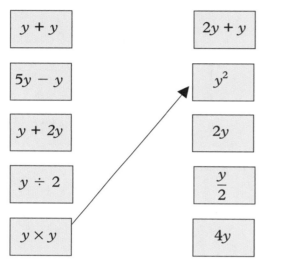

   (MEI, Spec)

3 Solve these equations.
   a $7x = 21$
   b $4 = x + 9$
   c $4x + 7 = 17$

   (OCR, 2003)

4 **a** Simplify $3p + 7q + p - 5q + 5p$.
   b Solve
      i $\dfrac{w}{2} = 1.5$
      ii $2z - 3 = 4$

   (OCR, 2004)

5 Solve these equations.
   a $3x = 12$
   b $y + 7 = 13$
   c $8z - 5 = 11$
   d $3(w - 2) = 9$

   (AQA, 2003)

6 a Simplify
   i $c + c + c + c$ **ii** $p \times p \times p \times p$
   iii $3g + 5g$ **iv** $2r \times 5p$

b Expand
   $5(2y - 3)$

(Edexcel, 2003)

7 a Solve
   $5x - 1 = 18$

b Use the formula

   $T = 5x + 2y$

   to find the value of y when $T = 80$ and $x = 9$.

(OCR, 2003)

8 a Write the perimeter of the pentagon, in terms of a and b, in its simplest form.

b i Write the volume of the cuboid, in terms of x and y, in its simplest form.
   ii Find the total area of the faces of the cuboid, in terms of x and y, in its simplest form.

(CCEA)

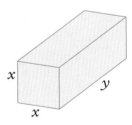

9 a Write, in symbols, the rule
   'To find y, multiply k by 3 and then subtract 1.'
b Work out the value of k when $y = 14$.

(Edexcel)

10 a Write as simply as possible an expression for the perimeter of this shape.
b Solve the equation $2x + 3 = 16$.
c When $y = 4x + 1$
   i find the value of y when $x = -2$
   ii find the value of x when $y = 19$.

(MEI, Spec)

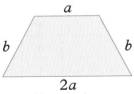

Not to scale

11 a Find the value of $3x + 5y$ when $x = -2$ and $y = 4$.
  b Find the value of $3a^2 + 5$ when $a = 4$.
  c k is an even number.

   Jo says that $\frac{1}{2}k + 1$ is always even.

   Give an example to show that Jo is wrong.

(AQA, 2004)

12 The total of each row is given at the side of the table.

$4x + 1$	$2(x + 5)$	20
$2x$	4	A

Find the values of x and A.

(AQA, 2004)

13 Tayub said, 'When $x = 3$, then the value of $4x^2$ is 144.'
Bryani said, 'When $x = 3$, then the value of $4x^2$ is 36.'
  a Who was right?
   Explain why.

  b Work out the value of $4(x + 1)^2$ when $x = 3$.

(Edexcel, 2003)

14 This word formula tells you how to find the cost in pounds
of hiring a minibus.

> Cost = Number of miles × 2, then add 10

  a Steve hires a minibus to travel 30 miles to get to Luton Airport.
   How much will it cost?
  b Sarah hires a minibus to travel forty-seven miles.
   How much will it cost?
  c For a different journey, Dave paid £42 to hire a minibus.
   How many miles was his journey?

(OCR, 2005)

15 In the table, the letters w, x, y and z represent different numbers.
The total of each row is given at the side of the table.

w	w	w	w	24
w	w	x	x	28
w	w	x	y	25
w	x	y	z	23

Find the values of w, x, y and z.

(AQA, 2004)

16 Simplify
   a $c + c + c + c$
   b $p \times p \times p \times p$
   c $2r \times 5p$

(Edexcel, 2003)

17 a The rule for working out the next number in a
     sequence is

> - If the number is even, divide by 2.
> - If the number is odd, multiple by 3 then add 1.

     i Using this rule, a sequence begins
        6, 3, 10, 5 ...
        Find the next two numbers.
     ii **Using the same rule**, another sequence begins
        2, 1, 4, ...
        Explain, in words, what happens as this sequence
        continues.
   b Using a **different** rule, the nth term of a sequence
     is $n^2 - 1$.
     Write the first three terms of this sequence.

(OCR, 2004)

18 Eggs are sold in boxes.
   A small box holds 6 eggs.
   Hina buys x small boxes of eggs.
   a Write, in terms of x, the total number of eggs in these
     small boxes.

   A large box holds 12 eggs.
   Hina buys 4 less of the large boxes of eggs than the
   small boxes.
   b Write, in terms of x, the number of large boxes she buys.
   c Find, in terms of x, the total number of eggs in the **large**
     boxes that Hina buys.
   d Find, in terms of x, the total number of eggs that Hina
     buys.
     Give your answer in its simplest form.

(Edexcel, 2004)

3 Shape and space 1

In this unit you will:
- revise plotting coordinates
- revise drawing, estimating and constructing angles and shapes
- learn about the properties of shapes, including congruency and symmetry
- revise area and perimeter
- learn to draw shapes as scale drawings and isometric drawings.

3.1 Coordinates

3.1.1 *x*- and *y*-coordinates

- To get to the point P on this grid you go **across** 1 and **up** 3 from the bottom corner.

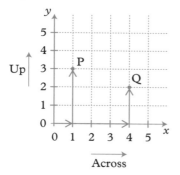

> Notice also that the **lines** are numbered, **not** the squares.

The position of P is $(1, 3)$.
The numbers 1 and 3 are called the **coordinates** of P.
The coordinates of Q are $(4, 2)$.
The **origin** is at $(0, 0)$.

- The first coordinate is the *x*-coordinate and the second coordinate is the *y*-coordinate.

- The **across** coordinate is always **first** and the **up** coordinate is **second**.

Remember: 'Along the corridor and up the stairs'.

Exercise 1 C

1 Write the coordinates of all the points
A to K like this: A(5, 1) B(1, 4)
Don't forget the brackets.

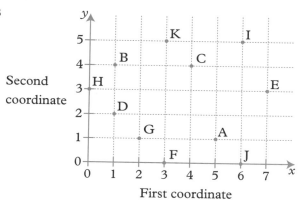

2 Write the coordinates of the points
which make up the '2' and the 'S'
in the diagram. You must give the
points in the correct order.

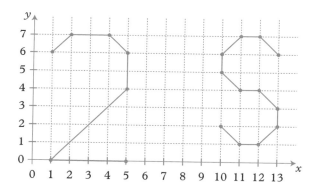

In questions **3** to **6** plot the points and join them up in order.
You will get a picture in each case.

3 Draw x- and y-axes from 0 to 10.
A: $(3, 2)$, $(4, 2)$, $(5, 3)$, $(3, 5)$, $(3, 6)$, $(2, 7)$, $(1, 6)$, $(1, 8)$,
$(2, 9)$, $(3, 9)$, $(5, 7)$, $(4, 6)$, $(4, 5)$, $(6, 4)$, $(8, 4)$, $(8, 5)$,
$(6, 7)$, $(5, 7)$.
B: $(7, 4)$, $(9, 2)$, $(8, 1)$, $(7, 3)$, $(5, 3)$.
C: $(1, 6)$, $(2, 8)$, $(2, 9)$, $(2, 7)$.
D: Draw a dot at $(3, 8)$.

4 Draw x- and y-axes from 0 to 10.
A: $(6, 5)$, $(7, 6)$, $(9, 5)$, $(10, 3)$, $(9, 1)$, $(1, 1)$, $(3, 3)$, $(3, 4)$,
$(4, 5)$, $(5, 4)$, $(4, 3)$, $(6, 4)$, $(8, 4)$, $(9, 3)$.
B: $(8, 3)$, $(8, 2)$, $(7, 1)$.
C: $(6, 3)$, $(6, 2)$, $(5, 1)$.
D: $(5, 2)$, $(4, 1)$.
E: Draw a dot at $(3, 2)$.

5 Draw x- and y-axes from 0 to 8.
A: $(6, 6)$, $(1, 6)$, $(2, 7)$, $(7, 7)$, $(6, 6)$, $(6, 1)$, $(7, 2)$, $(7, 7)$.
B: $(1, 6)$, $(1, 1)$, $(6, 1)$.
C: $(3, 5)$, $(3, 3)$, $(2, 2)$, $(2, 5)$, $(5, 5)$, $(5, 2)$, $(2, 2)$, $(3, 3)$, $(5, 3)$.

6 Draw the x-axis from 0 to 8 and the y-axis from 0 to 4.
A: $(7, 1)$, $(8, 1)$, $(7, 2)$, $(6, 2)$, $(5, 3)$, $(3, 3)$, $(2, 2)$, $(6, 2)$,
  $(1, 2)$, $(1, 1)$, $(2, 1)$.
B: $(3, 1)$, $(6, 1)$.
C: $(3, 3)$, $(3, 2)$.
D: $(4, 3)$, $(4, 2)$.
E: $(5, 3)$, $(5, 2)$.
F: Draw a circle of radius $\frac{1}{2}$ unit with centre at $\left(2\frac{1}{2}, 1\right)$.

G: Draw a circle of radius $\frac{1}{2}$ unit with centre at $\left(6\frac{1}{2}, 1\right)$.

7 Use the grid to work out the joke written in coordinates.
Work down each column.

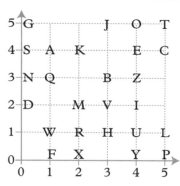

$(1, 1)$	$(4, 0)$	$(1, 4)$	$(1, 1)$	$(0, 4)$	$(0, 2)$
$(3, 1)$	$(4, 5)$		$(4, 2)$	$(5, 0)$	$(4, 5)$
$(1, 4)$	$(4, 1)$	$(2, 2)$	$(5, 5)$	$(1, 4)$	$(4, 1)$
$(5, 5)$		$(1, 4)$	$(3, 1)$	$(0, 2)$	$(0, 5)$
	$(5, 4)$	$(0, 3)$		$(4, 4)$	
$(0, 2)$	$(1, 4)$		$(1, 4)$		
$(4, 5)$	$(5, 1)$				
	$(5, 1)$				

Exercise 2 C

In this exercise some coordinates will have negative numbers.

1 Write the coordinates of all the points A to I
like this:
A$(3, -2)$ B$(4, 3)$.

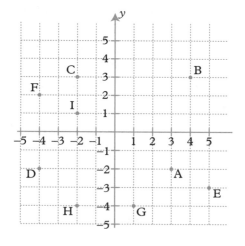

2 Plot the points and join them up in order.
You should produce a picture.
Draw axes with x from -4 to $+4$ and y
from 0 to 10.

A: (3, 5), (2, 7), (0, 8), (−1, 8), (−2, 7),
 (−3, 7), (−4, 8), (−2, 9), (0, 9), (2, 8), (3, 7),
 (3, 2), (1, 1), (0, 3), (−2, 2), (−2, 4), (−3, 4),
 (−2, 6), (−1, 6), (−1, 5), (−2, 6), (−2, 7).
B: (−1, 3), (−2, 3).
C: (1, 3), (0, 3).

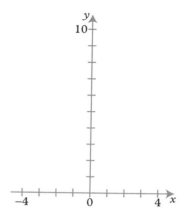

3 The graph shows several incomplete quadrilaterals. Copy the
diagram and complete the shapes.

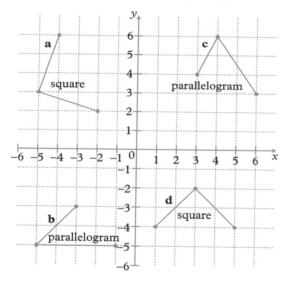

a Write the coordinates of the fourth vertex of each
shape.
b Write the coordinates of the centre of each shape.

A vertex is a
corner of a shape.

4 Copy this graph.
 a A, B and C are three corners of a square.
  Write the coordinates of the other corner.
 b C, A and D are three corners of another square.
  Write the coordinates of the other corner.
 c B, D and E are three corners of a rectangle.
  Write the coordinates of the other corner.
 d C, F and G are three vertices of a parallelogram.
  Write the coordinates of the other vertex.

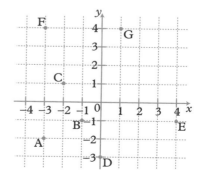

- A 'straight line' has infinite length. It goes on for ever!

- A 'line segment' has finite length.

For example the line segment PQ has end points P and Q.

P   Q

EXAMPLE

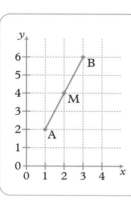

The diagram shows the line segment AB joining $A(1, 2)$ and $B(3, 6)$.

You find the coordinates of the midpoint, M, by adding the x- and y-coordinates of A and B and then dividing by 2.

So M is the point $\left(\dfrac{1+3}{2}, \dfrac{2+6}{2}\right)$.

M has coordinates $(2, 4)$.

*5 Find the coordinates of the midpoint of the line joining $C(2, 5)$ and $D(6, 7)$.

*6 Use the method in question **5** to find the midpoints of these line segments.

   a $A(2, 1)$ and $B(8, 3)$
   b $C(1, 3)$ and $D(5, 0)$
   c $E(2, 3)$ and $F(10, 7)$
   d $G(0, 8)$ and $H(4, 2)$
   e $I(2, -2)$ and $J(6, 4)$
   f $K(4, -3)$ and $L(0, 7)$

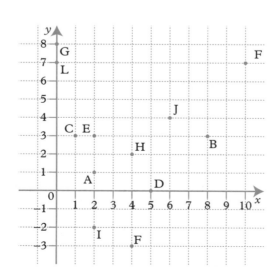

3.2 Using angles

3.2.1 Estimating angles

● Angles are measured in **degrees**.

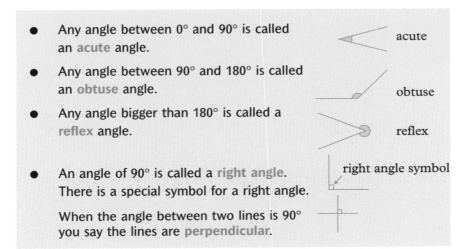

- ● Any angle between 0° and 90° is called an acute angle.
- ● Any angle between 90° and 180° is called an obtuse angle.
- ● Any angle bigger than 180° is called a reflex angle.

- ● An angle of 90° is called a right angle. There is a special symbol for a right angle.

  When the angle between two lines is 90° you say the lines are perpendicular.

acute

obtuse

reflex

right angle symbol

A full turn = 360°
A half turn = 180°
A quarter turn = 90°

Exercise 3 C

Write whether these angles are correctly or incorrectly labelled. Do **not** measure the angles – estimate! Where the angles are clearly incorrect, write an estimate for the correct angle.

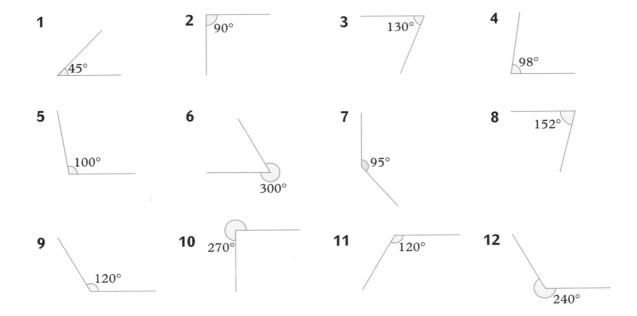

1. 45°
2. 90°
3. 130°
4. 98°
5. 100°
6. 300°
7. 95°
8. 152°
9. 120°
10. 270°
11. 120°
12. 240°

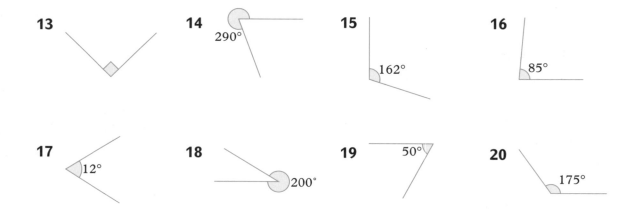

13 **14** 290° **15** 162° **16** 85°

17 12° **18** 200° **19** 50° **20** 175°

Exercise 4 C

For each angle in Exercise **3**, write whether the angle **marked** is acute, obtuse, reflex or a right angle.

3.2.2 Angle facts

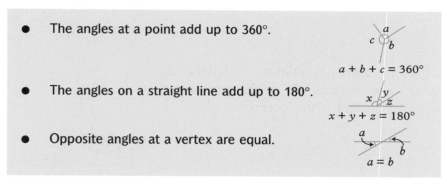

- The angles at a point add up to 360°.

  $a + b + c = 360°$

- The angles on a straight line add up to 180°.

  $x + y + z = 180°$

- Opposite angles at a vertex are equal.

  $a = b$

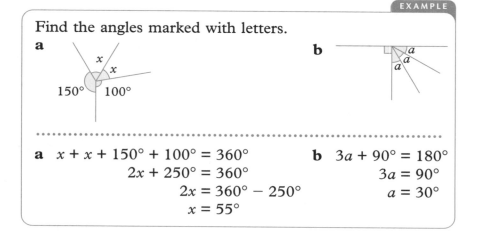

EXAMPLE

Find the angles marked with letters.

a x x 150° 100°

b a a a

....

a $x + x + 150° + 100° = 360°$
$2x + 250° = 360°$
$2x = 360° - 250°$
$x = 55°$

b $3a + 90° = 180°$
$3a = 90°$
$a = 30°$

Exercise 5 C

Find the angles marked with letters. The line segments AB and CD in the diagrams from question **9** onwards are straight.

1
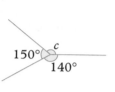
150°   c   140°

2
120°   x   140°

3
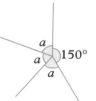
40°   a   90°   160°

4
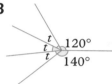
40°   y   60°   160°

5
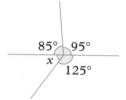
85°   95°   x   125°

6
70°   x   x   150°

7
a   150°   a   a

8
t   t   t   120°   140°

9

A   x   150°   B

10
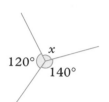
A   y   145°   B

11
A   25°   z   B

12
A   a   55°   B

13
A   71°   y   65°   B

14
A   58°   e   42°   B

15
A   100°   x   x   B

16
A   h   84°   h   B

17
A   60°   y   y   y   B

18
A   x   2x   75°   B

19

D   a   B   A   40°   b   C

20
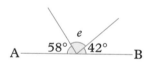
A   72°   x   D   C   y   B

3.2.3 Angles in a triangle

Draw a triangle of any shape on a piece of card and cut it out accurately. Now tear off the three corners as shown.

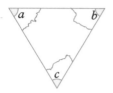

When the angles *a*, *b* and *c* are placed together they form a straight line.

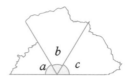

> This is a **demonstration** that the angles add up to 180°. It is not a mathematical **proof**. A proof of this result is on page 115.

● The angles in a triangle add up to 180°.

Isosceles and equilateral triangles

An **isosceles** triangle has two equal sides and two equal angles. It has a line of symmetry down the middle.
The sides AB and AC are equal (marked with a dash) and angles B and C are also equal.

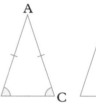

An **equilateral** triangle has three equal sides and three equal angles (all 60°).

Find the angles marked with letters.

a

b

a $y + 80° + 40° = 180°$
$\quad\quad y + 120° = 180°$
$\quad\quad\quad\quad y = 60°$

b $a = 180° - 150° = 30°$
The triangle is isosceles, so $\quad 2x + 30° = 180°$
$\quad\quad\quad\quad\quad\quad\quad\quad\quad\quad\quad 2x = 150°$
$\quad\quad\quad\quad\quad\quad\quad\quad\quad\quad\quad\quad x = 75°$

Exercise 6 C

Find the angles marked with letters. For the more difficult questions it is helpful to draw a diagram.

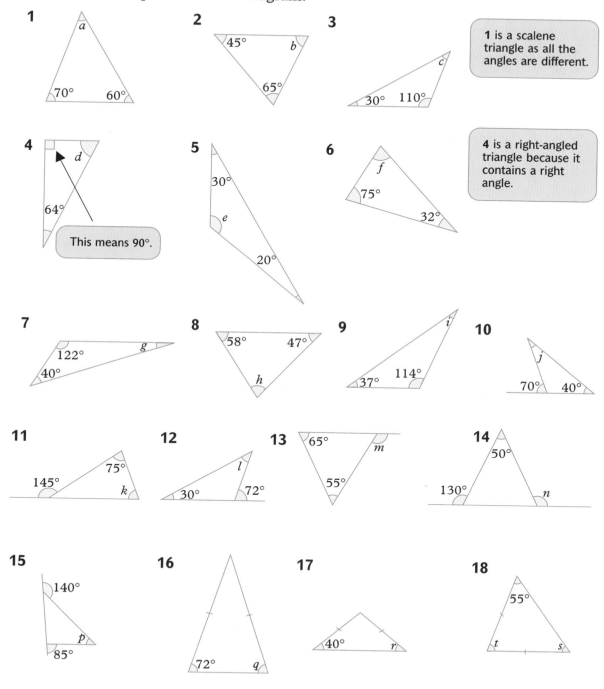

1 is a scalene triangle as all the angles are different.

4 is a right-angled triangle because it contains a right angle.

This means 90°.

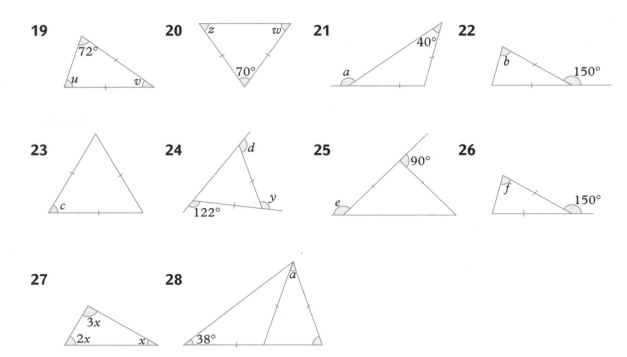

19

72°

u   v

20

z   w

70°

21

40°

a

22

b

150°

23

c

24

d

122°   y

25

90°

e

26

f

150°

27

3x

2x   x

28

a

38°

3.2.4   Parallel lines

- Parallel lines are always the same distance apart. They never meet.

- When a line cuts a pair of parallel lines all the acute angles are equal and all the obtuse angles are equal.

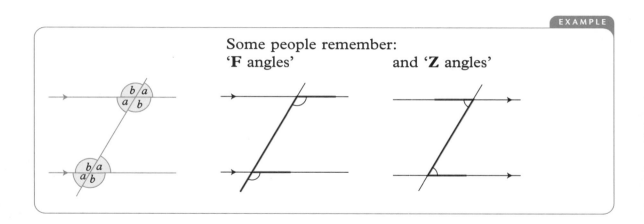

EXAMPLE

Some people remember:
'**F** angles'   and '**Z** angles'

b a
a b

b a
a b

Exercise 7Ⓔ

Find the angles marked with letters.

1

2

3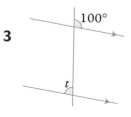

4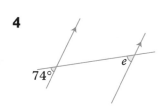

5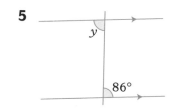

6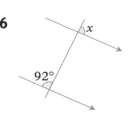

7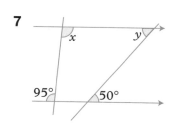

8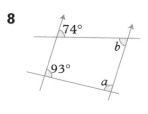

9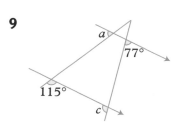

10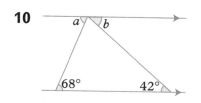

11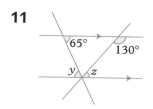

12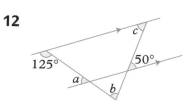

Exercise 8Ⓔ (Mixed questions)

The next exercise contains questions that use all the angle facts
you have learnt.

Find the angles marked with letters.

1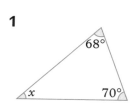

2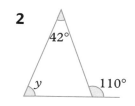

3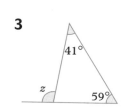

4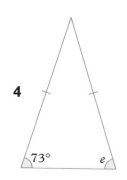

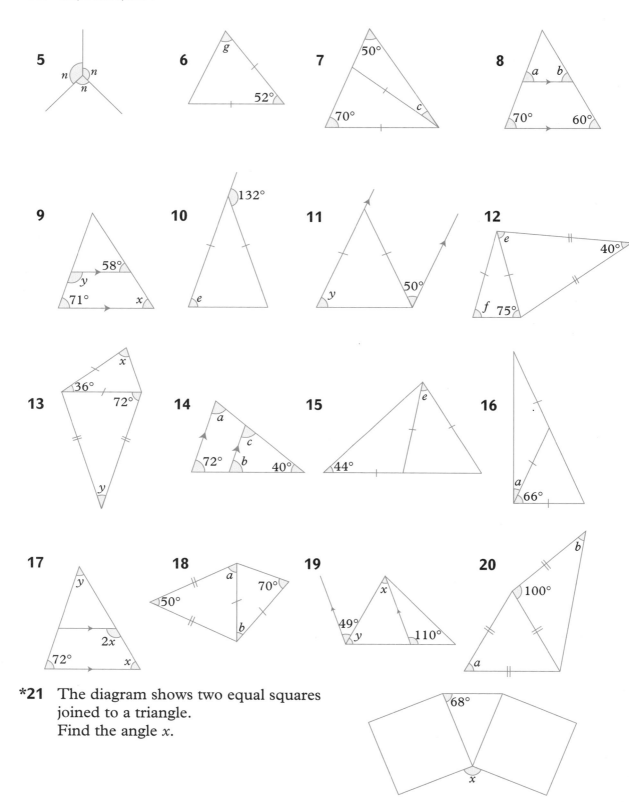

5

6 g 52°

7 50° 70° c

8 a b 70° 60°

9 132°

10 58° y 71° x e

11 50° y

12 e 40° f 75°

13 x 36° 72° y

14 a c 72° b 40°

15 e 44°

16 a 66°

17 y 2x 72° x

18 a 70° 50° b

19 x 49° y 110°

20 b 100° a

*21 The diagram shows two equal squares
joined to a triangle.
Find the angle x.

68° x

*22 Find the angle *a* between the diagonals
of the parallelogram.

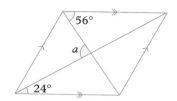

3.2.5 Proving results

On page 110, you **demonstrated** that the sum of the angles in
a triangle is 180°, by cutting out the angles and rearranging
them. A demonstration like this might not work for every
possible triangle.

When you **prove** results it means that the result is true for
every possible shape. You can often prove one simple result and
then use that result to prove further results.

- Here is a proof that the sum of the angles in a triangle is 180°.

Here is △ABC.

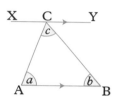

Draw line XCY
parallel to AB.

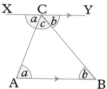

∠ABC = ∠YCB (alternate angles)
∠BAC = ∠ACX (alternate angles)
$a + b + c = 180°$ (angles on a
straight line)
So, angles in a triangle: $a + b + c = 180°$

> △ is a symbol that
> means triangle.

This proves that the sum of the angles in a triangle is 180°.

Exercise 9Ⓔ

Teacher's note: Proof is not an easy topic for most students.
Some teachers may choose to go through these proofs on the
board so that students can copy them down. It is desirable that
students should have an **understanding** of the proof rather
than the ability to reproduce it (which is not required in
examinations).

1 Copy and complete this proof that the sum of the angles
in a quadrilateral is 360°.
Draw any quadrilateral ABCD with diagonal BD.
Now $a + b + c = $ ☐ (angles in a △)
and $d + e + f = $ ☐ (angles in a △)
So, $a + b + c + d + e + f = $ ☐
This proves the result.

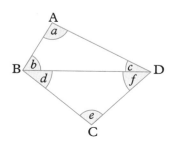

2 The exterior angle of a triangle is equal to the sum of the two interior opposite angles.

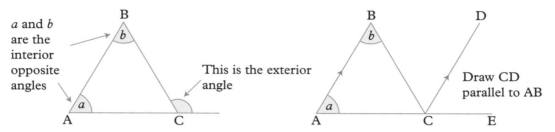

a and *b* are the interior opposite angles

This is the exterior angle

Draw CD parallel to AB

Copy and complete this proof:

∠BAC = ∠DCE (corresponding angles) ('F' angles)

∠ABC = ∠⬚ (alternate angles)   ('Z' angles)

So, ⬚ = ⬚ + ⬚

3 Prove that opposite angles of a parallelogram are equal. (Use alternate and corresponding angles.)

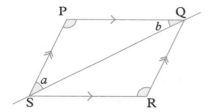

3.3 Accurate drawing

3.3.1 Using a protractor and a pair of compasses

Some questions involving bearings or irregular shapes are easy to solve if you draw an accurate diagram.

To improve the accuracy of your work, follow these guidelines.

- Use a **sharp** HB pencil.
- Don't press too hard.
- If you are drawing an **acute** angle make sure your angle is less than 90°.
- If you use a pair of compasses make sure they are fairly stiff so the radius does not change accidentally.

Using a protractor
You use a protractor to measure angles accurately.

> Remember: When you measure an **acute** angle the answer must be less than 90°. When you measure an **obtuse** angle the answer must be more than 90°.

Exercise 10 C

Measure these angles.

1

2

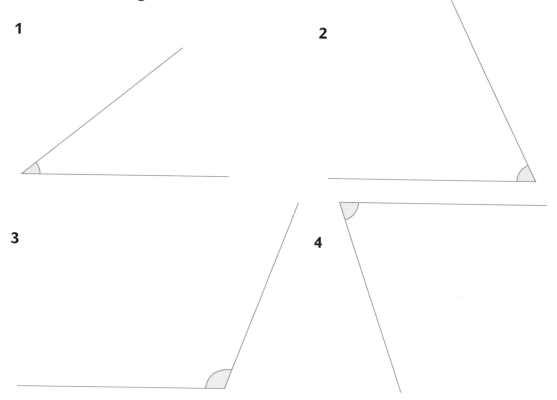

3

4

In questions **5, 6, 7** and **8** measure all the angles and all the sides in each triangle.

5

6

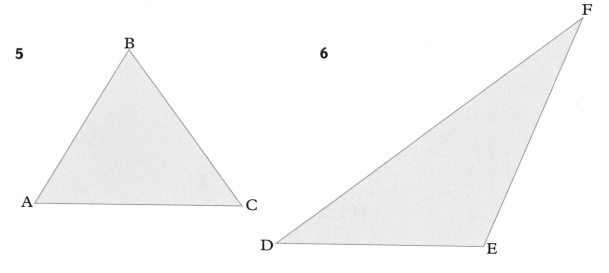

7

8

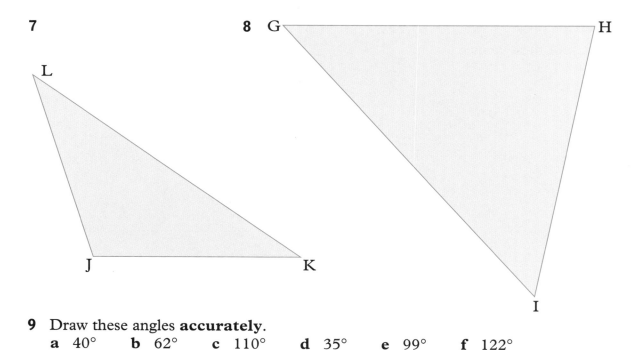

9 Draw these angles **accurately**.
   a 40°   **b** 62°   **c** 110°   **d** 35°   **e** 99°   **f** 122°

3.3.2 Constructing triangles

EXAMPLE

Draw the triangle ABC full size and measure the length x.

a Draw a base line **longer than 8·5 cm**.
b Put the centre of the protractor on A and measure an angle of 64°. Draw line AP.
c Similarly draw line BQ at an angle of 40° to AB.
d You have drawn the triangle!
   Measure $x = 5·6$ cm

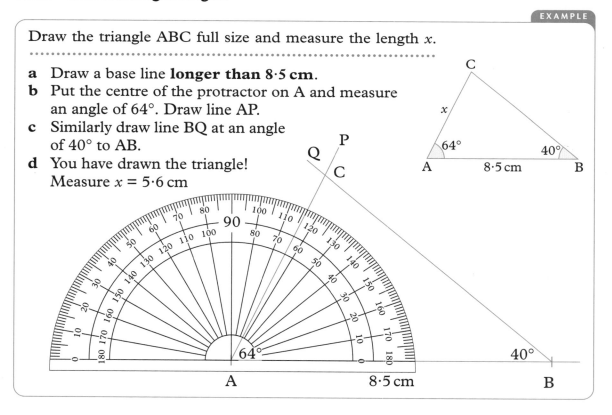

Exercise 11 Ⓒ

Use a protractor and ruler to draw full size diagrams and
measure the sides marked with letters.

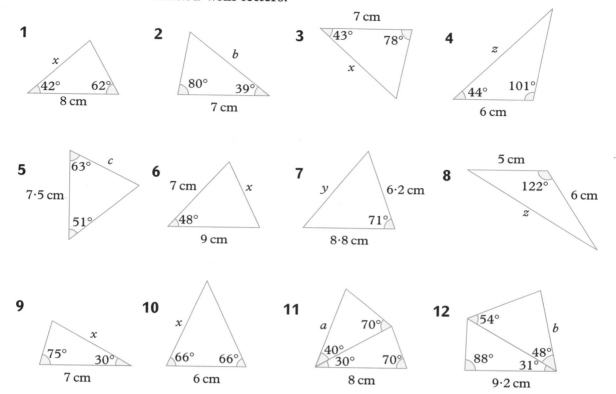

1 x, 42°, 62°, 8 cm

2 b, 80°, 39°, 7 cm

3   7 cm, 43°, 78°, x

4 z, 44°, 101°, 6 cm

5   63°, c, 7·5 cm, 51°

6   7 cm, x, 48°, 9 cm

7 y, 6·2 cm, 71°, 8·8 cm

8   5 cm, 122°, 6 cm, z

9 x, 75°, 30°, 7 cm

10 x, 66°, 66°, 6 cm

11 a, 70°, 40°, 30°, 70°, 8 cm

12   54°, b, 88°, 48°, 31°, 9·2 cm

Exercise 12 Ⓒ

This exercise will show you how to draw a triangle, if you
have three sides.

1   Follow these steps to draw a triangle with sides 7 cm,
5 cm, 6 cm.

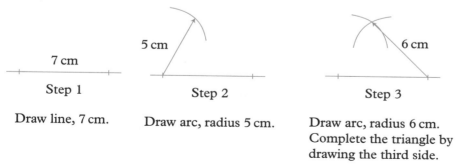

Step 1	Step 2	Step 3
Draw line, 7 cm.	Draw arc, radius 5 cm.	Draw arc, radius 6 cm. Complete the triangle by drawing the third side.

2 a Construct an equilateral triangle with sides of 6 cm.
  b Construct an equilateral triangle with sides of 7·5 cm.

In questions **3** to **6** construct the triangles using a pair of compasses.
Measure the angles marked with letters.

3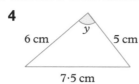
6 cm *x* 9 cm
8 cm

4
6 cm *y* 5 cm
7·5 cm

5
7·2 cm
9 cm *z*
8·1 cm

6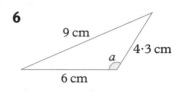
9 cm
a 4·3 cm
6 cm

Constructing unique triangles

You can describe a triangle according to the information you have.

You use: S when you know a side,
A when you know an angle,
R when you know there is a right angle,
H when you know the hypotenuse of a right-angled triangle.

● Here are five examples.

1
4 cm   5 cm
6 cm
SSS

All 3 sides

2
5 cm
65°
7 cm
SAS

2 sides and the included angle

3
60°   40°
8 cm
ASA

2 angles and one side

4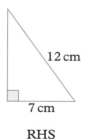
12 cm
7 cm
RHS

Right angle, hypotenuse and side

5
6 cm
40°
8 cm
SSA

2 sides and an angle (not included)

Exercise 13 Ⓔ

1 Using a ruler, protractor and a pair of compasses construct each of the triangles **1**, **2**, **3** and **4**.
Label the triangles SSS, SAS, ASA, RHS.

2 Construct triangle **5** and, using a pair of compasses, show that it is possible to construct two different triangles with the sides and angle given.

3 Construct the triangle in the diagram.
You are given SSA.
Show that you can construct two different triangles with the sides and angle given.

4·5 cm
35°
7 cm

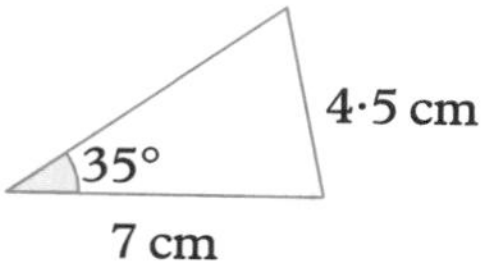

4 Copy and complete these two sentences.
'When you know SSS, SAS, [   ] or [   ] the constructed triangle is unique.

When you know [   ] the triangle is not unique and it is sometimes possible to construct two different triangles.'

> These are the conditions for triangles to be congruent (see page 125).

3.3.3 Nets

If the cube here was made of cardboard, and you cut along some of the edges and laid it out flat, you would have the **net** of the cube.

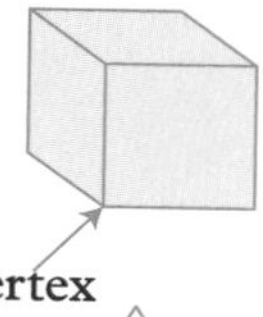

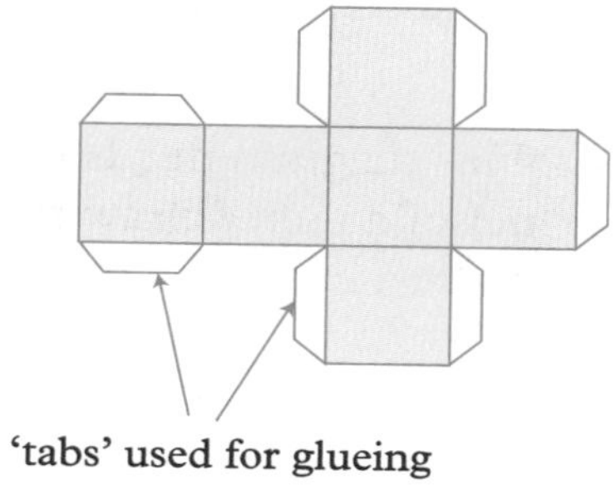

A cube has: 8 vertices (corners)
6 faces
12 edges.

vertex   'tabs' used for glueing

Here is the net for a square-based pyramid.
This pyramid has: 5 vertices
5 faces
8 edges.

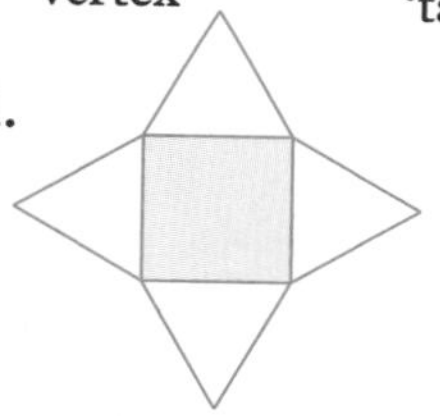

Exercise 14Ⓔ

You will need pencil, ruler, scissors and either glue or sticky tape.

1 Which of these nets could you use to make a cube?

a

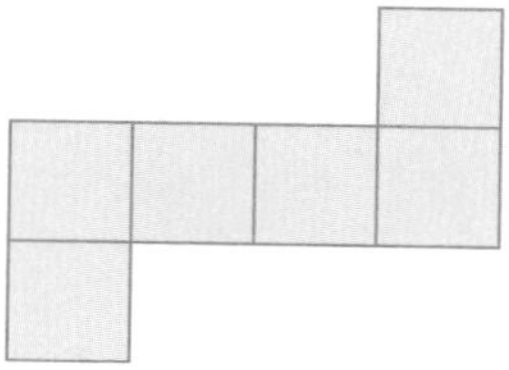

b

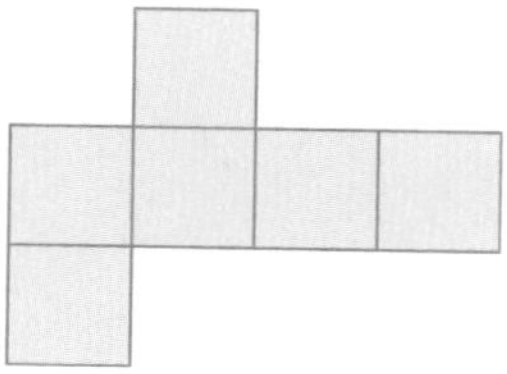

> If you want to create a 3-D shape you need to add tabs to your net to glue your shape together.

c

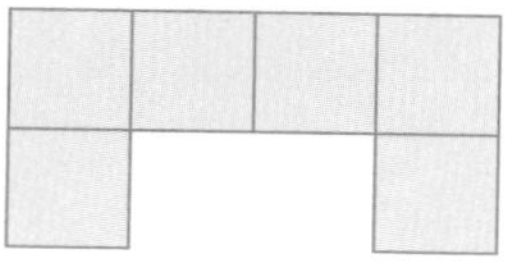

d

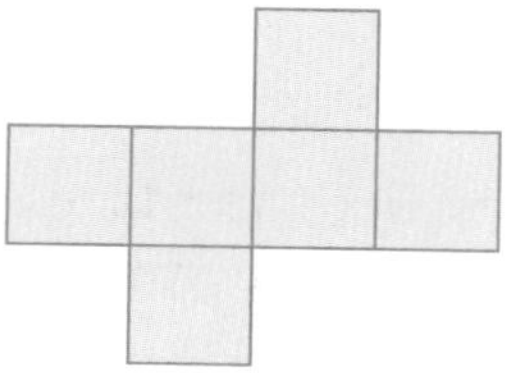

2 The numbers on opposite faces of a dice add up to 7. Take one of the possible nets for a cube from question **1** and show the number of dots on each face.

3 Here is the start of the net of a cuboid (a closed rectangular box) measuring 4 cm × 3 cm × 1 cm. Copy and then complete the net.

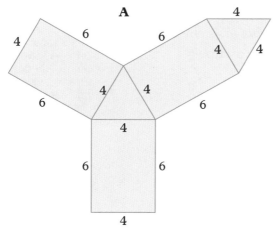

4 This diagram needs one more square to complete the net of a cube. Draw the **four** possible nets which would make a cube.

*5 You can make a solid from each of these nets.

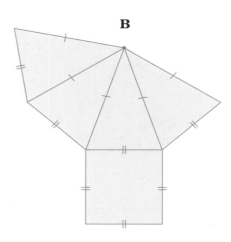

   a State the number of vertices and faces for each solid.
   b One object is a square-based pyramid and the other is a triangular-based prism. Which one is which?

6 Some interesting objects can be made using triangle
dotty paper. The basic shape for the nets is an equilateral
triangle. With the paper as shown the triangles are
easy to draw.

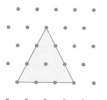

Make the sides of the triangles 3 cm long so that the solids
are easy to make. Here is the net of a regular tetrahedron.
Draw it and then cut it out.

3.3.4 Constructions with a ruler and compasses

1 Perpendicular bisector of a line segment AB

> An arc is a section of
> a circle.

With centres A and B draw two arcs with
your compass. Keep the same radius. Join the
points where the arcs intersect (the broken line).

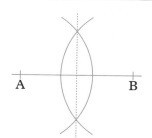

This line is the perpendicular bisector of AB.

2 Perpendicular from point P to a line

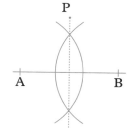

With centre P draw an arc
to cut the line at A and B.

Construct the perpendicular
bisector of AB. Use a smaller radius
on your compasses.

3 Bisector of an angle

With centre A draw arc PQ.
With centres at P and Q draw two more arcs.
Keep the same radius.
Join the point of intersection of the two arcs to A.
This line is the bisector of angle A.

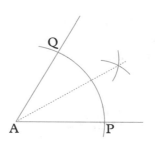

Exercise 15Ⓔ

Use only plain unlined paper, a pencil, a straight edge and a pair of compasses.

> A straight edge is a ruler with no markings on it.

1 Draw a line AB of length 6 cm. Construct the perpendicular bisector of AB.

2 Draw a line CD of length 8 cm. Construct the perpendicular bisector of CD.

3 Draw a line and a point P about 4 cm from the line. Construct the line which passes through P and is perpendicular to the line.

> See Construction 2 on page 123.

4 a Using a set square, draw a right-angled triangle ABC as shown. For greater accuracy draw lines slightly longer than 8 cm and 6 cm and **then** mark the points A, B and C.
  b **Construct** the perpendicular bisector of AB.
  c Construct the perpendicular bisector of AC.
  d If you do it accurately, your two lines from **b** and **c** should cross exactly on the line BC.

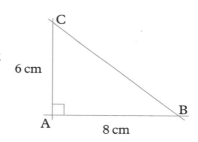

5 This is the construction of a perpendicular from a point P on a line, using ruler and compasses.
  a With centre P, draw arcs to cut the line at A and B.

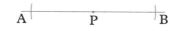

  b Now construct the perpendicular bisector of AB. (Use a larger radius, and join the points where the arcs meet.)

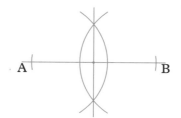

6 Draw an angle of about 60°. Construct the bisector of the angle.

7 Draw an angle of about 80°. Construct the bisector of the angle.

*8 Investigation

See Construction 3 on page 124.

Draw any triangle ABC and then construct the bisectors of angles A, B and C. If you do it accurately the three bisectors should all pass through one point.

If they do **not** pass through one point (or very nearly) do this question again with a new triangle ABC.

Does this work with every type of triangle, for example scalene, isosceles?

3.4 Congruent and similar shapes

3.4.1 What are congruent and similar shapes?

● **Congruent** shapes are exactly the same in shape and size. Shapes are congruent if one shape fits exactly over the other.

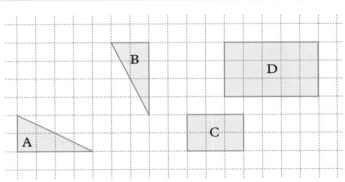

A and B are congruent   C and D are not congruent

● Shapes which are mathematically similar have the same shape, but different sizes. All corresponding angles in similar shapes are equal and corresponding lengths are in the same ratio.

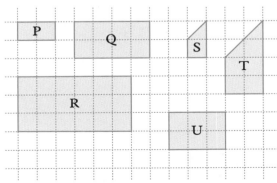

U is **not** similar to any of the shapes P, Q and R.

P, Q and R are similar.   S and T are similar.

If two shapes are similar one shape is an **enlargement** of the other.
Note that all **circles** are similar to each other and all **squares** are similar to each other.

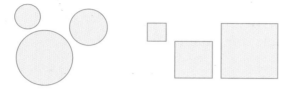

Exercise 16 C

1 Write the pairs of shapes that are congruent.

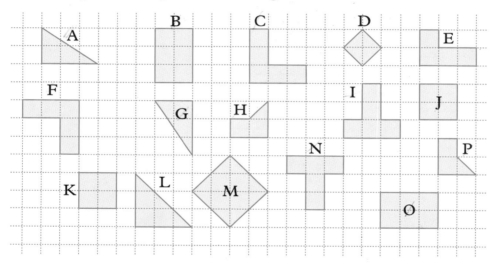

2 Copy the diagram onto squared paper and colour in the congruent shapes with the same colour.

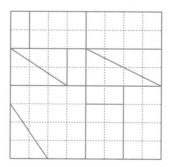

3 Copy shape X onto squared paper. Draw another shape which is congruent to shape X but turned into a different position.

4 Make a list of pairs of shapes which are similar.

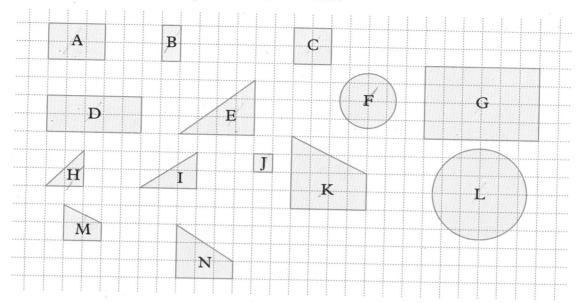

5 Draw shape I from question **4** on squared paper.
Now draw a shape which is similar to shape I.

6 Here are two rectangles. Explain why they are
not similar.

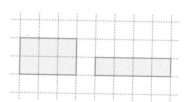

3.5 Symmetry

3.5.1 Line and rotational symmetry

Line symmetry
The letter M has one line of symmetry,
shown dotted.

> A line of symmetry
> is like a mirror line.

Rotational symmetry
The shape may be turned about its
centre into three identical positions.
It has rotational symmetry of order
three.

Exercise 17 C

For each shape write
a the number of lines of symmetry
b the order of rotational symmetry.

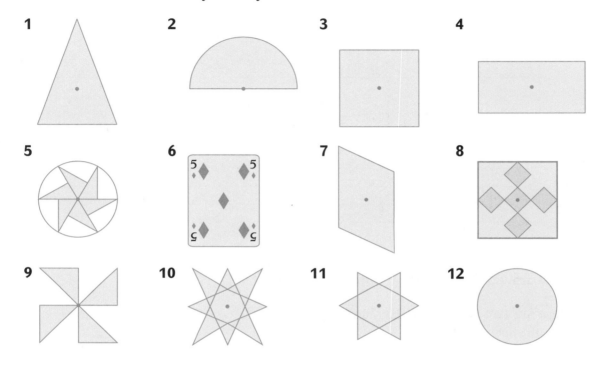

1   2   3   4

5   6   7   8

9   10   11   12

13 Here is a shape made using four squares.
  a Copy the shape and add one square so that the new shape has one line of symmetry. Do this in two different ways.
  b Copy the shape again and add one square so that the new shape has rotational symmetry of order 2.

14 a Copy this shape and shade one more triangle so that the new shape has one line of symmetry.
  b Copy the shape again and shade two more triangles so that the new shape has rotational symmetry of order 3.

15 Look at this shape.
  a Copy the shape and add one square so that the shape has rotational symmetry of order 2.
  b Copy the shape again and add one square so that the shape has one line of symmetry.

Exercise 18 C

In questions **1** to **8**, the broken lines are lines of symmetry.
In each diagram only **part of the shape** is given.
Copy the part shapes onto square grid paper and then
carefully complete them.

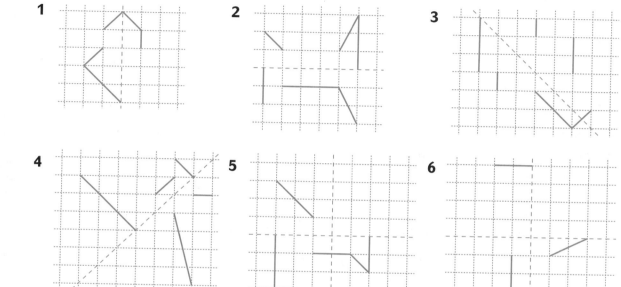

1   2   3

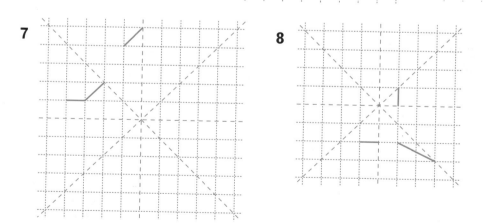

4   5   6

7   8

9 Fold a piece of paper twice and cut out any shape from the
corner. Open the shape and stick it in your book stating the
number of lines of symmetry and the order of rotational
symmetry.

cut here

3.5.2 Planes of symmetry

● A **plane of symmetry** divides a 3-D shape into two congruent shapes. One shape must be a mirror image of the other shape.

The diagrams show two planes of symmetry of a cube.

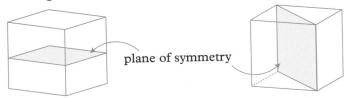

plane of symmetry

Exercise 19Ⓔ

1 How many planes of symmetry does this cuboid have?

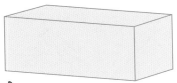

2 How many planes of symmetry do these shapes have?

a 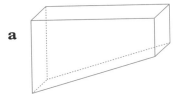   **b**    **c**

3 a Draw a diagram of a cube like the one at the top of this page and draw a different plane of symmetry.
  b How many planes of symmetry does a cube have?

4 Draw a pyramid with a square base so that the point of the pyramid is vertically above the centre of the square base. Show any planes of symmetry by shading.

5 The diagrams show the plan view and the side view of an object.

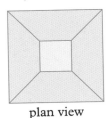   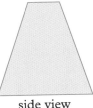

plan view   side view

How many planes of symmetry does this object have?

> The plan view is the view looking down on the object.

> The side view is the view from one side.

3.6 Quadrilaterals

3.6.1 Properties of quadrilaterals

- The angles in a quadrilateral always add up to 360°.

Title	Properties	
Square	Four equal sides All angles 90° Four lines of symmetry Rotational symmetry of order 4	
Rectangle (not square)	Two pairs of equal and parallel sides All angles 90° Two lines of symmetry Rotational symmetry of order 2	
Rhombus	Four equal sides Opposite sides parallel Diagonals bisect at right angles Diagonals bisect angles of rhombus Two lines of symmetry Rotational symmetry of order 4	
Parallelogram	Two pairs of equal and parallel sides Opposite angles equal No lines of symmetry Rotational symmetry of order 2	
Trapezium	One pair of parallel sides Rotational symmetry of order 1	
Kite	AB = AD, CB = CD Diagonals meet at 90° One line of symmetry Rotational symmetry of order 1	

Exercise 20 C

1 Write the correct names for these five quadrilaterals.

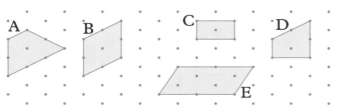

> These are drawn on isometric paper.

2 Copy the table and fill all the boxes with either ticks or crosses.

	Diagonals always equal	Diagonals always perpendicular	Diagonals always bisect the angles	Diagonals always bisect each other
Square	✓			
Rectangle				
Parallelogram				
Rhombus				
Trapezium				

3 Find the angle x.

a

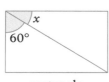

rectangle

b

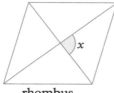

rhombus

c

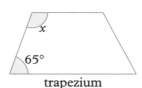

trapezium

4 Copy each diagram onto square grid paper and mark the fourth vertex with a cross.

a

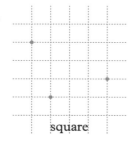

square

b

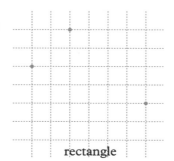

rectangle

c

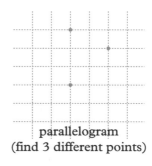

parallelogram
(find 3 different points)

5 What is the name for all shapes with four sides?

6 What four-sided shape has all sides the same length and all angles equal?

7 What quadrilateral has two pairs of sides the same length and all angles equal?

8 Which quadrilateral has only one pair of parallel sides?

9 Name the triangle with three equal sides and angles.

10 What is the name of the triangle with two equal angles?

11 Which quadrilateral has all four sides the same length but only opposite angles equal?

12 True or false: 'All squares are rectangles.'

13 True or false: 'Any quadrilateral can be cut into two equal triangles.'

14 Name the shapes made by joining these points.
   a A B F G
   b C E F I
   c A B E H
   d A B D I

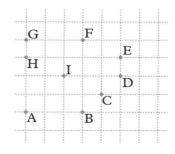

15 Name the shapes made by joining these points.
   a B I G E
   b A B E H
   c B C D F
   d C J G D
   e C J E

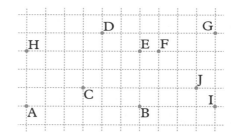

16 On square grid paper draw a quadrilateral with just two right angles and only one pair of parallel sides.

*17 ABCD is a rhombus whose diagonals intersect at M.
Find the coordinates of C and D, the other two vertices of the rhombus.

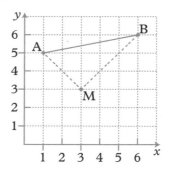

Exercise 21Ⓔ

In these questions, begin by drawing a diagram and remember to put the letters around the shape in alphabetical order.

1 In a rectangle KLMN, ∠LNM = 34°.
Calculate
  a ∠KLN
  b ∠KML.

2 In a trapezium ABCD, ∠ABD = 35°, ∠BAD = 110° and AB is parallel to DC.
Calculate
  a ∠ADB   **b** ∠BDC.

3 In a parallelogram WXYZ, ∠WXY = 72°, ∠ZWY = 80°.
Calculate
  a ∠WZY   **b** ∠XWZ   **c** ∠WYX.

4 In a kite ABCD, AB = AD, BC = CD, ∠CAD = 40° and ∠CBD = 60°. Calculate
  a ∠BAC   **b** ∠BCA   **c** ∠ADC.

5 In a rhombus ABCD, ∠ABC = 64°.
Calculate
  a ∠BCD
  b ∠ADB
  c ∠BAC.

6 In a rectangle WXYZ, M is the midpoint of WX and ∠ZMY = 70°. Calculate
  a ∠MZY   **b** ∠YMX.

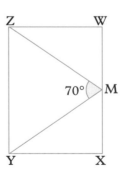

7 In a trapezium ABCD, AB is parallel to DC, AB = AD, BD = DC and ∠BAD = 128°.
Find
  a ∠ABD
  b ∠BDC
  c ∠BCD.

3.7 Isometric drawing

3.7.1 Drawing cuboids and solids made from cuboids

Here are two pictures of the same cuboid, measuring $4 \times 3 \times 2$ units.

> When you draw a solid on paper you make a 2-D representation of a 3-D object.

1 On ordinary square paper

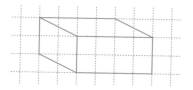

2 On isometric paper (a grid of equilateral triangles)

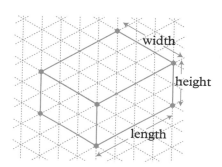

The dimensions of the cuboid cannot be taken from the first picture but they can be taken from the picture drawn on isometric paper. Instead of isometric paper you can also use 'triangular dotty' paper like this. Be careful to use it the right way round (as shown here).

Exercise 22Ⓔ

In questions **1** to **3** the solids are made from 1cm cubes joined together. Draw each solid on isometric paper (or 'triangular dotty' paper). Questions **1** and **2** are already drawn on isometric paper.

1

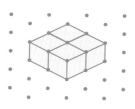

2

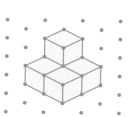

3

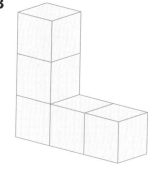

4 The diagram shows one edge of a cuboid.
Complete the drawings of **two** possible cuboids, each with
a volume of 12 cm³. Start with this edge both times.

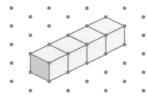

5 Here are two shapes made using four multilink cubes.

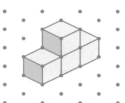

Using four cubes, make and then draw four more shapes
that are different from the two above.

6 The shape shown falls over on to the shaded face.
Draw the shape after it has fallen over.

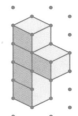

7 You need 16 small cubes.
Make the two solids in the diagram
and then arrange them into a
4 × 4 × 1 cuboid by adding
a third solid, which you have to find.
Draw the third solid on isometric paper.
There are **two** possible solids.

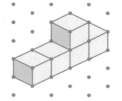

8 Make the letters of your initials from cubes and then draw
them on isometric paper.

9 The side view and plan view (from above) of
object A are shown.

> The side view is
> sometimes called
> the **'side elevation.'**

plan view ↓

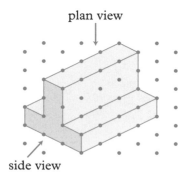

side view

A

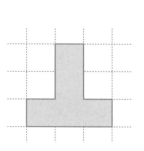

side view

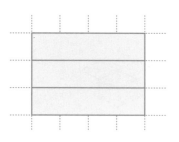

plan view

Draw the side view and plan view of objects B and C.

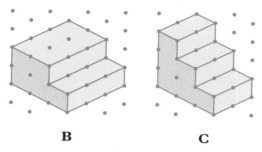

B   **C**

3.8 Area

3.8.1 Counting squares

● Area describes how much **surface** a shape has.

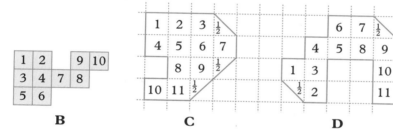

B   **C**   **D**

B contains 10 squares.   C has an area of $12\frac{1}{2}$ squares.

B has an area of 10 squares.   D has an area of 12 squares.

● A square one centimetre by one centimetre has
an area of one square centimetre. You write this
as $1\,cm^2$.

Exercise 23 ⓒ

In the diagrams each square represents $1\,cm^2$.
Copy each shape and find its area by counting squares.

1   **2**   **3**   **4**

   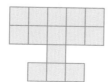

5 **6** **7**  **8**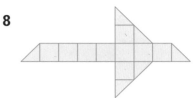

9 You can **estimate** the area of a curved shape like this.

a Count the whole squares inside the shape and mark them ✓.

b Count the parts where there is about half a square or more and mark them ✓.

c Ignore the parts where there is less than half a square inside the shape.

Each square represents 1 cm².

Some of the squares on the shape shown have been ticked according to the instructions **a**, **b** and **c**.
Use a photocopy of this page to find an estimate for the area of the shape.

10 Use the method in question **9** to estimate the area of each of these shapes.
(Ask your teacher for a photocopy of this page.)

a 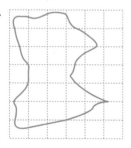 **b** 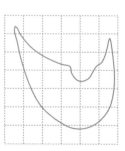 **c** 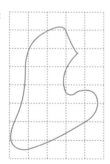 **d**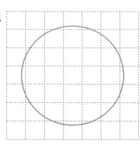

e 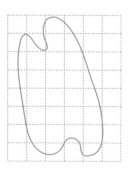 **f** 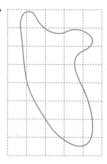 **g** 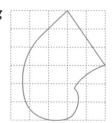 **h**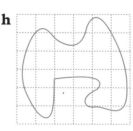

3.8.2 Areas of rectangles

You can find the area of a rectangle by counting squares. This rectangle has an area of 30 squares. If each square is 1 cm^2, this rectangle has an area of 30 cm^2.

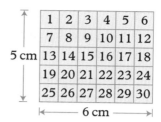

It is easier to multiply the length by the width of the rectangle than to count squares.

Area of rectangle = length × width
$$= (6 \times 5) \text{ cm}^2$$
$$= 30 \text{ cm}^2$$

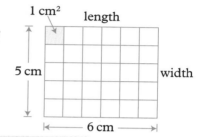

• Area of a rectangle = length × width

A square is a special rectangle in which the length and width are equal.

• Area of a square = length × length

Exercise 24 C

Calculate the areas of these rectangles. All lengths are in centimetres.

1
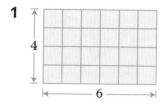
4
6

2
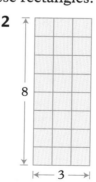
8
3

3
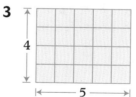
4
5

4
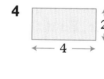
2
4

5
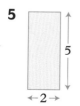
5
2

6
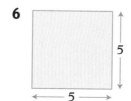
5
5

7
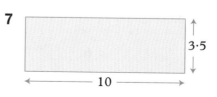
3·5
10

8 Find the area of a rectangular table measuring 60 cm by 100 cm.

9 A farmer's field is a square of side 200 m. Find the area of the field in m².

10 Here is Kelly's passport photo.
Measure the height and width and then work out its area in cm².

11 Work out the area of the shaded rectangle.

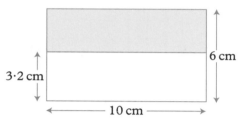

6 cm

3·2 cm

10 cm

Exercise 25 C

All lengths are in cm.

1 This shape is made from two rectangles joined together.
Find the total area of the shape.

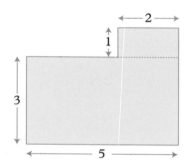

2

1

3

5

Start by finding the area of each rectangle then add them together.

The shapes in questions **2** to **12** are made of rectangles joined together. Find the area of each shape.

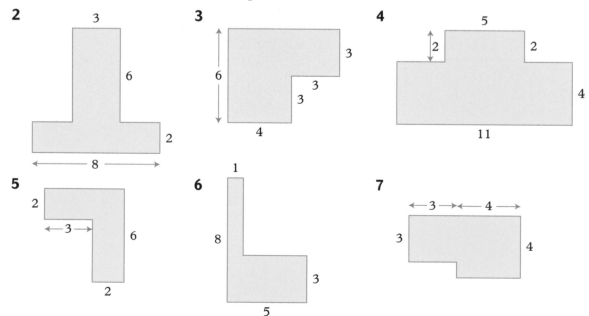

2

3

6

2

8

3

6

3

3

3

4

4

5

2

2

4

11

5

2

3

6

2

6

1

8

3

5

7

3

4

3

4

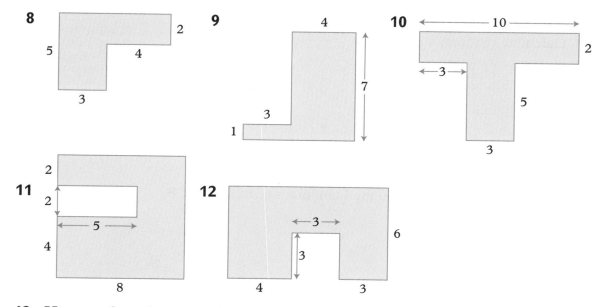

8

2

5   4

3

9

4

7

3

1

10

10

2

3

5

3

11

2

2

5

4

8

12

3

6

3

4   3

13 Here are four shapes made with centimetre squares.

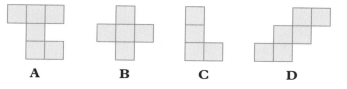

A   **B**   **C**   **D**

a Which shape has an area of 5 cm²?
b Find the areas of the other three shapes.
c Draw a shape using centimetre squares that has an area of 7 cm².

14 A rectangular pond, measuring 10 m by 6 m, is surrounded
by a path which is 1 m wide. Find the area of the path.

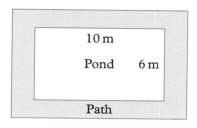

10 m

Pond   6 m

Path

> Find the area of
> the pond and path,
> then subtract the
> area of the pond.

15 Find the height of each rectangle.

a

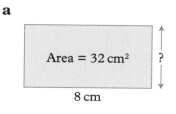

Area = 32 cm²   ?

8 cm

b

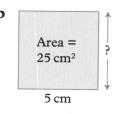

Area =
25 cm²   ?

5 cm

c

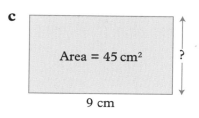

Area = 45 cm²   ?

9 cm

*16 A wooden cuboid has the dimensions shown.
Calculate the total surface area.

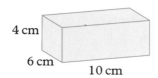

4 cm

6 cm

10 cm

> You find the surface area of a 3-D shape by adding the areas of all the faces.

3.8.3 Areas of triangles

This triangle has base 6 cm, height 4 cm and a right angle at B.

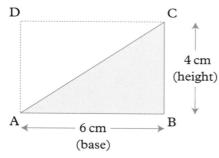

D

C

4 cm
(height)

A

6 cm
(base)

B

Area of rectangle ABCD $= (6 \times 4)$ cm^2
$$= 24 \text{ cm}^2$$
Area of triangle ABC $\quad=$ area of triangle ADC
Area of triangle ABC $\quad= 24 \div 2$
$$= 12 \text{ cm}^2$$

You can show that for any triangle
● Area $= \frac{1}{2}$ (base × height)   or   Area $= \dfrac{(\text{base} \times \text{height})}{2}$

> **Learn** this formula.

Exercise 26 Ⓒ

Find the area of each triangle. Lengths are in cm.

1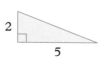

2

5

2

7

4

3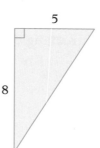

5

8

4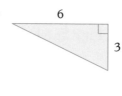

6

3

5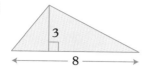

6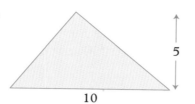

7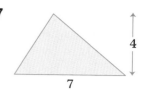

In questions **8** to **10** find the total area of each shape.

8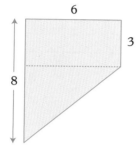

9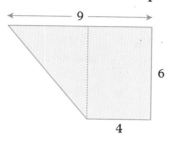

10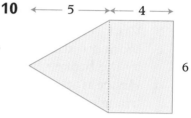

11 Find the total shaded area.

12 Find the total shaded area.

13 Find the shaded area.

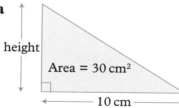

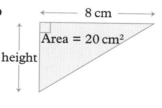

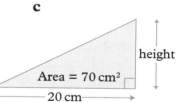

14 Find the height of each triangle.

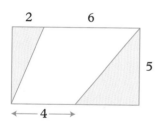
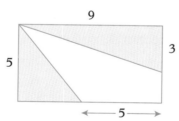

a height Area = 30 cm² 10 cm

b 8 cm Area = 20 cm² height

c height Area = 70 cm² 20 cm

15 The triangle on the right has an area of 2 cm².
 a On square dotty paper draw a triangle with an area of 3 cm².
 b Draw a triangle, different from the one shown, with an area of 2 cm².

16 Joe said 'There are $100\,cm^2$ in $1\,m^2$.' Mark said 'That's not right because one square metre is $100\,cm$ by $100\,cm$ so there are $10\,000\,cm^2$ in $1\,m^2$.' Who is right?

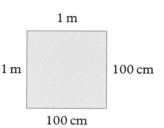

1 m

1 m   100 cm

100 cm

Exercise 27Ⓔ

1 a Copy the diagram.
  b Work out the areas of triangles A, B and C.
  c Work out the area of the square enclosed by the broken lines.
  d Hence work out the area of the shaded triangle.
    Give the answer in square units.

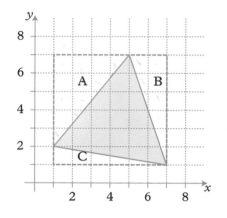

2 a Copy the diagram.
  b Work out the areas of triangles A, B and C.
  c Work out the area of the rectangle enclosed by the broken lines.
  d Hence work out the area of the shaded triangle.
    Give the answer in square units.

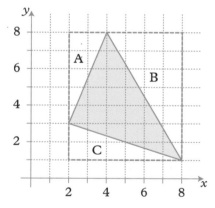

For questions **3** to **7**, draw a pair of axes similar to those in questions **1** and **2**. Plot the points in the order given and find the area of the shape they enclose.

3 $(1, 4), (6, 8), (4, 1)$

4 $(1, 7), (8, 5), (4, 2)$

5 $(2, 4), (6, 1), (8, 7), (4, 8), (2, 4)$

6 $(1, 4), (5, 1), (7, 6), (4, 8), (1, 4)$

7 $(1, 6), (2, 2), (8, 6), (6, 8), (1, 6)$

3.9 Perimeter

3.9.1 The perimeter of a shape

● The **perimeter** of a shape is the total length of its boundary.

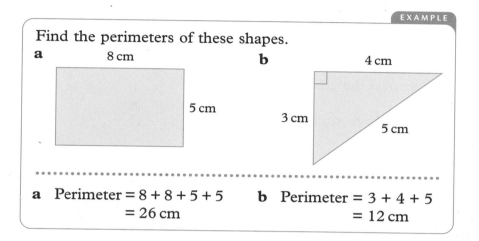

EXAMPLE

Find the perimeters of these shapes.

a   8 cm   5 cm

b   4 cm   3 cm   5 cm

a   Perimeter = 8 + 8 + 5 + 5
         = 26 cm

b   Perimeter = 3 + 4 + 5
         = 12 cm

Exercise 28 C

1 Find the perimeter of each shape. All lengths are in cm.

a   6   4

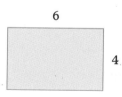

b   13   12   5

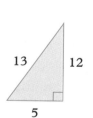

c   11   9   5   5   8

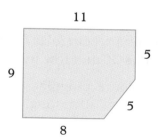

d   square   7

2 Measure the sides of these shapes and work out the perimeter of each one.

a

b

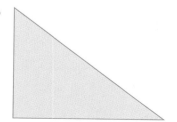

c

3 Here are three shapes made with centimetre squares.

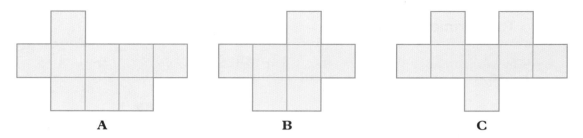

A   B   C

a Find the perimeter of each shape.
b Draw a shape of your own design with a perimeter
   of 14 cm.
   Ask another student to check your shape.

4 Use a ruler to find the perimeter of each picture.

a    **b** 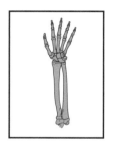   **c**

5 The perimeter of a rectangular field is 160 m. The shorter
   side of the field is 30 m. How long is the longer side?

6 The perimeter of a rectangular garden is 56 m. The two
   longer sides are each 20 m long. How long are the shorter
   sides of the garden?

In questions **7** to **14** find the perimeter of each shape.
All lengths are in cm.

7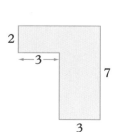

8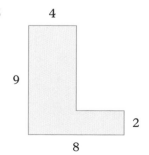

9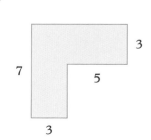

10

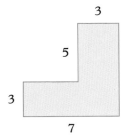

11

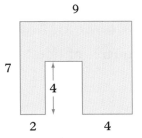

12

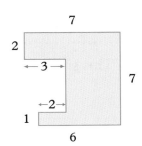

13

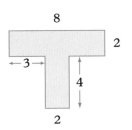

14

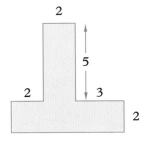

Test yourself

1 This shape is made up of squares of side 1 cm.

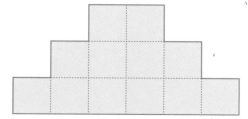

   a Find the perimeter of the shape.
   b Write the area of the shape.
   c Copy the shape and shade in $\frac{1}{4}$ of it.

(OCR, 2003)

2 a Copy the diagram and draw a line from the point C perpendicular to the line *AB*.

   b Sketch a cylinder in your book.

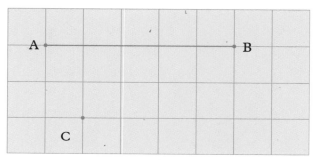

(Edexcel, 2004)

3 a The diagram shows three angles on a straight line AB.

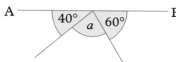

   Work out the value of *a*.

   b Work out the values of *b* and *c* in this diagram.

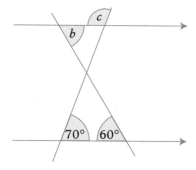

(AQA, 2004)

4 ABCD is a quadrilateral.
Work out the size of the largest angle in the quadrilateral.

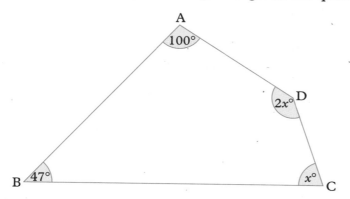

(Edexcel, 2004)

5 Here is a map of Morris Island.

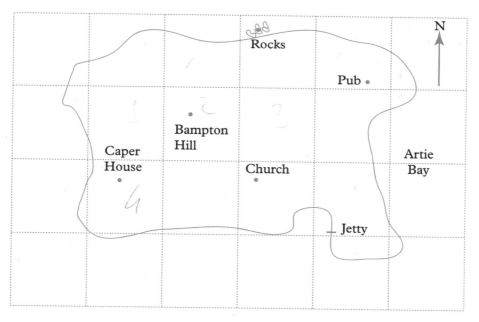

a Roy walks from the Rocks to the Church.
In what direction is he walking?
b From the church he walks north-west.
What place is he walking towards?

The side of each square on the map represents 1 km.
c How far is it from Caper House to the Church?
d Estimate the area of Morris island.

(OCR, 2005)

6 Here is a cuboid.

  a Write

    i the number of edges of this cuboid,

    ii the number of vertices of the cuboid.

  b Draw an accurate net for the cuboid.

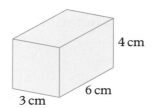

Not drawn accurately

(Edexcel, 2003)

7 A new country is designing its flag.
It makes the design in the shape of a triangle.
The sides of the triangle are

60 cm, 80 cm and 1 metre.

Using a scale of **1 cm to 10 cm**
construct an accurate drawing of the design.
Show your construction lines.

(OCR, 2004)

8 The diagram shows a solid shape
made from 4 one-centimetre cubes.

What is the surface area of the solid shape?

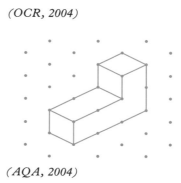

(AQA, 2004)

9 **a** Draw a circle of radius 4 cm.

  b Write the length of the diameter of the circle.

  c On your diagram draw a tangent to the circle.

  d On your diagram draw a chord of length 6 cm inside the circle.

(AQA, 2003)

10 The triangle ABC is shown on
a centimetre grid.

  a Find the coordinates of the
midpoint of BC.

  b Find the area of the triangle ABC.

  c Show how two triangles congruent
to triangle ABC can be put together
to form an isosceles triangle.

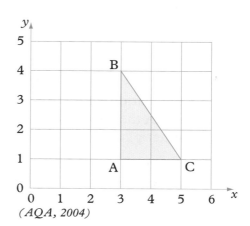

(AQA, 2004)

11 The diagram shows part of a design.
Only the first quarter has been completed.
The complete design must have
rotational symmetry of order 4.

Copy and complete the design by
shading twelve more squares.

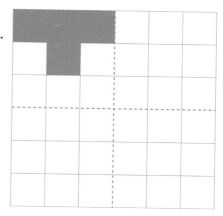

(OCR, 2004)

12 The length of a rectangle is 10·8 cm.
The perimeter of the rectangle is 28·8 cm.
Calculate the width of the rectangle.

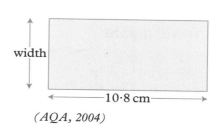

(AQA, 2004)

13 Look at this diagram.

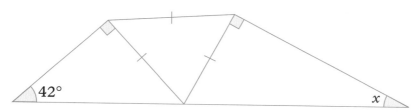

Work out the value of x.

(MEI, Spec)

14 The diagram shows a solid object.

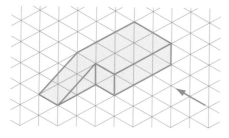

   a Sketch the front elevation from the direction marked
   with an arrow.
   b Sketch the plan of the solid object.

(Edexcel, 2004)

4 Handling data 1

In this unit you will:
- learn how to interpret and discuss graphs
- learn how to find the mean, median, mode and range for discrete, continuous and grouped data
- learn how to draw and interpret graphs, charts and diagrams.

4.1 Interpreting graphs

4.1.1 Travel graphs

- A distance–time graph illustrates a journey by showing how far you have travelled during a period of time.
- A sloping line shows movement — the steeper the line, the faster the movement.
- A horizontal line shows there is no movement.

Exercise 1 C

Look at each graph and then answer the questions.

1 This graph shows a car journey from Newbury to Tonbridge.
   a Find the time when the car is
      i   at Reigate
      ii at Basingstoke
      iii at Tonbridge.
   b Where is the car
      i   at 09:00
      ii at 09:45
      iii after travelling for two hours?

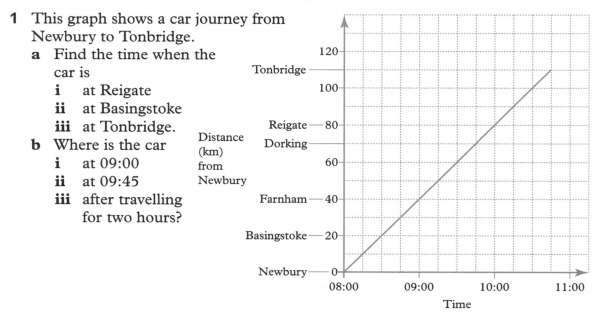

2 This graph shows a coach journey from Bristol to Portsmouth.

  a Find the time when
    the coach is
    i at Salisbury
    ii at Bath
    iii at Newton.
  b Where is the coach
    i at 14:00
    ii at 15:45
    iii at 13:00
    iv at 16:15?
  c How far is it from
    Bristol to Westbury?
  d How long does it take to travel
    from Bristol to Westbury?
  e At what speed does the coach
    travel from Bristol to Westbury?

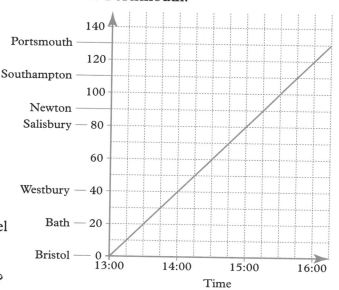

You can use the formula

$$\text{Speed} = \frac{\text{Distance travelled}}{\text{Time taken}}$$

3 This graph shows a bicycle journey from Royston to Harlow.

  a Find the time
    when the bicycle is
    i at Harlow
    ii at Cambridge.
  b Where is the bicycle
    i at 11:45
    ii at 12:45?
  c How far is it
    from Royston
    to Cambridge?
  d How long does
    it take to travel from
    Royston to
    Cambridge?
  e At what speed does
    the cyclist travel from
    Royston to Cambridge?

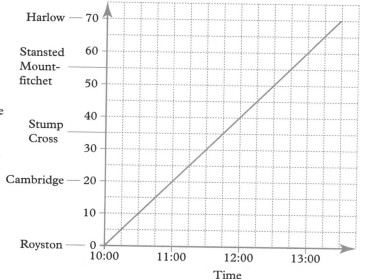

4 This graph shows a bus journey from Dover to Burgess Hill.

a When is the bus
  i at Throwley Forstal
  ii at Burgess Hill
  iii (more difficult) at Folkestone?

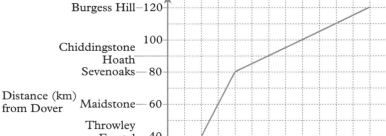
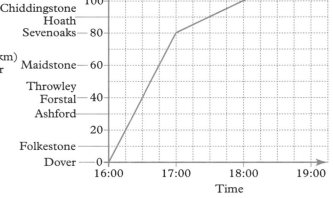

b Where is the bus
  i at 17:00
  ii at 17:30
  iii at 16:45?

c How far does the bus travel between 16:00 and 17:00?

d At what speed does the bus travel from Dover to Sevenoaks?

e How far does the bus travel between 17:00 and 19:00?

f At what speed does the bus travel from Sevenoaks to Burgess Hill?

5 Mike makes a journey partly on foot and partly on bicycle.

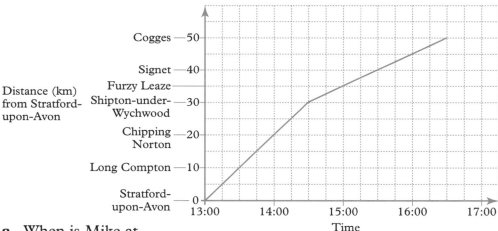

a When is Mike at
  i Signet   **ii** Cogges   **iii** Shipton-under-Wychwood?

b Where is he at
  i 14:00   **ii** 15:00   **iii** 13:30?

c How far does he travel between 13:00 and 14:30?

d At what speed does he travel from Stratford-upon-Avon to Shipton-under-Wychwood?

e How far does he travel between 14:30 and 16:30?

f At what speed does he complete the second part of the journey?

6 This graph shows Alex's journey cycling and running from
Claydon to Bury St. Edmunds.

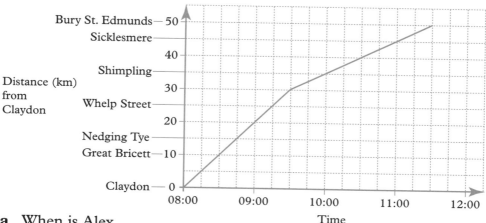

a When is Alex
  i at Nedging Tye   **ii** at Whelp Street   **iii** at Sicklesmere?
b Where is she
  i at 08:30   **ii** at 11:30   **iii** at 10:00?
c How far does she cycle between 08:00 and 09:30?
d At what speed does she cycle from 08:00 until 09:30?
e How far does she run between 09:30 and 11:30?
f At what speed does she run from 09:30 until 11:30?
g What is Alex's **average** speed on the complete journey?

> Use the formula
> Average speed =
> Total distance
> Time taken

4.1.2 Broken journeys

Exercise 2 C

Look at each graph and then answer the questions.
1 This graph shows a bus journey
from York to Skipton and
back again.
  a How long does the bus stop at
    Skipton?
  b At what speed does
    the bus travel
    i   from York to
      Harrogate
    ii from Harrogate to Skipton
    iii from Skipton back to Harrogate?
  c When does the bus
    i   return to York
    ii arrive at Skipton
    iii leave Harrogate?

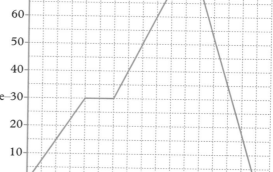

2 This graph shows a car journey from Albridge to Chington and back again.

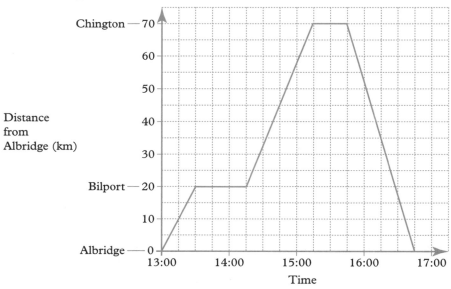

a How long does the car stop at Bilport?
b When does the car arrive at Chington?
c When does the car leave Bilport?
d At what speed does the car travel
   i from Albridge to Bilport
   ii from Bilport to Chington
   iii from Chington back to Albridge?
e At what time on the return journey is the car exactly halfway between Chington and Albridge?

3 The graph shows a return journey by car from Leeds to Scarborough.

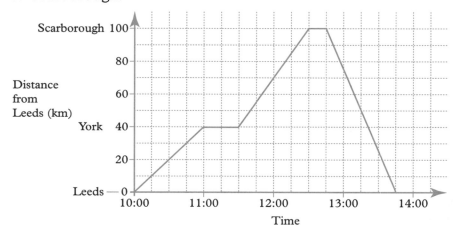

a How far is it from Leeds to York?
b How far is it from York to Scarborough?
c At which two places does the car stop?
d How long does the car stop at Scarborough?
e When does the car
    i   arrive in York
    ii   arrive back in Leeds?
f What is the speed of the car
    i   from Leeds to York
    ii   from York to Scarborough
    iii from Scarborough to Leeds?

> Speed =
> $\dfrac{\text{Distance travelled}}{\text{Time taken}}$

4 Steve cycles to a friend's house, but on the way his bike
gets a puncture and he has to walk the remaining distance.
At his friend's house, he repairs the puncture, has a game of
snooker and then returns home. On the way back, he stops
at a shop to buy a book on how to play snooker.

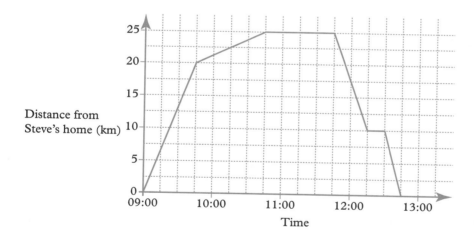

a How far is it to his friend's house?
b How far is it from his friend's house to the shop?
c At what time did his bike get a puncture?
d How long did he stay at his friend's house?
e At what speed did he travel
    i   from home until he had the puncture
    ii   after the puncture to his friend's house
    iii from his friend's house to the shop
    iv from the shop back to his own home?

5 Mr Berol and Mr Hale use the same road to travel between Aston and Borton.
This graph shows both journeys on one set of axes.

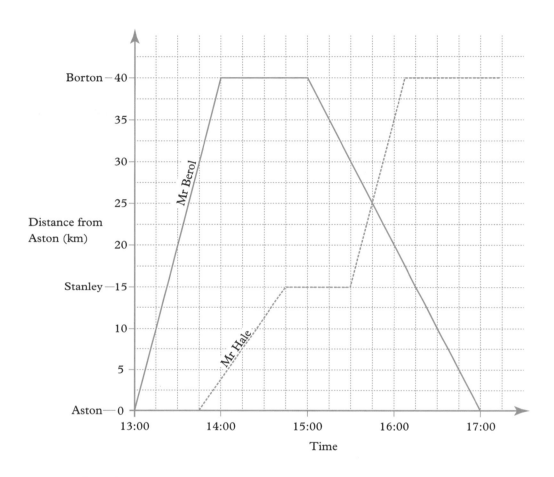

a At what time did
   i Mr Berol arrive in Borton
   ii Mr Hale leave Aston?

b **i** When did Mr Berol and Mr Hale pass each other?
   ii In which direction was Mr Berol travelling?

c Find the speeds of
   i Mr Hale from Aston to Stanley
   ii Mr Berol from Aston to Borton
   iii Mr Hale from Stanley to Borton
   iv Mr Berol from Borton back to Aston.

6 The graph shows the journeys made by a van and a car
   starting at York, travelling to Durham and returning to York.

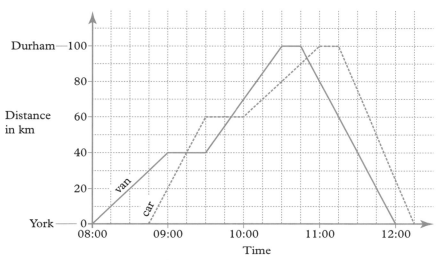

a For how long was the van stationary during the journey?
b At what time did the car first overtake the van?
c At what speed was the van travelling between 09:30
   and 10:00?
d What was the greatest speed of the car during the entire
   journey?
e What was the average speed of the car over its entire
   journey?

> Average speed =
> $\dfrac{\text{Distance travelled}}{\text{Total time taken}}$

4.1.3 Real-life graphs

You can use real-life graphs to find solutions to practical problems.

EXAMPLE

This graph converts miles
into kilometres.
Read off the axes to find
20 miles = 32 km
64 km = 40 miles.

Exercise 3 **C**

1 The graph converts pounds into euros.
  a Convert into pounds.
    i €4
    ii €3·30
  b Convert into euros.
    i £3
    ii £1·40

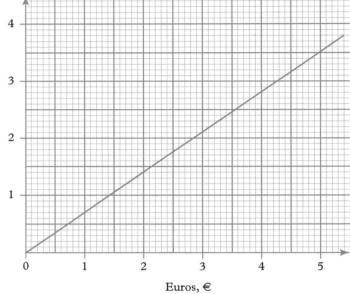

Pounds, £

Euros, €

  c Henri, the highway robber, held up the Paris–Berlin stagecoach and got away with €50 000. How much was that in pounds?

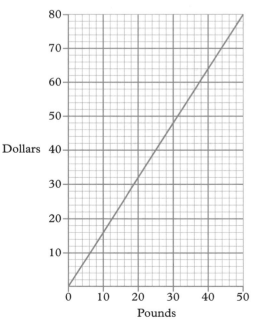

2 This graph converts US dollars to pounds.
  a Convert into pounds.
    i   40 dollars   **ii**   70 dollars
    iii 48 dollars   **iv**   64 dollars
    v   16 dollars   **vi**   24 dollars
  b Convert into dollars.
    i   £20   **ii** £50   **iii** £6
    iv £16   **v**   £36   **vi** £34
  c Before flying home from New York the Wilson family bought gifts for $72. How much did they spend in pounds?

Dollars

Pounds

3 The rupee is the currency used in India. This graph converts rupees into pounds.

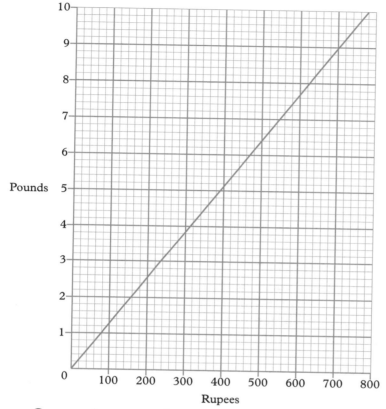

a Convert into pounds.
  i 280 rupees   **ii** 660 rupees   **iii** 420 rupees
  iv 380 rupees   **v** 600 rupees   **vi** 40 rupees
b Convert into rupees.
  i £8·00   **ii** £7·40   **iii** £4·60
  iv £2·00   **v** £10·00   **vi** £6·60
c On holiday in India, Stanley bought snake-charming lessons worth £9. How much did he pay in rupees?

4 This line graph shows the depth of water in a stream during one year.
  a How deep was the stream in March?
  b In which two months was the stream 8 cm deep?
  c Which month saw the largest increase in depth?
    Why do you think this happened?

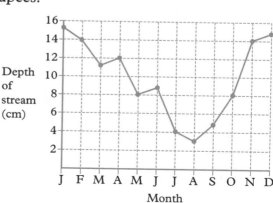

5 This line graph shows the average daily temperature in Sweden.

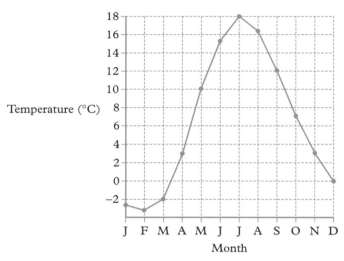

a What was the average temperature in June?
b In which month was the average temperature 7 °C?
c In which two months was the average temperature 3 °C?
d Which month saw the largest fall in temperature?
e The range is the difference between the highest and lowest temperatures.
What was the range of temperature over the year?

*6 A car travels along a motorway and the amount of petrol in its tank is monitored as shown on the graph.

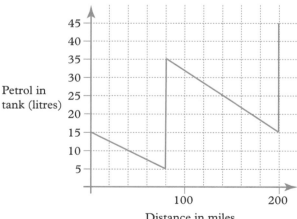

a How much petrol was bought at the first stop?
b What was the petrol consumption in miles per litre
  i before the first stop
  ii between the two stops?

*7 Kendal Motors hires out vans.

a Copy and complete the table where x is the number of miles travelled and C is the total cost in pounds.

x	0	50	100	150	200	250	300
C	35			65			95

b Draw a graph of C against x, using scales of 2 cm for 50 miles on the x-axis and 1 cm for £10 on the C-axis.

c Use the graph to find the number of miles travelled when the total cost was £71.

*8 Jeff sets up his own business as a plumber.

a Copy and complete the table where C stands for his total charge and h stands for the number of hours he works.

h	0	1	2	3
C		33		

b Draw a graph with h across the page and C up the page. Use scales of 2 cm to 1 hour for h and 2 cm to £10 for C.

c Use your graph to find how long Jeff worked if his charge was £55·50.

4.1.4 Sketch graphs

You can sketch a graph to see the shape of a general trend. Sketch graphs have no detailed labels on the axes.

Exercise 4(E)

1 Match each of these statements to one of the graphs.

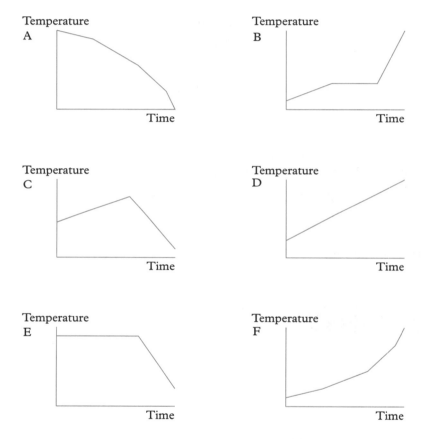

a The temperature rises steadily.
b The temperature begins to rise and then falls quickly.
c The temperature rises more and more quickly.
d The temperature stays the same and then falls quickly.
e The temperature falls faster and faster.
f The temperature rises, stays the same and then rises quickly.

2 Which of the graphs A to D best fits the statement
'Unemployment is still rising but by less each month'.

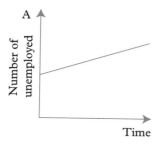

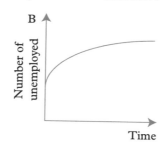

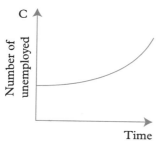

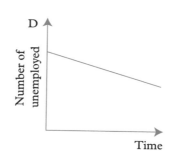

3 Which of the graphs A to D best fits the statement
'The price of oil was rising more rapidly in 2005 than at any
time in the previous ten years'.

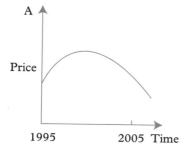

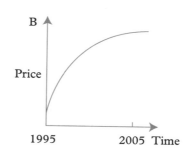

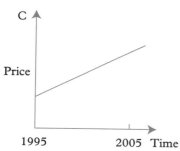

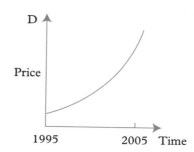

4 Which of the graphs A to D best fits each of the statements **a–d**?

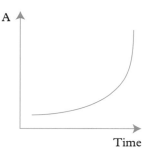

A

Time

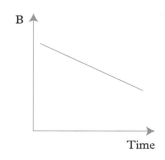

B

Time

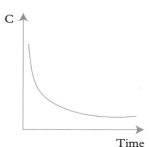

C

Time

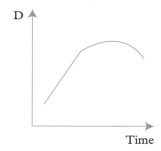

D

Time

a The birthrate was falling but is now steady.
b Unemployment, which rose slowly until 2004, is now rising rapidly.
c Inflation, which has been rising steadily, is now beginning to fall.
d The price of gold has fallen steadily over the last year.

*5 The graph shows the motion of three cars A, B and C along the same road. Answer these questions giving estimates where necessary.

a Which car is in front after
  i 10 s **ii** 20 s?
b When is B in the front?
c Which car is going fastest after 5 s?
d Which car starts slowly and then goes faster and faster?

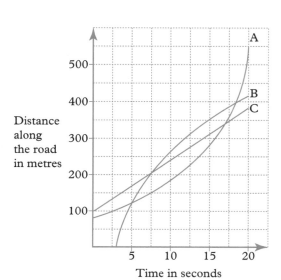

Distance along the road in metres

Time in seconds

*6 Water is poured at a constant rate into each of the
containers A, B and C.
The graphs X, Y and Z show how the water level rises.

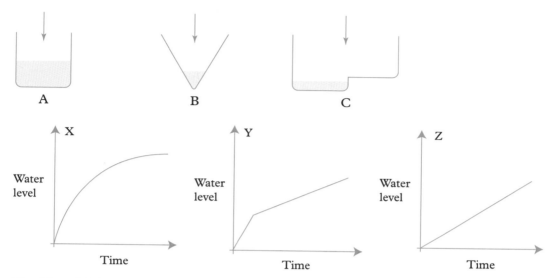

Decide which graph fits each container.

*7 The diagrams show three containers P, Q and R with
water coming out as they empty.
Sketch three graphs similar to those in question **6** to show
how the water level falls against time.

4.2 Averages and spread

If you have a set of data, say exam marks or heights, and
are told to find the 'average', just what are you trying to
find? The answer is: a single number which can be used to
represent the entire set of data.

The 'average' value can be found in three different ways:
the **median**, the **mode** and the **mean**.

4.2.1 The median

● The data are arranged in order from the smallest to the largest; the middle number is then selected. This is called the median.

If there are two 'middle' numbers, the median is in the middle of these two numbers.

Median

Seven people are lined up in order of height. The median height is the height of the person in the middle.

Find the median of these sets of numbers
a 8   9   2   4   13   10   5
b 5   6   4   5   10   8

a Arrange the numbers in order of size.
   2   4   5   8   9   10   13   The median is **8**.
                  ↑

b Arrange the numbers in order of size.
   4   5   5   6   8   10   The median is **5·5**. This number
                  ↑   is midway between 5 and 6.

Exercise 5 C

1 Find the median for each set of numbers.
  a 5, 6, 7, 3, 11, 9, 14
  b 5, 2, 4, 5, 7
  c 18, 4, 17, 10, 11, 16, 15
  d 7, 8, 4, 6, 4, 5, 9, 12
  e −1, 4, −4, −2, 5
  f 10, 8, 9, 10, 3, 2, 6, 7, 7
  g 2, −2, 6, 8, 4, 0, −1, 5

2 The table shows the age, height and weight of seven students.

	Mike	Steve	Dora	Sam	Pat	Rayyan	Gary
Age (years)	16	15	17	15	16	15	16
Height (cm)	169	180	170	175	172	163	164
Weight (kg)	50	50	52	44	51	41	48

  a What was the median age?
  b What was the median height?
  c What was the median weight?

3 **a** Find the median of   4,   5,   7,   9,   1,   9,   2
  b **Use your answer** to find the median of
                  14, 15, 17, 19, 11, 19, 12

4 Sally throws a dice eight times and her father gives her 20p if the median score is more than 3.
The dice scores are   6, 1, 2, 6, 4, 1, 3, 6.
Find the median score and find out if Sally wins 20p.

5 **a** Write five numbers so that the median is 7.
  b Write seven numbers so that the median is 6.
  c Write six numbers so that the median is 8.

6 Twenty-one people were asked to estimate the weight of a stork (in kg). Their estimates were

  7 30 16 30 17 21   9 24 18   9 25
21 14   8 21 12   7 10 27 20 11

What was the median estimate?

4.2.2 The mode

● The **mode** of a set of numbers is the number which occurs most often.

The mode is the easiest average to find and you can also find it for data which are not just numbers. You could, for example, find the modal colour of T-shirts worn by a group of children.

EXAMPLE

Find the mode of this set of data.
3 4 5 3 6 4 6 5 3 5 6 4 3 6 3

....

The number 3 occurs most often so the mode is 3.

You can order data using a tally chart and frequency table to make it easier to find the mode.

EXAMPLE

Find the mode of this set of data.
5 6 7 5 7 6 5 7 6 5

....

Number	Tally	Frequency
5	||||	④
6	|||	3
7	|||	3

5 is also called the modal number.

The mode is 5, as it has the highest frequency.

Exercise 6 Ⓒ

1 Find the mode of each set of data.
   a 1, 2, 4, 1, 3, 1, 2, 3
   b 3, 4, 5, 6, 6, 3, 6, 5, 4, 6, 6, 3, 4, 5, 6, 4
   c 2, 4, 6, 8, 4, 6, 2, 4, 6, 4, 8, 2, 8, 4, 8, 6
   d 2, 2, 2, 2, 3, 3, 3, 3, 4, 4, 4, 4, 4, 5, 5
   e red, red, blue, red, blue, white, white, red, blue, red
   f 10, 100, 10, 1000, 1000, 1000, 100, 10, 1000, 100

2 The frequency table shows the test results for a class of 30 students.

Mark	3	4	5	6	7	8
Frequency	2	5	4	7	6	6

What is the modal mark?

3 The temperature in a room is measured at midday every day for a month. The results are shown on the chart.
  a On how many days was the temperature 17 °C?
  b What was the modal temperature?
  c How many days were there in that month?

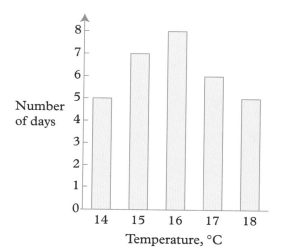

4 A dice was thrown 14 times.

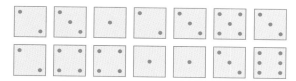

  a What was the modal score?
  b What was the median score?

5 a Write six numbers so that the mode is 2.
  b Write five numbers so that the median is 6 and the mode is 4.
  c Write seven numbers so that the median is 8 and the mode is 5.

6 The numbers of chicks in 32 nests were counted.

```
2 4 3 4 1 4 4 2
3 0 5 3 4 3 4 2
4 1 4 2 0 1 4 3
1 3 2 5 3 5 2 6
```

Make a tally/frequency chart and use it to find the modal number of chicks in a nest.

Number of chicks	Tally	Frequency
0	||	2
1		
2		
3		
4		
5		
6		

4.2.3 The mean

> ● To find the **mean**, add all the data and divide the total by the number of items.

a Find the mean of 3, 9, 6, 6, 8, 3, 9, 4
b Find the mean of 15, 7, 14, 17, 11, 16

....

a There are 8 numbers

$$\text{mean} = \frac{(3 + 9 + 6 + 6 + 8 + 3 + 9 + 4)}{8} = \frac{48}{8} = 6$$

b There are 6 numbers

$$\text{mean} = \frac{(15 + 7 + 14 + 17 + 11 + 16)}{6} = \frac{80}{6} = 13\cdot333\ldots$$

or mean = 13·3 correct to 1 decimal place (1 dp)

Exercise 7 Ⓒ

1 Find the mean of each set of data.
  a 3, 6, 9, 5, 8, 11
  b 10, 12, 10, 7, 8, 10, 13
  c 3, 3, 6, 7, 1, 4, 5, 5, 2
  d 9, 13, 4, 7, 6
  e 1·3, 0·6, 1·4, 3·7, 0·4, 1·6, 0·9, 0·5
  f 4, 4, 4, 4, 4, 4, 4, 4, 4, 4, 4, 4, 4, 4, 4

2 In a test the girls' marks were   8, 6, 7, 9, 5
       and the boys' marks were   6, 7, 9, 5, 8, 5.
  Find the mean mark for the girls and the mean mark
  for the boys.

3 Mr and Mrs Amrit and their three children weigh 84 kg,
  55 kg, 48 kg, 25 kg, and 9 kg.
  Find the mean weight of the Amrit family.

4 a Write three numbers so that the mean is 5.
  b Write five numbers so that the mean is 8.
  c Write five numbers so that the mean is 6 and the
    median is 7.

5 Copy and complete.

   a The mean of   3, 5, 6 and ☐ is 6.

   b The mean of   7, 8, 4 and ☐ is 8.

6 **a** Calculate the mean of   2, 5, 7, 7, 4.
   b **Hence** find the mean of   32, 35, 37, 37, 34.
      (Try to do this **without** adding 32 + 35 + 37 + 37 + 34.)

7 Oliver's hobby is collecting crabs. As an experiment
   he decides to weigh each member of his collection as it
   crawls across a scale pan.
   Their weights (in grams) were

   5·2, 6·9, 2·7, 10·1, 3·6, 8·7, 2·7, 7·5

   Find the mean weight of the crabs, correct to 1 decimal
   place.

8 The total mass of five men was 350 kg. Calculate the mean
   mass of the men.

9 The total height of 7 children is 1127 cm. Calculate the
   mean height of the children.

4.2.4   The range

● The range is not an average. It is the difference between the
  largest value and the smallest value.
  For example: the range of the numbers 3, 7, 11, 12, 16 is 16 − 3 = 13.

The range is a measure of how spread out the data is. It is
useful when comparing sets of data.

Exercise 8 C

1 The marks in a test were   2, 2, 3, 5, 7, 7, 7, 10.
   Copy and complete this sentence.
   The range of the marks is 10 − ☐ = ☐

2 Find the range for each set of data.
   a 4, 7, 8, 14, 20, 30
   b 1, 2, 2, 2, 4, 40, 50
   c 11, 14, 16, 25, 32, 44, 52

3 The temperature in a greenhouse was measured at midnight every day for a week. The results (in °C) were

1, −2, 4, 0, 5, 2, −1.

What was the range of the temperatures?

4 a Write six numbers so that the range is 10.
  b Write five numbers so that the median is 7 and the range is 12.

5 There were 9 children at a party, including one up in the air. The mean age of the children was 15 and the range of their ages was 5.
Write each sentence below and write next to it whether it is **True**, **Possible** or **Impossible**.

  a Every child was currently 15 years old.
  b All the children were at least 13 years old.
  c The oldest child was 5 years older than the youngest child.

6 There were six people living in a house. The **median** age of the people was 20 and the range of their ages was 3.
Write each sentence below and write next to it whether it is **True**, **Possible** or **Impossible**.
  a Every person was either 19 or 20 years old.
  b The oldest person in the house was 23 years old.
  c All six people in the house could speak French.

7 The bar chart shows the scores for some children in a spelling test. The scores are collected together in groups.
  a How many children took the test?
  b What is the modal group?
  c Explain why it is not possible to find the exact range for the scores in this test.

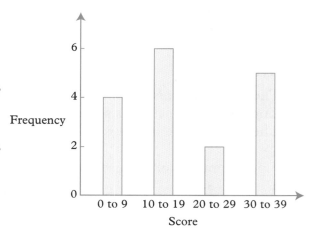

Which is the 'best' average to use?

● The **median** is fairly easy to find and is not really affected by very large or very small values that occur at the ends of the distribution.

Look at these exam marks:

20, 21, 21, 22, 23, 23, 25, 27, 27, 27, 29, 98, 98
                            ↑

The median (25) gives a truer picture of the centre of the distribution than the mean (35·7).

● The **mode** of the exam marks is 27. It is easy to find and it takes away some of the effects of extreme values. However, it does have disadvantages, particularly in data which has two 'most popular' values, and it is not widely used.

● The **mean** takes account of all of the data and is the 'average' which most people first think of. It does, of course, take a little longer to calculate than either the mode or the median.

Exercise 9 ⓒ (Mixed questions)

1 Find the mean, median and mode of these sets of numbers.

a 3, 12, 4, 6, 8, 5, 4
b 7, 21, 2, 17, 3, 13, 7, 4, 9, 7, 9
c 12, 1, 10, 1, 9, 3, 4, 9, 7, 9
d 8, 0, 3, 3, 1, 7, 4, 1, 4, 4
e Which is the best average to use in part **a**? Give your reason.

2 The temperature in °C on 17 days was:

1, 0, 2, 2, 0, 4, 1, 3, 2, 1, 2, 3, 4, 5, 4, 5, 5

a What was the modal temperature?
b What was the range?

3 For this set of numbers, find the mean and the median.

2, 2, 3, 3, 3, 5, 101

Which average best describes the set of numbers?

4 Find
  a the range
  b the mode of these data.

4	2	5	4	5	12	4	1	3	3
3	5	5	3	4	5	2	7	4	2
4	12	1	7	12	1	3	8	10	4

5 The range for the eight numbers on these cards is 40.

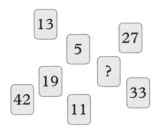

Find the **two** possible values of the missing number.

6 The mean weight of ten people in a lift is 70 kg. The weight limit for the lift is 1000 kg. Roughly how many more people can get into the lift?

7 The bar chart shows the marks scored in a test. What was the modal mark?

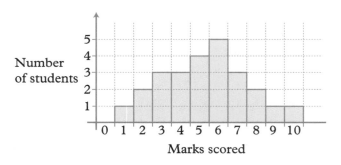

Number of students

Marks scored

8 Six boys have heights of 1·53 m, 1·49 m, 1·60 m, 1·65 m, 1·90 m and 1·43 m.
  a Find the mean height of the six boys.
  b Find the mean height of the remaining five boys when the shortest boy leaves.

9 In a maths test the marks for the boys were 9, 7, 8, 7, 5 and the marks for the girls were 6, 3, 9, 8, 2, 2.
  a Find the mean mark for the boys.
  b Find the mean mark for the girls.
  c Find the mean mark for the whole class.

10 Five different drinks have an average
(mean) price of £1·50.
When a sixth drink is added the new
average price is £1·60.
How much did the sixth drink cost?

11 Write five numbers so that
the mean is 6
the median is 5
the mode is 4.

12 There were ten cowboys in a saloon. The mean age
of the men was 25 and the range of their ages was 6.
Write each statement below and then write next to it
whether it is 'true', 'possible' or 'false'.
a The youngest man was 18 years old.
b All the men were at least 20 years old.
c The oldest person was 4 years older than the youngest.
d Every man was between 20 and 26 years old.

13 These are the salaries of 5 employees in a small
business.
Mr A : £22 500   Mr B : £17 900   Mr C : £21 400
Mr D : £22 500   Mr E : £155 300
a Find the mean and the median of their salaries.
b Which does **not** give a fair 'average'? Explain why in
one sentence.

14 A farmer has 32 cattle to sell.
The weights of the cattle in kg are
81	81	82	82	83	84	84	85
85	86	86	87	87	88	89	91
91	92	93	94	96	150	152	153
154	320	370	375	376	380	381	390
(Total weight = 5028 kg)

On the telephone to a potential buyer, the
farmer describes the cattle and says the
'average' weight is 'over 157 kg'.
a Find the mean weight and the median
weight.
b Which 'average' has the farmer used to describe his
animals?
Does this average describe the cattle fairly?

4.2.5 Calculating the mean from a frequency table

The frequency tables shows the weights of the eggs bought in a supermarket.
Find the mean, median and modal weight.

Weight	58 g	59 g	60 g	61 g	62 g	63 g
Frequency	3	7	11	9	8	2

Mean weight of eggs

$$= \frac{(58 \times 3) + (59 \times 7) + (60 \times 11) + (61 \times 9) + (62 \times 8) + (63 \times 2)}{(3 + 7 + 11 + 9 + 8 + 2)} = \frac{2418}{40}$$

$$= 60 \cdot 45 \text{ g}$$

There are 40 eggs so the median weight is the number between the 20th and 21st numbers. You can see that both the 20th and 21st numbers are 60 g.
So, median weight = 60 g

The modal weight is the weight with the highest frequency.
The modal weight = 60 g

Exercise 10Ⓔ

1 The frequency table shows the weights of the 40 apples sold in a shop.

Weight	70 g	80 g	90 g	100 g	110 g	120 g
Frequency	2	7	9	11	8	3

Calculate the mean weight of the apples.

2 The frequency table shows the price of a packet of butter in 30 different shops.

Price	49p	50p	51p	52p	53p	54p
Frequency	2	3	5	10	6	4

Calculate the mean price of a packet of butter.

3 A box contains 50 nails of different lengths as shown in the frequency table.

Length of nail	2 cm	3 cm	4 cm	5 cm	6 cm	7 cm
Frequency	4	7	9	12	10	8

Calculate the mean length of the nails.

4 The tables give the distribution of marks obtained by two classes in a test. For each class, find the mean, median and mode.

a Class 1

Mark	0	1	2	3	4	5	6
Frequency	3	5	8	9	5	7	3

b Class 2

Mark	15	16	17	18	19	20
Frequency	1	3	7	1	5	3

5 A teacher conducted a mental arithmetic test for 26 students and the marks out of 10 are shown in the table.

Mark	3	4	5	6	7	8	9	10
Frequency	6	3	1	2	0	5	5	4

a Find the mean, median and mode.
b The teacher congratulated the class saying that 'over three-quarters were above average'. Which average must the teacher be using?

*6 The table shows the number of goals scored in a series of football matches.

Number of goals	1	2	3
Number of matches	8	8	x

a If the modal number of goals is 3, find the smallest possible value of x.
b If the median number of goals is 2, find the largest possible value of x.

4.2.6 Data in groups

EXAMPLE

The results of 51 students in a test are given in the frequency table.
Find the   **a** mean   **b** median   **c** mode.

Mark	30–39	40–49	50–59	60–69
Frequency	7	14	21	9

a To find the mean you approximate by saying each interval is represented by its midpoint. For the 30–39 interval the midpoint is $(30 + 39) \div 2 = 34 \cdot 5$.

$$\text{Mean} = \frac{(34 \cdot 5 \times 7) + (44 \cdot 5 \times 14) + (54 \cdot 5 \times 21) + (64 \cdot 5 \times 9)}{(7 + 14 + 21 + 9)}$$

$$= 50 \cdot 774\,509\,8$$

$$= 51 \ (2 \ \text{sf})$$

b The median is the 26th mark, which is in the interval 50–59. You cannot find the exact median.

c The **modal group** is 50–59. You cannot find an exact mode.

> Don't forget the mean is only an **estimate** because you do not have the raw data and you have made an assumption using the midpoint of each interval.

Exercise 11Ⓔ

1 The table gives the number of words in each sentence of a page in a book.
   a Copy and complete the table.
   b Work out an estimate for the mean number of words in a sentence.

Number of words	Frequency f	Midpoint x	fx
1–5	6	3	18
6–10	5	8	40
11–15	4		
16–20	2		
21–25	3		
Totals	20	—	

2 The results of 24 students in a test are given in the table.

Mark	40–54	55–69	70–84	85–99
Frequency	5	8	7	4

Find the midpoint of each group of marks and calculate an estimate of the mean mark.

3 The table shows the number of letters delivered to the 26 houses in a street.

Number of letters delivered	Number of houses (frequency)
0–2	10
3–4	8
5–7	5
8–12	3

Calculate an estimate of the mean number of letters delivered per house.

4.3   Pie charts and pictograms

Raw data are collected when you do a survey or experiment. This sort of information is often much easier to understand in a pie chart or a frequency diagram.

> Teacher's note: See page 374 for 'Pie charts and bar charts on a computer'.

4.3.1   Pie charts

- In a pie chart all the data are represented by a circle and the sectors of the circle represent the different items.
- The angle of each sector is proportional to the frequency of that item.

Exercise 12 C

The questions in this exercise will remind you how to cancel fractions and how to find a fraction of a number.

In questions **1** to **20** simplify the fractions.

1 $\dfrac{20}{24}$ **2** $\dfrac{36}{48}$ **3** $\dfrac{45}{90}$ **4** $\dfrac{32}{40}$ **5** $\dfrac{21}{35}$

6 $\dfrac{15}{18}$ **7** $\dfrac{24}{30}$ **8** $\dfrac{120}{360}$ **9** $\dfrac{90}{360}$ **10** $\dfrac{150}{360}$

11 $\dfrac{180}{360}$ **12** $\dfrac{240}{360}$ **13** $\dfrac{300}{360}$ **14** $\dfrac{60}{360}$ **15** $\dfrac{45}{360}$

16 $\dfrac{40}{360}$ **17** $\dfrac{80}{360}$ **18** $\dfrac{210}{360}$ **19** $\dfrac{35}{360}$ **20** $\dfrac{54}{360}$

> Reminder:
> $\dfrac{15}{20} = \dfrac{3}{4}$
> Divide 15 and 20 by 5.

Work out these.

21 $\frac{1}{4}$ of 48   **22** $\frac{3}{4}$ of 60   **23** $\frac{2}{3}$ of 60

24 $\frac{2}{5}$ of 50   **25** $\frac{1}{8}$ of 88   **26** $\frac{3}{7}$ of 84

27 $\frac{2}{3}$ of 360°   **28** $\frac{5}{6}$ of 360°   **29** $\frac{3}{8}$ of 360°

30 $\frac{1}{12}$ of 360°   **31** $\frac{7}{10}$ of 360°   **32** $\frac{5}{12}$ of 360°

> Reminder:
> $\frac{3}{4}$ of 80
> Find $\frac{1}{4}$ of 80 and then multiply by 3.

EXAMPLE

The pie chart shows how 600 people are spending their summer holiday. How many people are
a camping   **b** touring   **c** going to the seaside?

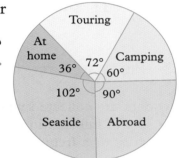

a   Number of people camping $= \frac{60}{360} \times 600$

$= \frac{1}{6} \times 600$

$= 100$

b   Number of people touring $= \frac{72}{360} \times 600$

$= \frac{1}{5} \times 600$

$= 120$

c   Number of people at seaside $= \frac{102}{360} \times 600$

$= 170$

Exercise 13 C

1 The pie chart shows the country of origin of 120 passengers on board an aircraft.

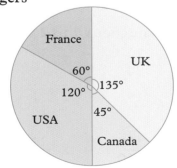

   a Simplify these fractions.

     i $\frac{60}{360}$ **ii** $\frac{120}{360}$ **iii** $\frac{45}{360}$ **iv** $\frac{135}{360}$

   b What **fraction** of the passengers came from

     i   France   **ii**   USA

     iii Canada   **iv** UK?

   c Work out these.

     i $\frac{1}{6}$ of 120   **ii** $\frac{1}{3}$ of 120   **iii** $\frac{1}{8}$ of 120

   d How many passengers came from

     i   France   **ii**   USA   **iii**   Canada   **iv**   UK?

2 The colour of each of the 36 cars in a car park was
noted and the results are shown in the pie chart.
  a Simplify these fractions.

$$\mathbf{i} \quad \frac{90}{360} \qquad \mathbf{ii} \quad \frac{60}{360} \qquad \mathbf{iii} \quad \frac{120}{360}$$

  b What **fraction** of the cars were
  i   red   **ii** green
  iii yellow   **iv** blue?

  c How many of the 36 cars were
  i   red   **ii** green
  iii yellow   **iv** blue?

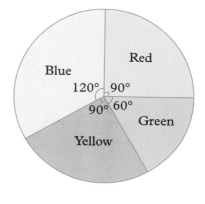

3 The pie chart shows how Simon spends £360 on
various items in one week.
  a What **fraction** of his money did Simon spend on
  i   food   **ii** fares
  iii rent   **iv** savings
  v   entertainment   **vi** clothes?
  b How much of the £360 did he spend on
  i   food   **ii** fares
  iii rent   **iv** savings?

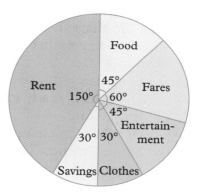

4 The total cost of a holiday was £900.
The pie chart shows how this cost was made up.

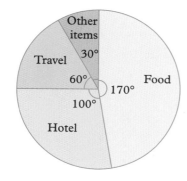

  a How much was spent on food?
  b How much was spent on travel?
  c How much was spent on the hotel?
  d How much was spent on other items?

5 Mr Choudry had an income of £60 000.
The pie chart shows how he used the money.
How much did he spend on
  a food
  b rent
  c savings
  d entertainment
  e travel?

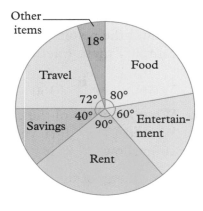

6 The pie chart shows how Sandra spends
her time in a maths lesson which lasts
60 minutes.
  a How much time does Sandra spend
   i   getting ready to work
   ii   talking
   iii sharpening a pencil?
  b Sandra spends 3 minutes working.
    What is the angle on the pie chart for the
    time spent working?

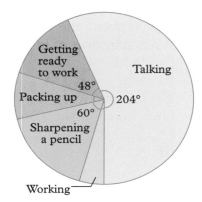

Exercise 14 C

1 At the semi-final stage of the F.A. Cup, 72 neutral referees
were asked to predict who they thought would win. Their
answers were

Spurs   9   Chelsea   40
Manchester United   22   York City   1

  a Work out

   i $\frac{9}{72}$ of 360°   **ii** $\frac{40}{72}$ of 360°

   iii $\frac{22}{72}$ of 360°   **iv** $\frac{1}{72}$ of 360°

  b Draw an accurate pie chart to display the predictions of
    the 72 referees.

2 A survey was carried out to find what 400 students did at the end of Year 11.

120 went into the sixth form
160 went into employment
80 went to F.E. colleges
40 were unemployed.

a Simplify these fractions: $\dfrac{120}{400}$; $\dfrac{160}{400}$; $\dfrac{80}{400}$; $\dfrac{40}{400}$.

b Draw an accurate pie chart to show the information above.

3 In a survey on washing powder 180 people were asked to state which brand they preferred.
45 chose Brand A.

If 30 people chose brand B and 105 chose Brand C, calculate the angles x and y.

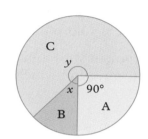

4 A packet of breakfast cereal weighing 600 g contains four ingredients.
Calculate the angles on the pie chart and draw an accurate diagram.

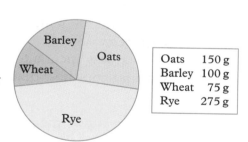

Oats	150 g
Barley	100 g
Wheat	75 g
Rye	275 g

5 The students at a school were asked to state their favourite colour. Here are the results.

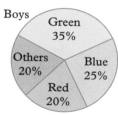

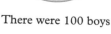

There were 100 boys

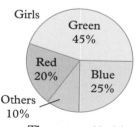

There were 60 girls

Tony says 'The same number of boys and girls chose red.'
Mel says 'More boys than girls chose blue.'
a Use both charts to explain whether or not Tony is right.
b Use both charts to explain whether or not Mel is right.

4.3.2 Pictograms

● In a pictogram you represent the frequency by a simple visual symbol that is repeated.

For example, this pictogram shows how many pizzas were sold on four days.

In the pictogram
    12 pizzas were sold on Monday
    10 pizzas were sold on Tuesday
     8 pizzas were sold on Wednesday
    17 pizzas were sold on Thursday.

Mon	○ ○ ○
Tues	○ ○ ◖
Wed	○ ○
Thur	○ ○ ○ ○ ◿

Key: ○ represents 4 pizzas

The main problem with a pictogram is showing fractions of the symbol which can sometimes only be approximate.

Exercise 15 Ⓒ

1 The pictogram shows the money spent at a school tuck shop by four students.
  a Who spent most?
  b How much was spent altogether?
  c How would you show that someone spent 50p?

Reena	£ £
Sharon	£ £ £ £
Tim	£ £
June	£ £ £

Key: £ represents £1

2 The pictogram shows the make of cars in a car park.
  a How many cars does the 🚗 represent?

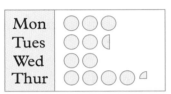

Make	Number of cars	
Ford	4	🚗 🚗
Renault		🚗 🚗 🚗
Toyota	6	
Audi		🚗 ◿

  b Copy and complete the pictogram.

3 The frequency table shows the number of letters posted to six houses one morning. Draw a pictogram to represent these data.

House 1	House 2	House 3	House 4	House 5	House 6
5	3	2	7	1	4

4.4   Data collection and frequency diagrams

4.4.1   Frequency diagrams and bar charts

EXAMPLE

The marks obtained by 36 students in a test were

1	3	2	3	4	2	1	3	0
5	3	0	1	4	0	4	4	3
3	4	3	1	3	4	3	1	2
1	3	4	0	4	3	2	5	3

Show the marks
a   on a tally chart
b   on a frequency diagram.

> A tally chart is a 'data collection sheet'.

a

Mark	Tally	Freq.
0	||||	4
1	L̶H̶T̶ |	6
2	||||	4
3	L̶H̶T̶ L̶H̶T̶ ||	12
4	L̶H̶T̶ |||	8
5	||	2

b

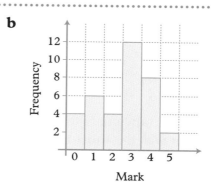

Exercise 16 Ⓒ

1   In a survey, the number of occupants in the cars passing a school was recorded. The bar chart shows the results.

   a   How many cars had 3 occupants?
   b   How many cars had fewer than 4 occupants?
   c   How many cars were in the survey?
   d   What was the total number of occupants in all the cars in the survey?
   e   What fraction of the cars had only one occupant?

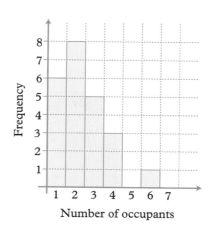

2 A dice was thrown 30 times. These were the scores.

```
1  2  5  5  3  6  1  5  2  4
3  6  1  5  6  5  6  5  1  4
1  3  5  5  1  4  6  3  2  2
```

Draw a tally chart, like the one in the example box on the previous page, and use it to draw a frequency diagram.

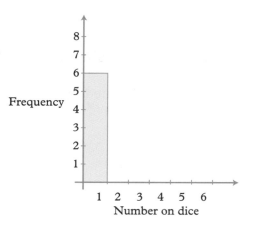

Frequency — y-axis labelled 1 to 8; x-axis labelled 1 2 3 4 5 6, Number on dice

3 The number of letters in each word on one page of a book was recorded and the results were

```
2   6  3  5  2  4  1  6  3  7  5  4  3   8
6   1  5  1  4  3  6  4  7  4  1  7  5   4
3   5  4  6  1  4  2  6  1  5  6  2  8   4
3   2  6  4  3  7  3  9  2  4  9  4  10  2
```

a Draw a tally chart for these results.
b What was the most frequent number of letters in a word?

4 Jake threw two dice sixty times and recorded the total score.

```
2    3   5   4   8   6   4   7   5   10
7    8   7   6   12  11  8   11  7   6
6    5   7   7   8   6   7   3   6   7
12   3   10  4   3   7   2   11  8   5
7    10  7   5   7   5   10  11  7   10
4    8   6   4   6   11  6   12  11  5
```

a Draw a tally chart to show the results of the experiment. The tally chart is started here.

Score	Tally marks	Frequency
2	||	2
3	||||	4
4		
.		
.		

b Draw a frequency graph to illustrate the results. Plot the frequency on the vertical axis.

5   The bar chart shows the profit/loss made by a toy shop
from September 2006 to April 2007.

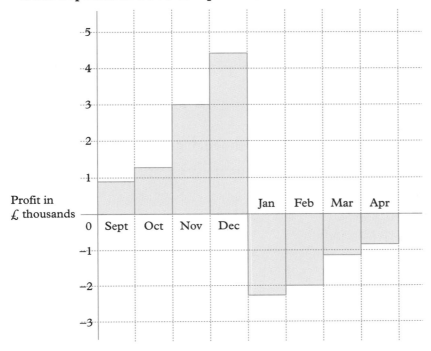

a   Estimate the total profit in this period.
b   Describe what is happening to the shop's profits in this period.
Try to think of an explanation for the shape of the bar chart.

4.4.2   Types of data

Discrete and continuous data
Data can be either **discrete** or **continuous**.
Discrete data can take only certain values, for example
● the number of peas in a pod
● the number of children in a class
● shoe size.

Continuous data comes from measuring and can take any
value, for example
● the height of a child
● the weight of an apple
● time taken to boil a kettle.

Grouped continuous data
Sometimes the data have a wide range of values. In such cases,
it is useful to put the data into groups before drawing a tally
chart and frequency diagram.

A group of 21 students measured their hand spans.

14·8	20·0	16·9	20·7	18·1	17·5	18·7
19·0	19·8	17·8	14·3	19·2	21·7	17·4
16·0	15·9	18·5	19·3	16·6	21·2	18·4

This is continuous data because it is measured.

Show the data grouped in
a a tally chart **b** a frequency diagram.

a

Class intervals	Tally
$14 \leqslant s < 16$	|||
$16 \leqslant s < 18$	⊔⊔⊤	
$18 \leqslant s < 20$	⊔⊔⊤ |||
$20 \leqslant s < 22$	||||

Note that '20·0' goes into the last group $20 \leqslant s < 22$.

$\leqslant$ means 'less than or equal to'
$<$ means 'less than'

b

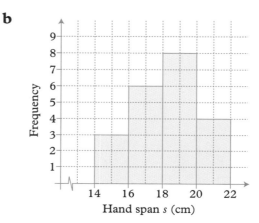

Hand span s (cm)

The diagram is a **histogram**.
The data is continuous so there are no gaps between the bars.
The data must be grouped into equal class intervals so that the
height of each bar represents the frequency.

Exercise 17(E)

1 The graph shows the heights of students
in a class.
  a How many students were over
    150 cm tall?
  b How many students had a
    height between 135 cm and
    155 cm?
  c How many students were
    in the class?

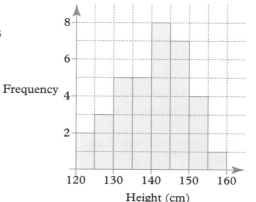

Height (cm)

2 In a survey, the heights of children aged 15 were
   measured in four countries around the world.
   A random sample of children was chosen by computer,
   not necessarily the same number from each country.

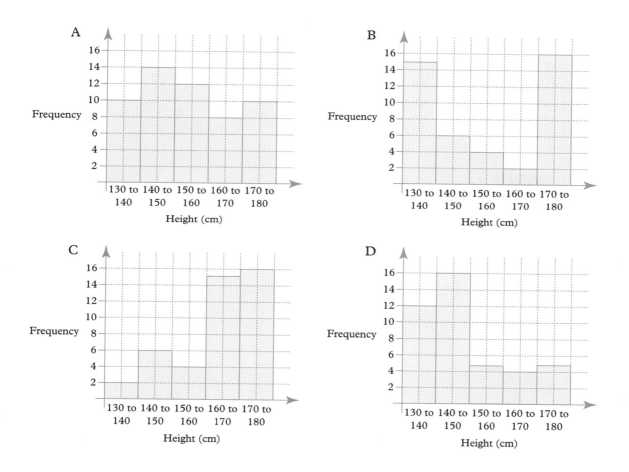

Use the graphs to identify the country in each of these statements.

a Country _____ is poor and the diet of children is not good.
   Two-thirds of the children were less than 150 cm tall.

b There were 54 children in the sample from Country _____.

c In Country _____ the heights were spread fairly evenly
   across the range 130 cm to 180 cm.

d The smallest sample of children came from Country _____.

e In Country _____ there were a similar number of very
   short and very tall children.

3 Scientists have developed a new fertiliser which is supposed to increase the size of carrots. A farmer grew carrots in two adjacent fields, A and B, and treated one of the fields with the new fertiliser. A random sample of 50 carrots was taken from each field and weighed. Here are the results for Field A (all in grams).

```
118   91    82  105    72    92  103    95    73  109
 63  111   102  116   101   104  107   119   111  108
112   97   100   75    85    94   76    67    93  112
 70  116   118  103    65   107   87    98   105  117
114  106    82   90    77    88   66    99    95  103
```

Make a tally chart using the groups given.

Weight	Tally	Frequency
$60 \leqslant w < 70$		
$70 \leqslant w < 80$		
$80 \leqslant w < 90$		
$90 \leqslant w < 100$		
$100 \leqslant w < 110$		
$110 \leqslant w < 120$		

This is the frequency graph for Field B.

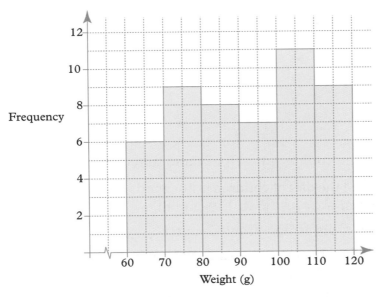

Draw a similar frequency graph for Field A.
Which field do you think was treated with the new fertiliser?

4 Karine and Jackie intend to go skiing in February. They have information about the expected snowfall in February for two possible places.

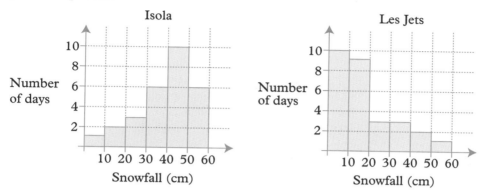

Isola

Les Jets

Decide where you think they should go. It doesn't matter where you decide, but you **must** say why, using the charts above to help you explain.

5 Some people think that children's IQs increase when they eat extra vitamins.
In an experiment, 52 children took an IQ test before and then after a course of vitamin pills. Here are the results.

Vitamin

Vitamins take Johnny to the top of the class!

Before

81	107	93	104	103	96	101	102	93	105	82	106	97
108	94	111	92	86	109	95	116	92	94	101	117	102
95	108	112	107	106	124	125	103	127	118	113	91	113
113	114	109	128	115	86	106	91	85	119	129	99	98

After

93	110	92	125	99	127	114	98	107	128	103	91	104
103	83	125	91	104	99	102	116	98	115	92	117	97
126	100	112	113	85	108	97	101	125	93	102	107	116
94	117	95	108	117	96	102	87	107	94	103	95	96

a Put the scores into convenient groups between 80 and 130.

b Draw two frequency graphs to display the results.

c Write a conclusion. Did the vitamin pills make a significant difference?

You could use
80–89, 90–99, ... or
80–84, 85–89,
90–94, ... It is up
to you!

4.5 Stem-and-leaf diagrams

4.5.1 Drawing a stem-and-leaf diagram

● You can display data in a stem-and-leaf diagram.

Here are the marks of 20 girls in a science test.

| 54 | 42 | 61 | 47 | 24 | 43 | 55 | 62 | 30 | 27 |
| 28 | 43 | 54 | 46 | 25 | 32 | 49 | 73 | 50 | 45 |

Put the marks into groups 20–29, 30–39, ... 70–79.
Choose the tens digit as the 'stem' and the units as the 'leaf'.

The first four marks are shown

	Stem (tens)	Leaf (units)
42 = 40 + 2	2	
	3	
54 = 50 + 4	4	2 7
	5	4
61 = 60 + 1	6	1
47 = 40 + 7	7	

This is the complete diagram ... and then with the leaves in numerical order:

Stem	Leaf		Stem	Leaf
2	4 7 8 5		2	4 5 7 8
3	0 2		3	0 2
4	2 7 3 3 6 9 5		4	2 3 3 5 6 7 9
5	4 5 4 0		5	0 4 4 5
6	1 2		6	1 2
7	3		7	3

The diagram shows the shape of the distribution.

It is also easy to find the mode, the median and the range from a stem-and-leaf diagram.

4.5.2 Back-to-back stem plots

Two sets of data can be compared using a **back-to-back stem plot**. The two sets of leaves share the same stem.

Here are the marks of 20 boys who took the same science test as the girls.

33	55	63	74	20	35	40	67	21	38
51	64	57	48	46	67	44	59	75	56

These marks are entered onto a back-to-back stem plot.

```
         Boys    | Stem |        Girls
             1 0 |  2   | 4  5  7  8
         8   5 0 |  3   | 0  2
     8  6  4 0   |  4   | 2  3  3  5  6  7  9
  9  7  6  5 1   |  5   | 0  4  4  5
     7  7  4 3   |  6   | 1  2
         5 4     |  7   | 3
```

It is helpful to have a key. →

Key (boys)
1 | 5 means 51.

Key (girls)
2 | 4 means 24.

Exercise 18 Ⓔ

1 The marks of 24 students in a test were

41 23 35 15 40 39 47 29
52 54 45 27 28 36 48 51
59 65 42 32 46 53 66 38

Copy and complete the stem-and-leaf diagram.

Stem	Leaf
1	
2	
3	
4	
5	
6	

2 Draw a stem-and-leaf diagram for each set of data.

a 24 52 31 55 40 37 58 61 25 46
44 67 68 75 73 28 20 59 65 39

b 30 41 53 22 72 54 35 47
44 67 46 38 59 29 47 28

Stem	Leaf
2	
3	
4	
5	
6	
7	

3 Here is the stem-and-leaf diagram showing the masses, in kg, of some people in a lift.
  a Write the range of the masses.
  b How many people were in the lift?
  c What is the median mass?

Stem (tens)	Leaf (units)
3	2 5
4	1 1 3 7 8
5	0 2 5 8
6	4 8
7	1
8	2

4 In this question the stem shows the units digit and the leaf shows the first digit after the decimal point.

Draw the stem-and-leaf diagram using these data.

2·4   3·1   5·2   4·7   1·4   6·2   4·5   3·3
4·0   6·3   3·7   6·7   4·6   4·9   5·1   5·5
1·8   3·8   4·5   2·4   5·8   3·3   4·6   2·8

  a What is the median?
  b What is the range?

Stem	Leaf
1	
2	
3	
4	
5	
6	

Key
3 | 7 means 3·7.

5 Here is a back-to-back stem plot showing the pulse rates of several people.

  a How many men were tested?
  b What was the median pulse rate for the women?
  c Write a sentence to describe the main features of the data.

Men		Women
5 1	4	
7 4 2	5	3
8 2 0	6	2 1
5 2	7	4 4 5 8 9
2 6	8	2 5 7
4	9	2 8

Key (men)
1 | 4 means 41.

Key (women)
5 | 3 means 35.

Test yourself

1 A shop has a sale.
The bar chart shows some information about the sale.

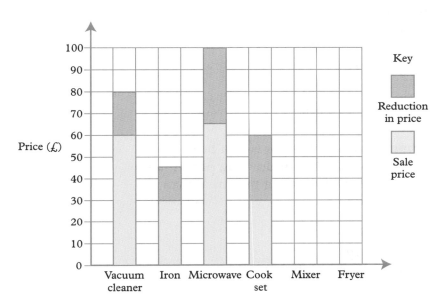

The normal price of a vacuum cleaner is £80.
The sale price of a vacuum cleaner is £60.
The price of a vacuum cleaner is reduced from £80 to £60.

a Write the sale price of a vacuum cleaner as a fraction of its normal price.
Give your answer in its simplest form.
b Find the reduction in the price of the iron.
c Which **two** items have the same sale price?

(Edexcel, 2004)

2 The sizes of the first eleven pairs of shoes sold in a shop one morning are

8   5   4   5   7   10   9   5   11   5   6

a What is the mode of the data?
b What is the median shoe size?
c Which of the mode or median would be more useful to the shopkeeper when he is ordering more shoes?
Explain your answer.

(AQA, 2003)

3   Rosie had 10 boxes of drawing pins.
She counted the number of drawing pins in each box.
The table gives information about her results.

Number of drawing pins	Frequency	
29	2	
30	5	
31	2	
32	1	

a   Copy the table.
b   Write the modal number of drawing pins in a box.
c   Work out the range of the number of drawing pins in a box.
d   Work out the mean number of drawing pins in a box.

(Edexcel, 2003)

4   Claire cycled from Ross to Slago.
She reached Slago at 3:30 p.m.
Her ride is represented by the graph below.

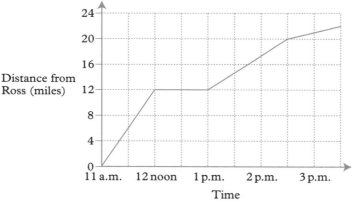

a   Between what times on her journey was she cycling at the fastest average speed?
b   She slowed down at the start of a long hill that goes up to Slago. How far from Slago does the hill start?
c   At 11:30 Miguel leaves Slago and cycles to Ross.
He cycles at a constant speed and arrives at Ross after 2 hours.
  i   Copy the graph and draw a line on it to show Miguel's journey.
  ii   At what time does Miguel pass Claire?

(OCR, 2004)

5 24 students were asked to name their favourite chocolates. The pie chart shows the results.

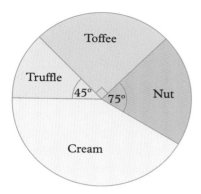

 a What fraction of the students chose Truffle?
 b Work out the size of the angle for Cream.
 c How many of the 24 students chose Toffee?

(OCR, 2004)

6 Sarah watched a water ride at a theme park.
She counted the number of people in each of 20 boats.
These numbers are shown below.

2	3	1	2	2	3	4	5	4	1
1	2	2	3	2	4	5	4	2	4

 a Copy and complete the frequency table.

Number of people in a boat	Tally	Frequency
1		
2		
3		
4		
5		

 b Write the mode of the number of people in a boat.

Emily asked 5 people the number of rides each of them had been on.
The numbers are shown below.

6   8   7   6   10

 c Work out the mean number of rides per person.

(Edexcel, 2004)

7 The stem-and-leaf diagram shows the ages, in years,
of 15 members of a badminton club.

```
2 | 7   8
3 | 0   2   4   8
4 | 1   2   3   3   4   6
5 | 3   6
6 | 2
```

Key: 2 | 7 means an age of 27 years.

a How many members are aged over 40?
b What is the median age of the members?
c What is the range of the ages?

(AQA, 2003)

8 Here are the times, in minutes, taken to change some tyres.

```
 5    10    15    12     8     7    20    35    24    15
20    33    15    25    10     8    10    20    16    10
```

a Draw a stem-and-leaf diagram to show these times.

The probability that a new tyre will be faulty is 0·05.
b Work out the probability that a new tyre will **not** be faulty.

(Edexcel, 2003)

9 Jenni recorded the time of each of the tracks in her CD
collection.
Her results are summarised below.

Time (t seconds)	Number of tracks
$120 < t \leqslant 150$	13
$150 < t \leqslant 180$	9
$180 < t \leqslant 210$	8
$210 < t \leqslant 240$	7
$240 < t \leqslant 270$	3

a Calculate an estimate of the mean time.
b Which class contains the median?
Explain how you found your answer.
c The random play on Jenni's CD player selects a track.
What is the probability it will last more than 240 seconds?

(OCR, 2003)

5 Number 2

In this unit you will:
- revise written multiplication and division
- learn how to use fractions and percentages in calculations
- learn about ratio and proportion
- revise mental methods of calculation
- learn about rounding and estimating.

5.1 Long multiplication and division

5.1.1 Long multiplication

- To work out 327×53 use the fact that
  $327 \times 53 = (327 \times 50) + (327 \times 3)$.
  Set out the working like this

$$
\begin{array}{r}
327 \\
53 \times \\
\hline
16\,350 \\
981 \\
\hline
17\,331 \\
\hline
\end{array}
$$

$16\,350 \rightarrow$ This is 327×50
$981 \rightarrow$ This is 327×3
$17\,331 \rightarrow$ This is 327×53

So $327 \times 53 = 17\,331$

- Here is another method using a grid.
  Many people prefer this method.
  Find 734×23.

  734×23
  $= 16\,882$

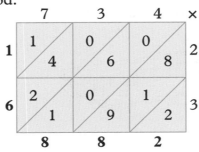

Add the numbers in the diagonals.

Exercise 1 ⑲

Work out these, without a calculator.

1 35×23	**2** 27×17	**3** 26×25
4 31×43	**5** 45×61	**6** 52×24
7 323×14	**8** 416×73	**9** 504×56
10 306×28	**11** 624×75	**12** 839×79
13 694×83	**14** 973×92	**15** 415×235

5.1.2 Long division

With ordinary 'short' division, you divide and find remainders.
The method for 'long' division is really the same but you set
it out so that the remainders are easier to find.

EXAMPLE

> Work out 736 ÷ 32
>
> ....
>
> ```
>        23
> 32)736          32 into 73 goes 2 times
>   -64↓          2 × 32 = 64
>    ‾‾‾
>    96           73 - 64 = 9
>   -96           'bring down' 6
>    ‾‾‾
>     0           32 into 96 goes 3 times
> ```

Exercise 2 C

1 672 ÷ 21	**2** 425 ÷ 17	**3** 576 ÷ 32
4 247 ÷ 19	**5** 875 ÷ 25	**6** 574 ÷ 26
7 806 ÷ 34	**8** 748 ÷ 41	**9** 666 ÷ 24
10 707 ÷ 52	**11** 951 ÷ 27	**12** 806 ÷ 34
13 2917 ÷ 45	**14** 2735 ÷ 18	**15** 56 274 ÷ 19

Exercise 3 C

1 A shop owner buys 56 tins of paint at 84p each. How much
does he spend altogether?

2 Eggs are packed eighteen to a box.

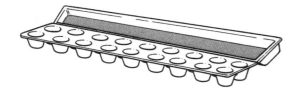 

How many boxes are needed for 828 eggs?

3 On average a man smokes 146 cigarettes a week. How many
does he smoke in a year?

4 Sally wants to buy as many 32p stamps as possible. She has
£5 to buy them. How many can she buy and how much
change is left?

5 How many 49-seater coaches will be needed for a school trip for a party of 366?

6 It costs £7905 to hire a plane for a day. A trip is organised for 93 people. How much does each person pay?

7 A lottery prize of £238 million was won by a syndicate of 17 people who shared the prize equally between them. How much did each person receive?

8 An office building has 24 windows on each of 8 floors. A window cleaner charges 42p for each window. How much is he paid for the whole building?

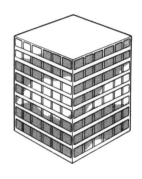

5.2 Fractions

5.2.1 What is a fraction?

- A fraction is a number that is less than a whole one. When a whole one is divided into equal parts each of the parts is a fraction of the whole one.

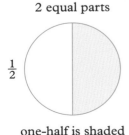

2 equal parts

$\frac{1}{2}$

one-half is shaded

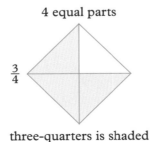

4 equal parts

$\frac{3}{4}$

three-quarters is shaded

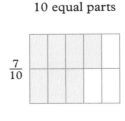

10 equal parts

$\frac{7}{10}$

seven-tenths is shaded

Exercise 4 C

1 What fraction of each diagram is shaded?
Write your answer in both words and figures.

a    b   c 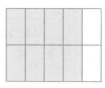   d

e 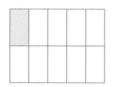   f   g   h

2 What fractions are shown on each of these number lines?

a    b 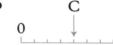   c

3 Draw three grids like this one and label them A, B, and C.

In grid A shade in $\frac{1}{4}$ of the squares.

In grid B shade in $\frac{1}{12}$ of the squares.

In grid C shade in $\frac{1}{3}$ of the squares.

5.2.2 Equivalent fractions

● **Equivalent** fractions are fractions that look different but are the same.

For example

one-half $\frac{1}{2}$ = $\frac{2}{4}$ two-quarters

four-fifths $\frac{4}{5}$ 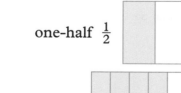 =  $\frac{8}{10}$ eight-tenths

- Changing a fraction into a simpler form is known as '**cancelling down**'.

This is also called simplifying.

Method: Find the highest number that divides exactly into both the numerator (top) and the denominator (bottom).

The fraction $\frac{15}{20}$ cancels down to $\frac{3}{4}$.

$$\frac{15 \div 5}{20 \div 5} = \frac{3}{4} \qquad \frac{4 \div 4}{12 \div 4} = \frac{1}{3}$$

Exercise 5 Ⓒ

Use the fraction charts to copy and complete the equivalent fractions by putting the correct number in the box.

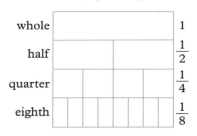
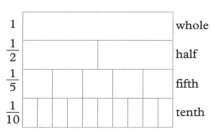

1 $\frac{1}{2} = \frac{\square}{4}$ **2** $\frac{1}{4} = \frac{\square}{8}$ **3** $\frac{1}{5} = \frac{\square}{10}$ **4** $\frac{3}{4} = \frac{\square}{8}$

5 $\frac{1}{2} = \frac{\square}{10}$ **6** $1 = \frac{\square}{4}$ **7** $\frac{3}{5} = \frac{\square}{10}$ **8** $\frac{4}{5} = \frac{\square}{10}$

9 Copy and complete these fraction chains.

   a $\frac{1}{2} = \frac{\square}{4} = \frac{\square}{8} = \frac{\square}{16}$ **b** $\frac{1}{2} = \frac{\square}{4} = \frac{\square}{6} = \frac{\square}{8} = \frac{\square}{10}$

Copy and cancel down these fractions.

10 $\frac{9}{12}$ **11** $\frac{6}{24}$ **12** $\frac{8}{10}$ **13** $\frac{8}{20}$ **14** $\frac{9}{36}$

15 $\frac{8}{12}$ **16** $\frac{9}{15}$ **17** $\frac{6}{18}$ **18** $\frac{7}{21}$ **19** $\frac{32}{36}$

20 $\frac{24}{30}$ **21** $\frac{4}{12}$ **22** $\frac{4}{18}$ **23** $\frac{20}{30}$ **24** $\frac{14}{42}$

25 $\frac{6}{15}$ **26** $\frac{27}{45}$ **27** $\frac{56}{64}$ **28** $\frac{18}{30}$ **29** $\frac{28}{36}$

30 $\frac{18}{63}$ **31** $\frac{44}{55}$ **32** $\frac{24}{60}$ **33** $\frac{18}{72}$ **34** $\frac{75}{100}$

35 What fraction of £1 is 1p?

36 What fraction of £1 is 50p?

37 What fraction of one minute is 15 seconds?

38 What fraction of one hour is 10 minutes?

39 What fraction of £10 is £3?

40 What fraction of the months of the year begin with the letters J or A?

41 In a class of 30 children, 23 are right-handed. What fraction are left-handed?

42 What fraction of the numbers from 1 to 30 contain the digit 9?

43 What fraction of these numbers are greater than ten?

44 Here are some number cards.

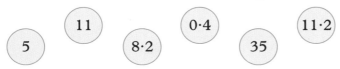

  a Use two cards to make a fraction which is equal to $\frac{1}{2}$.

  b Use three of the cards to make the smallest possible fraction.

*45 **a** Copy and complete. $\frac{2}{3} = \frac{\square}{12}$ $\frac{1}{2} = \frac{\square}{12}$ $\frac{3}{4} = \frac{\square}{12}$.

  b Hence write $\frac{2}{3}, \frac{1}{2}, \frac{3}{4}$ in order of size, smallest first.

*46 **a** Copy and complete $\frac{5}{6} = \frac{\square}{12}$, $\frac{2}{3} = \frac{\square}{12}$, $\frac{1}{4} = \frac{\square}{12}$.

  b Hence write $\frac{5}{6}, \frac{2}{3}, \frac{1}{4}$ in order of size, smallest first.

In questions **47** to **50** write the fractions in order of size.

*47 $\frac{3}{5}, \frac{7}{10}, \frac{1}{2}$ ***48** $\frac{7}{12}, \frac{5}{6}, \frac{5}{4}$

*49 $\frac{1}{2}, \frac{3}{8}, \frac{3}{4}$ ***50** $\frac{7}{15}, \frac{1}{3}, \frac{2}{15}$

5.2.3   Proper and improper fractions

- A proper fraction is one in which the numerator (top number) is less than the denominator (bottom number).

The fractions $\frac{1}{2}, \frac{2}{3}, \frac{3}{4}$ and $\frac{99}{100}$ are all examples of proper fractions.

$$\frac{1}{2} \quad \text{numerator} \atop \text{denominator}$$

- An improper fraction is one in which the numerator is larger than the denominator. They are sometimes called 'top-heavy' fractions.

The fractions $\frac{2}{2}, \frac{4}{3}, \frac{8}{5}$ and $\frac{100}{33}$ are all examples of improper fractions.

- A mixed number is one that contains both a whole number and a fraction. Improper fractions can be changed into mixed numbers and vice versa.

$$\frac{3}{2} = 1\frac{1}{2}$$

EXAMPLE

a Change $\frac{3}{2}$ and $\frac{16}{3}$ into mixed numbers.

b Change $2\frac{1}{2}$ and $3\frac{5}{6}$ into improper fractions.

....

a $\frac{3}{2} = 1\frac{1}{2}$ 2 into 3 goes once, giving the whole number 1.

The remainder is 1 which is written as $\frac{1}{2}$.

$\frac{16}{3} = 5\frac{1}{3}$ 3 into 16 goes five times, giving the whole number 5.

The remainder is 1, which is written $\frac{1}{3}$.

b $2\frac{1}{2} = \frac{5}{2}$ 2 times 2, gives 4 (4 halves).

Add 1 for the $\frac{1}{2}$ to 4 giving 5.

There are 5 halves altogether.

$3\frac{5}{6} = \frac{23}{6}$ 3 times 6 is 18 (18 sixths).

There are 5 sixths to add to 18 giving 23.

There are 23 sixths altogether.

Exercise 6(E)

Change these improper fractions to mixed numbers or whole numbers.

1 $\dfrac{7}{2}$ 2 $\dfrac{5}{3}$ 3 $\dfrac{7}{3}$ 4 $\dfrac{5}{4}$ 5 $\dfrac{8}{3}$

6 $\dfrac{8}{6}$ 7 $\dfrac{9}{3}$ 8 $\dfrac{9}{2}$ 9 $\dfrac{9}{4}$ 10 $\dfrac{10}{2}$

11 $\dfrac{10}{6}$ 12 $\dfrac{10}{7}$ 13 $\dfrac{13}{8}$ 14 $\dfrac{35}{15}$ 15 $\dfrac{42}{21}$

16 $\dfrac{120}{10}$ 17 $\dfrac{22}{7}$ 18 $\dfrac{15}{9}$ 19 $\dfrac{12}{5}$ 20 $\dfrac{150}{100}$

In questions **21** to **35** change the mixed numbers to improper fractions.

21 $1\dfrac{1}{4}$ 22 $1\dfrac{1}{3}$ 23 $2\dfrac{1}{4}$ 24 $2\dfrac{2}{3}$ 25 $1\dfrac{7}{8}$

26 $1\dfrac{2}{3}$ 27 $3\dfrac{1}{7}$ 28 $2\dfrac{1}{6}$ 29 $4\dfrac{3}{4}$ 30 $7\dfrac{1}{2}$

31 $3\dfrac{5}{8}$ 32 $4\dfrac{2}{5}$ 33 $3\dfrac{2}{5}$ 34 $8\dfrac{1}{4}$ 35 $1\dfrac{3}{10}$

5.2.4 Fraction of a number

- If four people share a lottery prize of £100 000 equally, each person receives $\dfrac{1}{4}$ of £100 000.

  That is £100 000 ÷ 4 = £25 000 each.

> To find one **half** of a number, divide the number by **two**.
> To find one **third** of a number, divide the number by **three**.

Exercise 7(C)

1 Copy and complete this table.

	Fraction of quantity required	Divide the quantity by
a	$\dfrac{1}{2}$	2
b	$\dfrac{1}{3}$	
c	one-quarter	
d		10
e	$\dfrac{1}{5}$	

In questions **2** to **19** copy and complete each calculation.

2 $\frac{1}{2}$ of £12 = ☐

3 $\frac{1}{4}$ of £40 = ☐

4 $\frac{1}{3}$ of 15 litres = ☐

5 $\frac{1}{4}$ of 20 kg = ☐

6 $\frac{1}{2}$ of 48 cm = ☐

7 $\frac{1}{5}$ of £20 = ☐

8 $\frac{1}{2}$ of 150 miles = ☐

9 $\frac{1}{10}$ of £100 = ☐

10 $\frac{1}{4}$ of 280 kg = ☐

11 $\frac{1}{5}$ of 30 litres = ☐

12 $\frac{1}{8}$ of 24 kg = ☐

13 $\frac{1}{100}$ of £2000 = ☐

14 $\frac{1}{11}$ of 66 kg = ☐

15 $\frac{1}{7}$ of 490 eggs = ☐

16 $\frac{1}{9}$ of 990 hens = ☐

17 $\frac{1}{8}$ of 888 miles = ☐

18 $\frac{1}{5}$ of £60 = ☐

19 $\frac{1}{8}$ of 320 pages = ☐

20 Here are calculations with letters.
Put the answers in order of size, smallest first.
Write the letters to make a word.

P
$\frac{1}{3}$ of 9

L
$\frac{1}{7}$ of 42

P
$\frac{1}{8}$ of 40

A
$\frac{1}{4}$ of 8

E
$\frac{1}{9}$ of 81

EXAMPLE

A petrol tank in a car holds 56 litres when full.
How much petrol is in the tank when it is $\frac{3}{4}$ full?

....

You need to work out $\frac{3}{4}$ of 56.

$\frac{1}{4}$ of 56 = 56 ÷ 4
$\quad$ = 14

So $\frac{3}{4}$ of 56 = 14 × 3
$\quad$ = 42

There are 42 litres in the tank when it is $\frac{3}{4}$ full.

Working:
$$\begin{array}{r} 1\ 4 \\ 4\overline{)5^16} \end{array}$$

Because $\frac{3}{4}$ of 56
is 3 times as much
as $\frac{1}{4}$ of 56.

Work out

a $\frac{2}{5}$ of £55

b $\frac{3}{4}$ of 52 weeks

....

a $\frac{1}{5}$ of 55 = 55 ÷ 5

           = 11

So $\frac{2}{5}$ of 55 = 11 × 2

           = 22

$\frac{2}{5}$ of £55 = £22

b $\frac{1}{4}$ of 52 = 52 ÷ 4

           = 13

So $\frac{3}{4}$ of 52 = 13 × 3

           = 39

$\frac{3}{4}$ of 52 weeks = 39 weeks

Exercise 8 C

Work out these amounts.

1 $\frac{2}{3}$ of £69

2 $\frac{3}{4}$ of £64

3 $\frac{3}{8}$ of 24 kg

4 $\frac{4}{5}$ of £65

5 $\frac{4}{7}$ of 84 miles

6 $\frac{5}{9}$ of 108 cm

7 $\frac{2}{3}$ of £216

8 $\frac{3}{4}$ of 20 kg

9 $\frac{7}{10}$ of 50 hens

10 $\frac{5}{8}$ of 480 cm

11 $\frac{3}{5}$ of 80 pence

12 $\frac{2}{3}$ of 600 miles

13 $\frac{4}{7}$ of 49p

14 $\frac{5}{8}$ of £4000

15 $\frac{3}{10}$ of 90p

16 $\frac{5}{9}$ of £144

17 $\frac{2}{11}$ of 88 kg

18 $\frac{3}{4}$ of 804 km

19 The total mark for a science test was 72. How many marks did Sandy get if she got $\frac{3}{4}$ of the full marks?

20 There are 450 apples on a tree and there are maggots in $\frac{2}{5}$ of them.
How many apples have maggots in them?

21 A DVD that cost £18 was sold on eBay for $\frac{2}{9}$ of the original price. What was the selling price?

22 Mark has 216 houses on his paper round. One day his sister does $\frac{2}{3}$ of the houses. How many houses are left for Mark?

23 Here are calculations with letters. Put the answers in order of size, smallest first. Write the letters to make a word.

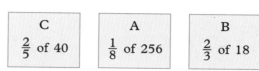

C	A	B
$\frac{2}{5}$ of 40	$\frac{1}{8}$ of 256	$\frac{2}{3}$ of 18

E	M	K	H
$\frac{1}{4}$ of 60	$\frac{2}{9}$ of 180	$\frac{3}{7}$ of 42	$\frac{5}{8}$ of 48

5.2.5 Adding and subtracting fractions

● You can add or subtract fractions when they have the same denominator.

For fractions with different denominators, first change them into equivalent fractions with the same denominator.

> **EXAMPLE**
>
> Find   **a** $\frac{3}{8} + \frac{1}{8}$ **b** $\frac{1}{2} + \frac{1}{5}$ **c** $\frac{3}{4} - \frac{1}{3}$
>
> ....
>
> **a** $\frac{3}{8} + \frac{1}{8} = \frac{4}{8} = \frac{1}{2}$
> Notice that you add the numerators but not the denominators!
>
>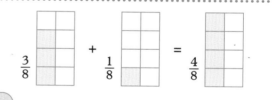
>
> **b** $\frac{1}{2} + \frac{1}{5}$
>
> The L.C.M. of 2 and 5 is 10.
>
> $\frac{5}{10} + \frac{2}{10} = \frac{7}{10}$
>
> **c** $\frac{3}{4} - \frac{1}{3}$
>
> The L.C.M. of 4 and 3 is 12.
>
> $\frac{9}{12} - \frac{4}{12} = \frac{5}{12}$

Exercise 9 Ⓔ

1 Work out

a $\frac{5}{7} + \frac{1}{7}$ **b** $\frac{4}{9} + \frac{1}{9}$ **c** $\frac{7}{10} - \frac{1}{10}$ **d** $\frac{3}{5} + \frac{2}{5}$

e $\frac{1}{8} + \frac{4}{8}$ **f** $\frac{2}{11} + \frac{3}{11}$ **g** $\frac{5}{12} - \frac{1}{12}$ **h** $\frac{7}{15} - \frac{3}{15}$

2 Copy and complete

a $\dfrac{1}{4} + \dfrac{1}{8} = \dfrac{\square}{8} + \dfrac{1}{8} =$ **b** $\dfrac{1}{2} + \dfrac{2}{5} = \dfrac{\square}{10} + \dfrac{\square}{10} =$ **c** $\dfrac{2}{5} + \dfrac{1}{3} = \dfrac{\square}{15} + \dfrac{\square}{15} =$

d $\dfrac{1}{2} + \dfrac{1}{5} = \dfrac{\square}{10} + \dfrac{\square}{10} =$ **e** $\dfrac{3}{4} + \dfrac{1}{8} = \dfrac{\square}{8} + \dfrac{1}{8} =$ **f** $\dfrac{1}{2} + \dfrac{1}{6} = \dfrac{\square}{6} + \dfrac{1}{6} =$

3 Work out

a $\dfrac{1}{2} + \dfrac{1}{3}$ **b** $\dfrac{1}{4} + \dfrac{1}{6}$ **c** $\dfrac{1}{3} + \dfrac{1}{4}$ **d** $\dfrac{1}{4} + \dfrac{1}{5}$

e $\dfrac{2}{5} + \dfrac{2}{3}$ **f** $\dfrac{1}{4} + \dfrac{2}{5}$ **g** $\dfrac{1}{2} + \dfrac{2}{7}$ **h** $1\dfrac{1}{4} + 2\dfrac{1}{2}$

4 Work out

a $\dfrac{2}{3} - \dfrac{1}{2}$ **b** $\dfrac{3}{4} - \dfrac{1}{3}$ **c** $\dfrac{4}{5} - \dfrac{1}{2}$ **d** $\dfrac{2}{3} - \dfrac{1}{4}$

e $\dfrac{1}{2} - \dfrac{2}{5}$ **f** $\dfrac{4}{5} - \dfrac{2}{3}$

5 Work out the perimeter of this drawing.

$\frac{1}{4}$ m

$\frac{1}{5}$ m

Perimeter = sum of the sides

6 Add together $\dfrac{1}{4}, \dfrac{1}{3}$ and $\dfrac{1}{6}$.

*7 Of the cars which failed MoTs

$\dfrac{1}{4}$ failed on brakes

$\dfrac{1}{3}$ failed on steering

$\dfrac{1}{6}$ failed on lights

the rest failed on worn tyres.

What fraction of the cars failed on worn tyres?

*8 In this equation, all the asterisks stand for the same number. What is the number?

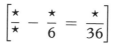

$$\left[\frac{\star}{\star} - \frac{\star}{6} = \frac{\star}{36} \right]$$

> Try different numbers until it works.

*9 The border round the shaded rectangle is $\frac{1}{16}$ inch wide. Find the outside length and width.

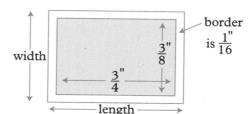

5.2.6 Multiplying and dividing fractions

> EXAMPLE

Work out

a $\frac{2}{5} \times \frac{6}{7}$

b $2\frac{2}{5} \times \frac{1}{5}$

. . . .

a $\frac{2}{5} \times \frac{6}{7} = \frac{2 \times 6}{5 \times 7}$

$= \frac{12}{35}$

b $2\frac{2}{5} \times \frac{1}{5}$

$= \frac{12}{5} \times \frac{1}{5}$

$= \frac{12 \times 1}{5 \times 5} = \frac{12}{25}$

> Change $2\frac{2}{5}$ to an improper fraction.

> EXAMPLE

Work out

a $\frac{5}{11} \div \frac{1}{2}$

b $\frac{3}{8} \div \frac{4}{5}$

c $\frac{3}{5} \times 4$

. . . .

a $\frac{5}{11} \div \frac{1}{2}$

$= \frac{5}{11} \times \frac{2}{1} = \frac{10}{11}$

b $\frac{3}{8} \div \frac{4}{5}$

$= \frac{3}{8} \times \frac{5}{4}$

$= \frac{15}{32}$

c $\frac{3}{5} \times 4$

$= \frac{3}{5} \times \frac{4}{1}$

$= \frac{12}{5}$

$= 2\frac{2}{5}$

> Invert $\frac{1}{2}$ and then multiply.

> Invert $\frac{4}{5}$ and then multiply.

> Write 4 as $\frac{4}{1}$.

Exercise 10 Ⓔ

1 Work out

a $\frac{2}{3} \times \frac{1}{5}$ **b** $\frac{3}{4} \times \frac{5}{7}$ **c** $\frac{4}{5} \times \frac{2}{3}$ **d** $\frac{5}{6} \times \frac{5}{7}$

e $\frac{5}{9} \times \frac{2}{3}$ **f** $\frac{4}{11} \times \frac{5}{6}$ **g** $\frac{7}{8} \times \frac{3}{4}$ **h** $\frac{8}{9} \times \frac{3}{4}$

2 a $\frac{1}{2} \times \frac{4}{5}$ **b** $\frac{1}{3} \times \frac{6}{7}$ **c** $\frac{2}{3} \times \frac{1}{4}$ **d** $\frac{7}{4} \times \frac{4}{9}$

e $\frac{9}{10} \times \frac{4}{3}$ **f** $\frac{5}{12} \times \frac{6}{7}$ **g** $\frac{3}{4} \times \frac{12}{13}$ **h** $1\frac{1}{2} \times \frac{2}{3}$

3 a $2\frac{1}{4} \times \frac{1}{3}$ **b** $1\frac{3}{4} \times \frac{1}{2}$ **c** $2\frac{1}{2} \times \frac{3}{10}$ **d** $3\frac{1}{2} \times \frac{1}{5}$

e $\frac{3}{4} \times 2$ **f** $\frac{3}{5} \times 2$ **g** $5 \times \frac{7}{10}$ **h** $8 \times \frac{3}{4}$

4 a $\frac{3}{4} \div \frac{1}{2}$ **b** $\frac{3}{5} \div \frac{2}{3}$ **c** $\frac{5}{6} \div \frac{1}{4}$ **d** $\frac{2}{3} \div \frac{3}{4}$

e $\frac{5}{6} \div \frac{3}{4}$ **f** $\frac{5}{7} \div \frac{3}{4}$ **g** $1\frac{1}{4} \div 2$ **h** $3\frac{2}{9} \div \frac{1}{3}$

5 Copy and complete this multiplication square.

×	$\frac{2}{3}$		
$\frac{1}{2}$		$\frac{3}{8}$	
		$\frac{3}{16}$	
$\frac{2}{5}$			$\frac{2}{25}$

6 Work out one-half of one-third of £320.

7 A rubber ball is dropped from a height of 300 cm. After each bounce, the ball rises to $\frac{4}{5}$ of its previous height. How high will it rise after the second bounce?

8 Steve Braindead spends his income as follows:

a $\frac{2}{5}$ of his income goes in tax

b $\frac{2}{3}$ of what is left goes on food, rent and transport

c he spends the rest on cigarettes, beer and betting.
What fraction of his income is spent on cigarettes, beer and betting?

9 Copy and fill in the missing numbers so that the answer is always $\frac{3}{8}$.

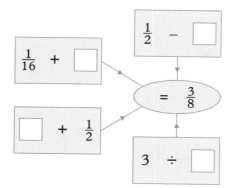

10 This question involves adding, subtracting, multiplying and dividing.

a $\frac{2}{3} \times \frac{3}{4}$ **b** $\frac{5}{6} - \frac{1}{2}$ **c** $\frac{4}{7} \div \frac{2}{3}$ **d** $\frac{7}{10} - \frac{2}{5}$

e $1\frac{1}{2} \div \frac{6}{7}$ **f** $\frac{5}{8} + \frac{1}{12}$ **g** $2\frac{3}{4} \times \frac{4}{5}$ **h** $\frac{2}{15} + \frac{1}{3}$

11 Work out

$$\frac{1}{3} \times \frac{2}{4} \times \frac{3}{5} \times \frac{4}{6} \times \frac{5}{7} \times \frac{6}{8} \times \frac{7}{9}$$

5.3 Percentages

5.3.1 What is a percentage?

• Percentages are equivalent to fractions with denominator (bottom number) equal to 100.

So 21% means $\frac{21}{100}$, 63% means $\frac{63}{100}$ and so on.

• Some percentages are used a lot and you should learn them.

$10\% = \frac{10}{100} = \frac{1}{10}$ $30\% = \frac{30}{100} = \frac{3}{10}$ $20\% = \frac{20}{100} = \frac{1}{5}$

$40\% = \frac{2}{5}$ $60\% = \frac{3}{5}$ $80\% = \frac{4}{5}$ $25\% = \frac{1}{4}$

$75\% = \frac{3}{4}$ $33\frac{1}{3}\% = \frac{1}{3}$ $66\frac{2}{3}\% = \frac{2}{3}$ $50\% = \frac{1}{2}$

Exercise 11 C

1 Draw each square and write underneath it
   a what fraction is shaded   **b** what percentage is shaded.

i   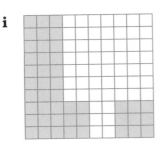   ii   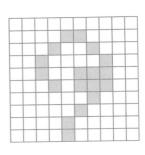   iii

2 If 40% of a square is shaded, what percentage of the square is not shaded?

3 If 73% of a square is shaded, what percentage of the square is not shaded?

4 Approximately 67% of the Earth's surface is covered with water.
   What percentage of the Earth's surface is land?

5 The diagram shows the percentage of people who took part in activities offered at a sports centre on a Friday night.

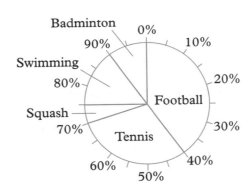

   a What percentage went swimming?
   b What percentage played squash?
   c What percentage played a racket sport?
   d What percentage did not play football?
   e What percentage played activities involving a ball?

6 Copy these and fill in the spaces.

   a $30\% = \dfrac{}{100} = \dfrac{}{10}$ **b** $\dfrac{3}{4} = \dfrac{}{100} = \quad\%$ **c** $\dfrac{1}{3} = \quad\%$

   d $1\% = \dfrac{}{100}$ **e** $80\% = \text{---}$ **f** $\dfrac{1}{10} = \quad\%$

7 For each shape write
   a what fraction is shaded   **b** what percentage is shaded.

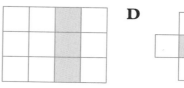

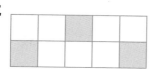

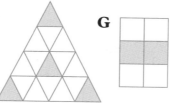

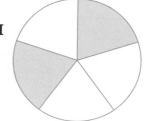

8 These pictures show how much petrol is in a car.
   E is Empty and F is Full.
   What percentage of a full tank is in each car?

   a E ────────────── F   **b** E ────────────── F

   c E ────────────── F   **d** E ────────────── F

9 Write the percentage that could be used in each
   sentence.
   a Three-quarters of the students skated to school.
   b Three out of five workers voted for a strike.
   c Lela got 15 out of 20 in the maths test.
   d One in three cats prefer 'Kattomeat'.
   e Half of the customers at a computer store thought
       that prices were going down.
   f One in four mothers think children are too tidy
       at home.

10 Draw three diagrams of your own design, like those in
    question **7** and shade in

    a 30%   **b** 75%   **c** $66\frac{2}{3}\%$

5.3.2 Changing fractions to percentages

● To change a fraction to a percentage you multiply by 100%.

a Change $\dfrac{7}{20}$ to a percentage.   **b** Change $\dfrac{3}{8}$ to a percentage.

a $\dfrac{7}{20} \times \dfrac{100\%}{1} = \dfrac{7}{\underset{1}{20}} \times \dfrac{\overset{5}{\cancel{100}}\%}{1} = 35\%$ **b** $\dfrac{3}{8} \times \dfrac{100\%}{1} = \dfrac{300\%}{8} = 37\cdot5\%$

Multiplying by 100% means multiplying by one whole, so you are not changing the value of the fraction.

Exercise 12 ⓒ

Change to percentages.

1 $\dfrac{3}{4}$ **2** $\dfrac{2}{5}$ **3** $\dfrac{3}{8}$ **4** $\dfrac{9}{10}$ **5** $\dfrac{17}{20}$

6 $\dfrac{5}{20}$ **7** $\dfrac{17}{25}$ **8** $\dfrac{3}{20}$ **9** $\dfrac{49}{50}$ **10** $\dfrac{7}{100}$

11 $\dfrac{15}{60}$ **12** $\dfrac{16}{50}$ **13** $\dfrac{27}{40}$ **14** $\dfrac{1}{8}$ **15** $\dfrac{235}{1000}$

16 16 marks out of 25.   **17** 18 marks out of 20.

18 12 marks out of 30.   **19** £45 out of £200

20 £17 out of £50   **21** £93 out of £100

22 **a** In a test Ann obtained 11 marks out of 25. What is her percentage mark?
  b In a second test she obtained 13 marks out of 20. What is her percentage mark?

23 A motorist has to drive a distance of 400 km. After an hour he has driven 84 km.
What percentage of his journey has he completed?

24 In a survey 1600 people were asked which
television channel they preferred.
   800 chose ITV1
   640 chose BBC1
   160 had no television set.
   a What percentage chose ITV1?
   b What percentage chose BBC1?
   c What percentage had no television?

25 Of the people in a room 11 are men and 14 are women.
   a How many people are in the room?
   b What percentage of the people are men?
   c What percentage of the people are women?

26 Three girls in different classes all had maths tests on the
same day.
   Jane scored 27 out of 50.
   Susan scored 28 out of 40.
   Jackie scored 39 out of 75.
Work out their marks as percentages and put them in order
with the highest first.

5.3.3 Working out percentages

EXAMPLE

a Work out 22% of £40.   **b** Work out 16% of £85.

$\dfrac{22}{100} \times \dfrac{40}{1} = \dfrac{880}{100}$ (Alternative method)

Since $16\% = \dfrac{16}{100}$ you can replace 16% by 0·16

22% of £40 = £8·80   So 16% of £85 = 0·16 × 85 = £13·60

Exercise 13 C

Work out

1 20% of £60   **2** 10% of £80   **3** 5% of £200
4 6% of £50   **5** 4% of £60   **6** 30% of £80
7 9% of £500   **8** 18% of £400   **9** 61% of £400
10 12% of £80   **11** 6% of $700   **12** 11% of $800
13 5% of 160 kg   **14** 20% of 60 kg   **15** 68% of 400 g
16 15% of 300 m   **17** 2% of 2000 km   **18** 71% of $1000
19 26% of 19 kg   **20** 1% of 6000 g   **21** 8·5% of €2400

22 Ayesha earns £25 for doing a paper round. How much **extra** does she earn when she gets a 20% rise?

23 Full marks in a maths test is 80. How many marks did Tim get if he got 60%?

24 There are 240 children at a school and 65% of them walk to school. How many children is that?

25 The normal price of a garden hoe is £15.
In a sale prices are reduced by 20%.
Find the sale price of the hoe.

26 Sam buys a car for £4200.
How much does he pay each month?

> **Star deal***
> - No interest
> - 40% deposit
> - 36 equal monthly payments

EXAMPLE

> A coat originally cost £24. Calculate the new price after a 5% reduction.
>
> Price reduction $= 5\%$ of £24
>
> $$= \frac{5}{100} \times \frac{24}{1} = £1\cdot20$$
>
> New price of coat $= £24 - £1\cdot20$
> $$= £22\cdot80$$

There is another way of finding the reduced price of the coat. The price, after a 5% reduction, will be 95% of the original price.
So, new price of coat = 95% of £24
$$= 0\cdot95 \times 24$$
$$= £22\cdot80$$

> For a 5% **increase** you multiply by 1·05.

> EXAMPLE
>
> A CD originally cost £11·60. Calculate the new price after a 7% increase.
> ....
> New price = 107% of £11·60
> $\qquad$ = 1·07 × 11·6
> $\qquad$ = £12·41 to the nearest penny.

Exercise 14 Ⓒ

1 Increase a price of £60 by 5%.
2 Increase a price of £800 by 8%.
3 Increase a price of £82·50 by 6%.
4 Increase a price of £65 by 60%.
5 Reduce a price of £2000 by 2%.
6 Increase a price of £440 by 80%.
7 Increase a price of £66 by 100%.
8 Reduce a price of £91·50 by 50%.
9 Increase a price of £88·24 by 25%.
10 Reduce a price of £63 by $33\frac{1}{3}$%.
11 Increase a price of £8·50 by 46%.
12 Increase a price of £240 by 11%.
13 Increase a price of £5·75 by 20%.
14 Reduce a price of £8500 by 4%.
15 Increase a price of £11·20 by 15%.
16 Reduce a price of £88 by 10%.

In the remaining questions give the answers to the nearest penny.
17 Increase a price of £28·20 by 13%.
18 Increase a price of £8·55 by 4%.
19 Reduce a price of £9·60 by 7%.
20 Increase a price of £12·80 by 11%.

Exercise 15 Ⓒ

1 In a closing-down sale a shop reduces all its prices by 20%. Find the sale price of a coat which previously cost £44.

2 The price of a car was £5400 but it is increased by 6%. What is the new price?

3 The price of a sideboard was £245 but, because the sideboard is scratched, the price is reduced by 30%. What is the new price?

4 A hi-fi shop offers a 7% discount for cash. How much does a cash-paying customer pay for an amplifier advertised at £95?

5 The insurance premium for a car is normally £90. With a 'no-claim bonus' the premium is reduced by 35%. What is the reduced premium?

6 Myxomatosis kills 92% of a colony of 300 rabbits. How many rabbits survive?

7 The population of a town increased by 32% between 1945 and 2005. If there were 45 000 people in 1945, what was the 2005 population?

8 A restaurant adds a 12% service charge onto the basic price of meals. How much do I pay for a meal with a basic price of £18·50?

*9 A new-born baby weighs 3·1 kg. Her weight increases by 8% over the next fortnight. What does she weigh then?

*10 A large snake normally weighs 12·2 kg. After swallowing a rat, the weight of the snake is increased by 7%. How much does it weigh after dinner?

*11 At the beginning of the year a car is valued at £3250. During the year its value falls by 15%. How much is it worth at the end of the year?

*12 The area of a square is 400 cm².
   a The area is increased by 20%. Find the new area.
   b The new area is then decreased by 20%. Find the final area.

400 cm²

Exercise 16Ⓔ

Find the total bill.

1 2 hammers at £5·30 each
50 screws at 25p for 10
5 bulbs at 38p each
1 tape measure at £1·15
VAT at 17·5% is added to the total cost.

2 5 litres of oil at 85p per litre
3 spanners at £1·25 each
2 manuals at £4·30 each
200 bolts at 90p for 10
VAT at 17·5% is added to the total cost.

3 12 rolls of wallpaper at £3·70 per roll
3 packets of paste at £0·55 per packet
2 brushes at £2·40 each
1 step ladder at £15·50
VAT at 17·5% is added to the total cost.

4 5 golf clubs at £12·45 each
48 golf balls at £15 per dozen
100 tees at 1p each
1 bag at £21·50
1 umbrella at £12·99
VAT at 17·5% is added to the total cost.

5.3.4   Income tax

The tax that employees pay on their income depends on
a how much they are paid
b their allowances
c the tax rate.

Tax is paid only on the 'taxable income'.

- Taxable income = Total income − Allowances

Employees find their allowances by looking at their 'Tax Code Number'.

- Allowances = (Tax Code Number) × 10

> Income is the amount of money a person is paid.

EXAMPLE

A man earns £65 000 per year.
If his Tax Code Number is 538, calculate his taxable income.

....

Allowances = 538 × 10 = £5380
Taxable income = £65 000 − £5380
         = £59 620

Exercise 17 **C**

Calculate the taxable income for each of these.

	Earnings	Tax Code Number		Earnings	Tax Code Number
1	£35 000 per year	213	2	£50 000 per year	274
3	£80 000 per year	315	4	£42 000 per year	289
5	£36 500 per year	265	6	£9800 per year	341
7	£8655 per year	286	8	£600 per month	412
9	£450 per month	263	10	£825 per month	311
11	£710 per month	278	12	£985 per month	415
13	£160 per week	342	14	£144 per week	214
15	£180 per week	289			

EXAMPLE

A woman earns £295 per week and her Tax Code Number is 215. Find the total amount of tax paid in a year when the tax rate is 30%.

Amount earned in year = £295 × 52   Allowances = 215 × 10

$\qquad$ = £15 340 $\qquad\qquad\qquad\qquad$ = £2150

Taxable income = £15 340 − £2150

$\qquad\qquad\quad$ = £13 190

Tax paid = 30% of £13 190

$$= \frac{30}{100} \times \frac{13\,190}{1} = £3957$$

> You multiply the taxable income by the tax rate to find the amount of tax you have to pay.

Exercise 18 Ⓒ

In all questions the tax rate is 30%.

1 Sue earns £210 per week and her Tax Code Number is 304. Find the total amount of tax she pays in a year.

2 Del earns £204 per week and his Tax Code Number is 361.
Find the total amount of tax he pays in a year.

3 Ann earns £265 per week. How much tax does she pay in a year if her Tax Code Number is 247?

4 John earns £348·50 per week. How much tax does he pay in a year if his Tax Code Number is 302?

5 David earns a salary of £1620 per month. How much tax does he pay in a year if his Tax Code Number is 342?

6 Louise earns £1850 per month and her Tax Code Number is 357. Find the total amount of tax she pays in a year.

7 Mr Tebbit's salary is £96 500 per year and his Tax Code Number is 465. Find the total amount of tax he pays in a year.

5.3.5 Percentage change

Price changes are sometimes easier to understand when expressed as a percentage of the original price. For example if the price of a car goes up from £7000 to £7070, this is only a 1% increase. If the price of a jacket went up from £100 to £170 this would be a 70% increase! In both cases the actual increase is the same: £70.

- Percentage increase $= \dfrac{\text{actual increase}}{\text{original value}} \times \dfrac{100}{1}$

- Percentage decrease $= \dfrac{\text{actual decrease}}{\text{original value}} \times \dfrac{100}{1}$

EXAMPLE

The price of a car is increased from £6400 to £6800. What percentage is that?

....

Percentage increase $= \dfrac{400}{6400} \times \dfrac{100\%}{1} = \dfrac{40\,000\%}{6400} = 6\dfrac{1}{4}\%$

Exercise 19 Ⓔ

In questions **1** to **10** calculate the percentage increase.

In questions **11** to **20** calculate the percentage decrease.

	Original price	Final price
1	£50	£54
2	£80	£88
3	£180	£225
4	£100	£102
5	£75	£78
6	£400	£410
7	€5000	€6000
8	£210	£315
9	£600	£690
10	$4000	$7200

	Original price	Final price
11	£800	£600
12	£50	£40
13	£120	£105
14	£420	£280
15	£6000	£1200
16	€880	€836
17	$15 000	$14 100
18	$7·50	$6·00
19	£8·20	£7·79
20	£16 000	£15 600

If the selling price is greater than the cost price, the seller has made a profit. If it is less, they have made a loss.

You can find this profit or loss as a percentage of the original price the seller paid (cost price).

EXAMPLE

John buys a coat for £16 and sells it again for £20. What is his percentage profit?

Percentage profit $= \dfrac{20-16}{16} \times \dfrac{100\%}{1} = \dfrac{4}{16} \times \dfrac{100\%}{1} = 25\%$

Exercise 20Ⓔ

Find the percentage profit/loss using one of these formulae.

percentage profit $= \dfrac{\text{actual profit}}{\text{cost price}} \times \dfrac{100\%}{1}$ or percentage loss $= \dfrac{\text{actual loss}}{\text{cost price}} \times \dfrac{100\%}{1}$

Give the answers correct to one decimal place.

	Cost price	Selling price		Cost price	Selling price
1	£11	£15	**11**	£20	£18·47
2	£21	£25	**12**	£17	£11
3	£36	£43	**13**	£13	£9
4	£41	£50	**14**	£211	£200
5	£411	£461	**15**	£8·15	£7
6	£5·32	£5·82	**16**	£2·62	£3
7	£6·14	£7·00	**17**	£1·52	£1·81
8	£2·13	£2·50	**18**	$13·50	$13·98
9	£6·11	£8·11	**19**	$3·05	$4·00
10	£18·15	£20	**20**	$1705	$1816

Exercise 21Ⓔ

1 The number of people employed by a firm increased from 250 to 280. Calculate the percentage increase in the workforce.

2 During the first four weeks of his life Samuel's weight increases from 3000 g to 3870 g. Calculate the percentage increase in his weight.

3 After cleaning out the drains, Peter's clothes went down in value from £80 to £56. Calculate the percentage decrease in the value of his clothes.

4 When cold, an iron rod is 200 cm long. After it is heated, the length of the rod increases to 200·5 cm. Calculate the percentage increase in the length.

5 A man buys a car for £4000 and sells it for £4600. Calculate his percentage profit.

6 A shopkeeper buys jumpers for £6·20 and sells them for £9·99. Calculate the percentage profit correct to one decimal place.

7 A grocer buys bananas at 20p per pound but after the fruit are spoiled he has to sell them at only 17p per pound. Calculate the percentage loss.

5.3.6   Compound interest

Suppose a bank pays a fixed interest of 10% a year on money in deposit accounts. A man puts £500 in the bank.

After one year he has
    500 + 10% of 500 = £550

After two years he has
    550 + 10% of 550 = £605
      (Check that this is $1·10^2 \times 500$)

After three years he has
    605 + 10% of 605 = £665·50
      (Check that this is $1·10^3 \times 500$)

In general after n years the money in the bank will be
£$(1·10^n \times 500)$.

> With **compound interest** the interest is added to the original amount and the whole sum earns interest the following year.

Exercise 22Ⓔ

1 A bank pays interest of 10% on money in deposit accounts. Mrs Wells puts £2000 in the bank. How much has she after
   a   one year   **b**   two years   **c**   three years?

2 A bank pays interest of 12%. Mr Olsen puts £5000 in the bank. How much has he after
   a   one year   **b**   two years   **c**   three years?

3 A computer operator is paid £10 000 a year. Assuming her
pay is increased by 7% each year, what will her salary be
in two years' time?

4 Mrs Bergkamp's salary in 2006 is £30 000 per year. Every
year her salary is increased by 5%.
In 2007 her salary will be 30 000 × 1·05   = £31 500
In 2008 her salary will be 30 000 × 1·05 × 1·05   = £33 075
In 2009 her salary will be 30 000 × 1·05 × 1·05 × 1·05 = £34 728·75
And so on.
  a What will her salary be in 2010?
  b What will her salary be in 2011?

5 The price of a house was £90 000 in 2005. At the end of each
year the price increases by 6%.
  a Find the price of the house after 1 year.
  b Find the price of the house after 2 years.
  c Find the price of the house after 4 years.

6 Assuming an average inflation rate of 8%, work out
the probable cost of the following items in 2 years' time.
  a car £6500   **b** TV £340   **c** house £500 000

7 A new bike is valued at £8000. At the end of each year its
value is reduced by 15% of its value at the start of the year.
What will it be worth after 3 years?

8 The population of an island is 25 000 and it increases by 10% each
year. After how many years will the original population be doubled?

9 A bank pays interest of 11% on £6000 in a deposit account.
After how many years will the money have trebled?

5.4 Ratio

5.4.1 What is a ratio?

● You use a ratio to compare the sizes of two or more quantities.

● Suppose in a classroom there are 9 girls and 12 boys.
The ratio of girls to boys is

girls : boys = 9 : 12

You can simplify the ratio by dividing both numbers by a
common factor, 3.

So girls : boys = 3 : 4.

You can use the ratio of girls to boys to find what fraction
of the class are girls and what fraction are boys.
Add 3 + 4 to give 7.

$\frac{3}{7}$ of the class are girls $\frac{4}{7}$ and are boys.

● You must use the same units in both parts of a ratio.

For example, you write the ratio of 22 cm to 1 m as 22 cm to 100 cm,
then divide both sides by 2 to get 22 : 100 = 11 : 50.

Exercise 23Ⓔ

1 Express each ratio in its simplest form.
   a 9 : 6   **b** 15 : 25   **c** 10 : 40   **d** 30 : 12   **e** 48 : 44
   f 18 to 24   **g** 40 to 25   **h** 21 to 49   **i** 60 : 42   **j** 16 : 22
   k 9 : 6 : 12 **l** 40 : 5 : 15 **m** 12 : 10 : 8 **n** 18 : 12 : 18

2 In a box there are 9 pencils and 12 pens. Find the ratio of
   pencils to pens. Give your answer in its simplest form.

3 In a hall there are 36 chairs and 9 tables. Find the ratio of
   chairs to table. Give your answer in its simplest form.

4 Express each ratio in its simplest form. Remember to use
   the same units in both parts.
   a 8 kg to 20 kg   **b** 20p to £1   **c** 20 cm to 1 m
   d 20 minutes to 1 hour   **e** 2 kg to 500 g   **f** 400 m to 2 km
   g 5 mm to 5 cm   **h** 25p to £1·50   **i** 40 cm to 1·2 m

5 In a minibus there are 5 women and 3 men. What fraction of
   the people are men?

6 In a room there are 36 plastic chairs and 9 wooden chairs.
   What fraction of the chairs are wooden?

7 In a class the ratio of girls to boys is 3 : 2.
   What fraction of the class are girls?

8 In a shop the ratio of apples to pears to oranges is $3 : 4 : 5$.
What fraction of the total fruit are
a apples   **b** pears   **c** oranges?

9 In an office there are twice as many men as women. What
fraction of the people in the office are men?

5.4.2   Dividing quantities in a given ratio

EXAMPLE

Share £60 in the ratio $2 : 3$.
....

Total number of shares = $2 + 3 = 5$
So one share = £60 ÷ 5 = £12
Two shares = £24 and three shares = £36.
The two amounts are £24 and £36.

Exercise 24Ⓔ

1 Share £30 in the ratio $1 : 2$.

2 Share £60 in the ratio $3 : 1$.

3 Divide 880 g of food between a cat and a dog in the
ratio $3 : 5$.

4 Divide $1080 between Sam and Chris in the ratio $4 : 5$.

5 Share 126 litres of petrol between Steven and Dave in the
ratio $2 : 5$.

6 Share £60 in the ratio $1 : 2 : 3$.

7 Alan, Brian and Dawn divided £560 between them in the
ratio $2 : 1 : 5$. How much did Brian receive?

8 A sum of £120 is divided in the ratio $3 : 4 : 5$. What is the largest share?

9 At an election 7800 people voted Labour, Conservative or
Liberal Democrat in the ratio $4 : 3 : 5$. How many people
voted Liberal Democrat?

10 Find the ratio, 'shaded area' : 'unshaded area' for each diagram.

a   **b**   **c**

In a class, the ratio of boys to girls is 3 : 4.
If there are 12 boys, how many girls are there?

....

Boys : Girls = 3 : 4
Multiply both parts by 4.
Boys : Girls = 12 : 16
So there are 12 boys and 16 girls.

Exercise 25Ⓔ

1 In a room, the ratio of boys to girls is 3 : 2. If there are
12 boys, how many girls are there?

2 In a room, the ratio of men to women is 4 : 1. If there are
20 men, how many women are there?

3 In a box, the ratio of nails to screws is 5 : 3. If there are
15 nails, how many screws are there?

4 An alloy consists of copper, zinc and tin in the ratio 1 : 3 : 4.
If there is 10 g of copper in the alloy, find the weights of
zinc and tin.

5 In a shop the ratio of oranges to apples is 2 : 5. If there are
60 apples, how many oranges are there?

6 A recipe for 5 people calls for 1·5 kg of meat. How
much meat is required if the recipe is adapted to feed
8 people?

7 A cake for 6 people requires 4 eggs. How many eggs are
needed to make a cake big enough for 9 people?

8 A photocopier enlarges the original in the ratio 2 : 3. The height of a tree is 12 cm on the original. How tall is the tree on the enlarged copy?

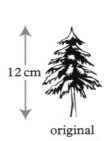

12 cm

original   enlarged copy

9 A photocopier enlarges copies in the ratio 4 : 5. The length of the headline 'BRIDGE COLLAPSES' is 18 cm on the original. How long is the headline on the enlarged copy?

10 A photocopier **reduces** in the ratio 5 : 3. The height of a church spire is 20 cm on the original. How high is the church spire on the reduced copy?

11 A cake weighing 550 g has three ingredients: flour, sugar and butter. There is twice as much flour as sugar and one and a half times as much sugar as butter. How much flour is there?

12 If $\frac{5}{8}$ of the children in a school are boys, what is the ratio of boys to girls?

13 A man and a woman share a bingo prize of £1000 between them in the ratio 1 : 4. The woman shares her part between herself, her mother and her dog in the ratio 2 : 1 : 1. How much does her dog receive?

14 The number of pages in a newspaper is increased from 36 to 54. The price is increased in the same ratio. If the old price was 28p, what will the new price be?

*15 Two friends bought a house for £220 000. Sam paid £140 000 and Joe paid the rest. Three years later they sold the house for £275 000. How much should Sam receive from the sale?

*16 Concrete is made from 1 part cement, 2 parts sand and 5 parts aggregate (by volume). How much cement is needed to make 2 m³ of concrete?

5.4.3   Map scales and ratio

The map below is drawn to a scale of 1 : 50 000.
1 cm on the map represents 50 000 cm on the land.

> You always draw a map to a scale.

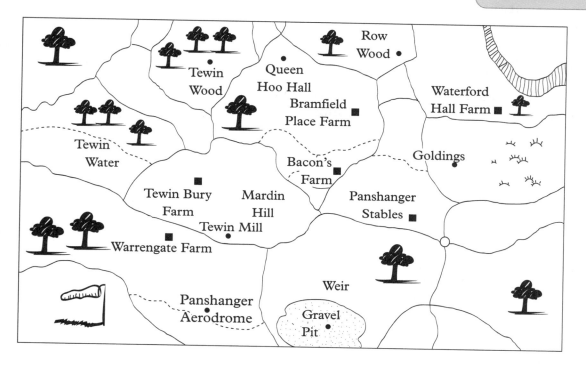

EXAMPLE

On a map of scale 1 : 25 000 two towns appear 10 cm apart. What is the actual distance between the towns in km?

....

  1 cm on map = 25 000 cm on land
10 cm on map = 250 000 cm on land
    250 000 cm = 2500 m
                      = 2·5 km
The towns are 2·5 km apart.

Exercise 26Ⓔ

1   The scale of a map is 1 : 1000. Find the actual length in metres represented on the map by 20 cm.

2   The scale of a map is 1 : 10 000. Find the actual length in metres represented on the map by 5 cm.

3 Copy and complete the table.

	Map scale	Length on map	Actual length on land
a	1 : 10 000	10 cm	1 km
b	1 : 2000	10 cm	m
c	1 : 25 000	4 cm	km
d	1 : 10 000	6 cm	km

4 Find the actual distance in metres between two points which are 6·3 cm apart on a map whose scale is 1 : 1000.

5 On a map of scale 1 : 300 000 the distance between York and Harrogate is 8 cm. What is the actual distance in km?

6 A builder's plan is drawn to a scale of 1 cm to 10 m. How long is a road which is 12 cm on the plan?

*7 The map on page 233 is drawn to a scale of 1 : 50 000. Make your own measurements to find the actual distance in km between
  a Goldings and Tewin Wood (marked ●)
  b Panshanger Aerodrome and Row Wood
  c Gravel Pit and Queen Hoo Hall.

EXAMPLE

The distance between two towns is 18 km.
How far apart will they be on a map of scale 1 : 50 000?

18 km = 1 800 000 cm

1 800 000 cm on land = $\frac{1}{50\,000}$ × 1 800 000 cm on map

Distance between towns on map = 36 cm

Exercise 27Ⓔ

1 The distance between two towns is 15 km. How far apart will they be on a map of scale 1 : 10 000?

2 The distance between two points is 25 km. How far apart will they be on a map of scale 1 : 20 000?

3 The length of a road is 2·8 km. How long will the road be on a map of scale 1 : 10 000?

4 The length of a reservoir is 5·9 km. How long will it be on a map of scale 1 : 100 000?

5 Copy and complete the table.

	Map scale	Actual length on land	Length on map
a	1 : 20 000	12 km	__cm
b	1 : 10 000	8·4 km	__cm
c	1 : 50 000	28 km	__cm
d	1 : 40 000	56 km	__cm
e	1 : 5000	5 km	__cm

6 The scale of a drawing is 1 cm to 10 m. The length of a wall is 25 m. What length will the wall be on the drawing?

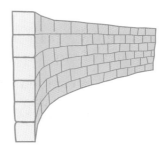

5.5   Proportion

5.5.1   Direct proportion

EXAMPLE

If 12 calculators cost £54, find the cost of 17 calculators.

···

12 calculators cost £54
So 1 calculator costs £54 ÷ 12
   = £4·50
17 calculators cost £4·50 × 17
   = £76·50

Exercise 28 C

1 If 5 books cost £15, find the cost of 8.

2 If 7 apples cost 63p, find the cost of 12.

3 If 4 batteries cost 180p, find the cost of 7.

4 Toy cars cost £3·36 for 8. Find the cost of
   a 3 toy cars   **b** 10 toy cars.

5 Crisps cost £1·32 for 12 packets. Find the cost of
   a 20 packets   **b** 200 packets.

6 Stair carpet costs £78 for 12 m. Find the cost of 15 m.

7 The total weight of 7 ceramic tiles is 1750 g. What do 11 tiles weigh?

8 A machine fills 2000 bottles in 10 minutes. How many bottles will it fill in 7 minutes?

9 The total contents of 8 cartons of fruit drink is 12 litres. How much fruit drink is there is 3 cartons?

10 Find the cost of 15 cakes if 9 cakes cost £2·07.

11 Find the cost of 7 screws if 20 screws cost £4·60.

12 How much would 7 cauliflowers cost, if 10 cauliflowers cost £5·70?

13 A machine takes 20 seconds to make 8 coins. How long does it take to make 50 coins?

14 A plane flies 50 km in 15 minutes. How long will it take to fly 300 km?

Exercise 29 C

1 If 8 pencils cost 56p, how many can you buy for 70p?

2 If you can buy 6 pineapples for £3·12, how many can you buy for £5·20?

3 If 20 m^2 of carpet costs £150, what area of carpet can you buy for £90?

4 Tins of cat food cost £2·88 for 12. How many tins can I buy for £2·16?

5 Twenty men produce 500 articles in 6 days. How many articles would 4 men produce in 6 days?

6 Forty people take 8 days to produce 400 articles. How many articles would 16 people produce in 8 days?

7 12 people produce 600 components in 12 hours. How many components would 9 people produce in 12 hours?

8 7 cycles cost £623.
   a What is the cost of 3 cycles?
   b How many cycles can be bought for £979?

9 11 CDs cost £93·50.
   a What is the cost of 15 CDs?
   b How many CDs can be bought for £17?

10 $2\frac{1}{2}$ m of metal tube cost £1·40. Find the cost of
   a 4 m   b $7\frac{1}{2}$ m.

11 $3\frac{1}{4}$ kg of sweets costs £2·60. Find the cost of
   a 2 kg   b $4\frac{1}{2}$ kg.

12 A car travels 210 km on 30 litres of petrol. How much petrol is needed for a journey of 245 km?

13 A light aircraft flies 375 km on 150 litres of fuel. How much fuel is needed for a journey of 500 km?

14 A tank travels 140 miles on 40 gallons of fuel. How much fuel is needed for a journey of 245 miles?

5.6 Mental arithmetic

Ideally these questions should be read out by a teacher or another student and you should not be looking at them. Each of the 30 questions should be repeated once and then the answer, and only the answer, should be written down.

Each test, including the recording of results, should take about 30 minutes.

If you do not have anyone to read out the questions for you, try to do the test without writing down any detailed working.

Test 1

1 Find the cost in pounds of ten books at 35 pence each.

2 Add together £4·20 and 75 pence.

3 What number divided by six gives an answer of four?

4 I spend £1·60 and pay with £2. My change consists of three coins. What are they?

5 Find the difference between $13\frac{1}{2}$ and 20.

6 Write one centimetre as a fraction of one metre.

7 How many ten pence coins are there in a pile worth £5?

8 Ten per cent of the students in a school play hockey, 15% play basketball and the rest play football. What percentage play football?

9 In a room of 20 people, three-quarters were women. What was the number of women?

10 Four lemons costing eleven pence each are bought with a one pound coin. What is the change?

11 Write the number that is thirteen less than one hundred.

12 What number is ten times as big as 0·6?

13 A hockey pitch measures 20 metres by 40 metres. Find the distance around the pitch.

14 What is twenty multiplied by ten?

15 The side of a square is four metres. What is the area of the square?

16 How many 2p coins are worth the same as ten 5p coins?

17 What number must be added to $1\frac{1}{4}$ to make $2\frac{1}{2}$?

18 Three books cost six pounds. How much will five books cost?

19 A rubber costs 20 pence. How many can be bought for £2?

20 What number is a hundred times as big as 0·15?

21 How many millimetres are there in 5 cm?

22 Find the mean of 12 and 20.

23 In the morning the temperature is minus three degrees Celsius. What will be the temperature after it rises eleven degrees?

24 A certain number multiplied by itself gives 81 as the answer. What is half of that number?

25 The difference between two numbers is 15. One of the numbers is 90. What is the other?

26 How many half-litre glasses can be filled from a vessel containing ten litres?

27 How much will a dozen oranges cost at 20 pence each?

28 On a coach forty-one out of fifty people are men. What percentage is this?

29 A prize of £400 000 is shared equally between one hundred people. How much does each person receive?

30 How many quarters make up two whole ones?

Test 2

1 What are 48 twos?

2 How many fives are there in ninety-five?

3 What is 6:30 a.m. on the 24-hour clock?

4 Add together £2·25 and 50 pence.

5 I go shopping with £2·80 and buy a magazine for ninety pence. How much money have I left?

6 Two angles of a triangle are 65° and 20°. What is the third angle?

7 Write in figures the number 'five million, eighteen thousand and one'.

8 How many 20 pence biros can be bought for £3?

9 Work out 1% of £600.

10 A packet of 10 small cakes costs 35 pence. How much does each cake cost?

11 Add ten to 9 fives.

12 A packet of flour weighing 2400 grams is divided into three equal parts. How heavy is each part?

13 Add together 7, 23 and 44.

14 A car does 7 miles per litre of petrol. How far does the car travel on 40 litres of petrol?

15 How many twenty pence coins are needed to make eight pounds?

16 A certain butterfly lives for just 96 hours. How many days is this?

17 What number is 25 more than 37?

18 Find the average of 2, 5 and 8.

19 Pears cost eleven pence each. How many can I buy for sixty pence?

20 How many minutes are there in eight hours?

21 Write seven-tenths as a decimal number.

22 One-third of a number is six. What is the number?

23 Write one-fifth as a decimal.

24 Which is the larger: 0·7 or 0·071?

25 If a woman earns £8·40 per hour, how much does she earn in ten hours?

26 A car costing £2500 is reduced by £45. What is the new price?

27 How many half-kilogram packets of sugar can be filled from a large bowl containing 32 kilograms?

28 My daily paper costs 60 pence and I buy the paper six days a week. What is my weekly bill?

29 A car journey of 110 miles took two hours. What was the average speed of the car?

30 How many days will there be in February 2011?

Test 3

1   What number is fifteen more than fifty-five?

2   What is a tenth of 2400?

3   What is twenty times forty-five?

4   Write in figures the number ten thousand, seven hundred and five.

5   A play lasting $2\frac{1}{4}$ hours starts at half-past eight. When does it finish?

6   What number is fifty-five less than 300?

7   What is one-quarter of twenty-eight?

8   A book costs £1·95. How much change do I receive from a five pound note?

9   What is half of half of sixty?

10 What four coins make 61 pence?

11 Work out $\frac{1}{2}$ plus $\frac{1}{4}$ and give the answer as a decimal.

12 A box holds 16 cans. How many boxes are needed for 80 cans?

13 If the 25th of December is a Tuesday, what day of the week is the first of January?

14 By how much is two kilos more than 500 g?

15 Write fifteen thousand and fifty pence in pounds and pence.

16 The sides of a square field measure 160 metres. Find the total distance around the field.

17 A quarter of my wages is taken in tax. What percentage have I got left?

18 A bingo prize of £150 000 is shared equally between five people. How much does each person receive?

19 Ice creams cost twenty-four pence each. How many can I buy with one pound?

20 A bag contains 22 five-pence coins. How much is in the bag?

21 What is two point three multiplied by ten?

22 A wine merchant puts 100 bottles in crates of 12. How many crates does he need?

23 Add together 73 and 18.

24 What is 5% of £120?

25 Peaches cost twenty pence each. How much do I pay for seven peaches?

26 A toy costs 54 pence. Find the change from a five-pound note.

27 A boy goes to and from school by bus and a ticket costs 33 pence each way. How much does he spend in a five-day week?

28 In your purse, you have two ten-pound notes, three five-pound notes and seven one-pound coins. How much have you got altogether?

29 What is double seventeen?

30 Fifty per cent of a number is thirty-two. What is the number?

Test 4

1 What is the change from a £10 note for goods costing £1·95?

2 Add 12 to 7 nines.

3 How many 20 pence coins are needed to make £5?

4 A pile of 100 sheets of paper is 10 cm thick. How thick is each sheet?

5 Lemons cost 7 pence each or 60 pence a dozen. How much is saved by buying a dozen instead of 12 separate lemons?

6 How many weeks are there in two years?

7 What is 1% of £40?

8 How much more than £92 is £110?

9 My watch reads five past 6. It is 15 minutes fast. What is the correct time?

10 How many degrees are there in three right angles?

11 I am facing south-west and the wind is hitting me on my back. What direction is the wind coming from?

12 If eight per cent of students of a school are absent, what percentage of students are present?

13 I go shopping with £5 and buy 3 items at 25 pence each. How much money have I left?

14 From one thousand and seven take away nine.

15 If I can cycle a mile in 3 minutes, how many miles can I cycle in one hour?

16 How many millimetres are there in 20 cm?

17 A metal rod 90 cm long is cut into four equal parts. How long is one part?

18 Find the cost of fifteen items at 5 pence each.

19 A 2 pence coin is about 2 mm thick. How many coins are in a pile which is 2 cm high?

20 Add up the first four odd numbers.

21 Add up the first four even numbers.

22 My daily paper costs 25 pence. I pay for it with a £10 note. What change do I receive?

*23 A film starts at 8:50 p.m. and finishes at 9:15 p.m. How long is the film?

24 School finishes at twenty to four. What is that on the 24-hour clock?

25 What is the sum of the numbers 1, 2, 3, 4 and 5?

26 What is 10% of £7?

27 How many 2 pence coins are needed to make £4?

28 35% of a class prefer BBC1 and 30% prefer ITV1. What percentage prefer the other channels?

29 How many minutes is it between 6:20 p.m. and 8:00 p.m.?

30 What is the cost of 1000 books at £2·50 each?

Test 5

1 What are eight twenties?

2 What number is nineteen more than eighty-seven?

3 Write in figures the number six-thousand and eleven.

4 What is the cost of six items at thirty-five pence each?

5 What is the sum of sixty-three and twenty-nine?

6 How many sevens are there in eighty-four?

7 A pair of shorts costs £8·99; how much change do you get from a £10 note?

8 Subtract forty-five centimetres from two metres giving your answer in metres as a decimal number.

9 If you have three thousand and eleven pennies, how much do you have in pounds and pence?

10 The perimeter of a square is sixteen centimetres. What is the length of one side of the square?

11 How many metres are there in 1·5 kilometres?

12 What is fifty per cent of fifty pounds?

13 How many sides has a heptagon?

14 I think of a number, double it and the answer is five. What was the number I thought of?

15 When playing darts you score double twenty and treble eight. What is your total score?

16 How many inches are there in one foot?

17 A petrol pump delivers 1 litre in 5 seconds. How many litres will it deliver in one minute?

18 If seventy-seven per cent of the students in a school are right-handed, what percentage are left-handed?

19 How much more than 119 is 150?

20 What is the cube root of 27?

21 What number is eight squared?

22 A rectangular lawn is 7 yards wide and 12 yards long. What area does it cover?

23 How many centimetres are there in 1 km?

24 A ship was due at noon on Tuesday, but arrived at 15:00 on Wednesday. How many hours late was it?

25 A litre of wine fills 9 glasses. How many litre bottles are needed to fill 50 glasses?

26 Work out 25% of £40.

27 How many grams are there in half a kilogram?

28 How many seconds are there in 3 minutes?

29 Write any multiple of nine.

30 If the eighth of May is a Monday, what day of the week is the thirteenth?

Test 6

1 What is the angle between the hands of a clock at two o'clock?

2 What is a half of a half of 60?

3 In a test Paul got 15 out of 20. What percentage is that?

4 Work out $2 \times 20 \times 200$.

5 Two friends share a bill for £52. How much does each person pay?

6 What number is halfway between six and sixteen?

7 A television programme starts at five minutes to seven and lasts thirty-five minutes. At what time does the programme finish?

8 What is $\frac{1}{5}$ as a percentage?

9 What is twenty-fifteen in twelve hour clock time?

10 Work out $0 \cdot 1$ plus $0 \cdot 01$.

11 What number should you subtract from fifty-one to get the answer twenty-four?

12 What is ten per cent of £25?

13 A car has a 1795 cc engine. What is that approximately in litres?

14 Write the number two thousand, one hundred and four in figures.

15 Find the cost of smoking 40 cigarettes a day for five days if a packet of 20 costs £4.

16 How many minutes are there in $2\frac{1}{2}$ hours?

17 A pie chart has a red sector representing 10% of the whole chart. What is the angle of the sector?

18 How many five-pence coins are needed to make £3?

19 What is the reflex angle between clock hands showing three o'clock?

20 A rectangular pane of glass is 3 feet long and 2 feet wide. Glass costs £1·50 per square foot. How much will the pane cost?

21 A car journey of 150 miles took 3 hours. What was the average speed?

22 Add 218 to 32.

23 Pencils cost 10 pence each. How many can I buy with £2·50?

24 Write the next prime number after 31.

25 A ruler costs 32 pence. What is the total cost of three rulers?

26 A salesman receives commission of 2% on sales. How much commission does he receive when he sells a computer for £1000?

27 How many edges does a cube have?

28 What four coins make seventy-six pence?

29 One angle in an isosceles triangle is one hundred and ten degrees. How large is each of the other two angles?

30 What number is one hundred times bigger than nought point two?

Test 7

1 Two angles of a triangle are 42° and 56°. What is the third angle?

2 Telephone charges are increased by 10%. What is the new charge for a call which previously cost 60p?

3 Write the number two thousand, one hundred and seven in figures.

4 What is nine hundred and fifty-eight to the nearest hundred?

5 In a survey three-quarters of people like football. What percentage of people like football?

6 What decimal number is twenty-three divided by one hundred?

7 How many twenty-pence coins make three pounds?

8 How many twelve-pence pencils can you buy for one pound?

9 One-third of a number is eight. What is the number?

10 Write nine-tenths as a decimal number.

11 What is the name of the quadrilateral which has only one pair of parallel sides?

12 What number is 10 less than ninety thousand?

13 A sphere is a prism. True or false?

14 What is the probability of scoring less than six on a fair dice?

15 I think of a number, divide it by three and the answer is seven. What number did I think of?

16 What is the area of a rectangle nine metres by seven metres?

17 How many quarters are there in one and a half?

18 How many millimetres are there in one metre?

19 How many hours of recording time are there on a two hundred and forty minute video tape?

20 What number is squared to produce eighty-one?

21 A packet of peanuts costs 65 pence. I buy two packets and pay with a ten-pound note. Find the change.

22 What is a half of a half of 22?

23 I bought three kilograms of flour and I used four hundred and fifty grams. How many grams of flour do I have left?

24 A bingo prize of two hundred thousand pounds is shared equally between five people. How much does each person receive?

25 What is the angle between the hands of a clock at 4 o'clock?

26 Write down a sensible estimate for eleven multiplied by ninety-nine.

27 How many 24p stamps can I buy for £2?

28 A milk crate has space for 24 bottles. How many crates are needed for 100 bottles?

29 Write nought point two five as a fraction.

30 What is two hundred and ten divided by one hundred?

Test 8	Test 9	Test 10	Test 11
1 $39 + 22$	**1** 65×2	**1** $7 + 77$	**1** 25% of 880
2 $60 - 21$	**2** $84 - 7$	**2** $330 - 295$	**2** $400 \div 20$
3 20% of 50	**3** $0 \cdot 7 + 0 \cdot 3$	**3** $(8 - 2)^2$	**3** 8×7
4 $0 \cdot 2 + 0 \cdot 62$	**4** 23×100	**4** $37 + 63$	**4** $3 \cdot 5 + 0 \cdot 35$
5 20×6	**5** 7^2	**5** 30×7	**5** 1% of 20 000
6 $200 - 145$	**6** £10 − 50p	**6** 25×200	**6** $301 - 102$
7 £5 − £1·20	**7** 50% of 684	**7** 12×7	**7** $(3 + 8)^2$
8 9×7	**8** $1 - 0 \cdot 2$	**8** $5^2 - 5$	**8** Half of 630
9 $14 + 140$	**9** 25×12	**9** $76 + 14$	**9** $1000 \times 0 \cdot 5$
10 50×22	**10** $8 + 9 + 10$	**10** 5% of 440	**10** $3 \times 4 \times 5$
11 5% of 300	**11** $210 \div 7$	**11** $500 - 85$	**11** 25×16
12 Half of 330	**12** 100×100	**12** $0 \cdot 6 + 0 \cdot 4$	**12** $54 \div 9$
13 $600 - 245$	**13** $5 \cdot 5 + 1 \cdot 5$	**13** 10^3	**13** 30×60
14 $2 \cdot 4 + 1 \cdot 7$	**14** $(2 + 3)^2$	**14** $425 - 198$	**14** $22 + 23 + 24$
15 $200 \div 5$	**15** $240 \div 6$	**15** 200×8	**15** $6^2 - 2^2$
16 8×25	**16** $8 - 2 \cdot 5$	**16** $420 \div 7$	**16** $1100 - 999$
17 60p + £1·50	**17** Half of 38·4	**17** $4^2 + 7$	**17** $11 - 0 \cdot 3$
18 $82 - 63$	**18** $400 \div 50$	**18** $9 - 0 \cdot 2$	**18** $200 \div 200$
19 $7 + 8 + 9$	**19** 15% of 300	**19** $2 \cdot 6 + 2 \cdot 6$	**19** 60×60
20 $2 \times 3 \times 4$	**20** $18 + 81$	**20** $2600 \div 100$	**20** $4 \times 5 \times 6$

5.7 Rounding

A car travels a distance of 158 miles in $3\frac{1}{2}$ hours. What is its average speed?

$$\text{Speed} = \frac{\text{Distance}}{\text{Time}} = \frac{158}{3\cdot5}$$

> You will meet this formula again in Section 9·2.

On a calculator the answer is 45·142 857 14 mph.
It is not sensible to give all these figures in the answer.
The distance and time may not be all that accurate. It would be reasonable to give the answer as '45 mph'. This is called 'rounding'.

> ● You can approximate in several ways:
> 1 you can round to **the nearest whole number**
> 2 you can round to one or more **significant figures**
> 3 you can round to one or more **decimal places**.

5.7.1 Rounding to the nearest whole number

> ● To round a number to the nearest whole number, look at the first digit after the decimal point to see if it is '**five or more**'. If that number is '**five or more**' you round **up**. Otherwise you round **down**.

Here are some examples.
a 13·82 = 14 to the nearest whole number
       ↑
b   6·2   = 6 to the nearest whole number
c 211·54 = 212 to the nearest whole number
d 0·971 = 1 to the nearest whole number

Exercise 30 C

Round these numbers to the nearest whole number.
1 18·32	**2** 22·8	**3** 41·51
4 3·24	**5** 224·9	**6** 36·11
7 8·07	**8** 56·52	**9** 3·911

10 0·87	**11** 18·421	**12** 111·1
13 17·62	**14** 5·52	**15** 712·89
16 62·66	**17** 3·333	**18** 5742·2
19 19·501	**20** 59·7	

Work these out on a calculator and then give the answer to the
nearest whole number.

21 $8·2 \times 1·7$	**22** $11·3 \times 11·4$	**23** $8·06 \times 19$
24 $6·9 \times 1·5$	**25** $0·71 \times 5·2$	**26** $16·4 \times 2$
27 $8·05 \div 5$	**28** $11·2 \div 3$	**29** $6·6 \times 6·6$
30 $27 \div 4·7$	**31** $10 \div 0·42$	**32** $624 \div 11$
33 $7084 \div 211$	**34** $6168 \div 217$	**35** $18·2^2$
36 $6·09^2$	**37** $6·84 + 11·471$	**38** $19 - 7·364$
39 $22·1 + 0·724$	**40** $16·3 - 7·82$	

5.7.2   Rounding to one significant figure

● To round a number to one significant figure, look at the place
value of the first non-zero digit and then round to this place value.

Here are some examples.

a   41 is 40 to one significant figure
  ↑

b   278 is 300 to one significant figure
   ↑

Look at the number
marked with an
arrow to see if it is
'five or more.'

c   4562 is 5000 to one significant figure
  ↑

d   84 796 is 80 000 to one significant figure
  ↑

Exercise 31 C

Round these numbers to one significant figure.

1 214	**2** 378	**3** 4911	**4** 6684
5 8209	**6** 4592	**7** 376	**8** 29
9 42	**10** 196	**11** 417	**12** 4211
13 685	**14** 701	**15** 6666	**16** 28
17 4192	**18** 16 234	**19** 8523	**20** 672

Work these out on a calculator and then give the answer to
one significant figure.

21 41×11	**22** 7×229	**23** 82×83	**24** 17×5
25 $3540 \div 15$	**26** $1426 \div 23$	**27** $682 \div 31$	**28** $1760 \div 32$
29 $4714 + 525$	**30** $6024 - 4111$	**31** $378 + 5972$	**32** $84 + 871 + 246$

33 The newspaper article contains several numbers in bold type.

    a For each number decide whether or not to replace the number with an approximate value to an **appropriate** degree of accuracy. ('appropriate' means 'sensible')

    b Some of the numbers should **not** be replaced. State which these are.

> The Olympic swimming pool in Athens contained **1493.2** m³ of water at a temperature of **23.41**°C. The crowd of **2108** cheered as Marisa won the **100** m butterfly in a new World Record time of **58.23** seconds. Altogether there were about **5173** swimmers taking part in the swimming events. The next Games take place in **2008** in ...

5.7.3 Rounding to 2 or 3 significant figures

EXAMPLE

Write these numbers correct to three significant figures (3 sf).

  a 2·6582   **b** 0·5142
  c 84 660   **d** 0·040 31

In each case look at the fourth significant figure, marked with an arrow, to see if it is 'five or more'.
Here you count figures from the left starting from the first **non-zero** figure.

  a 2·6582 = 2·66 to 3 sf   **b** 0·5142 = 0·514 to 3 sf   (Ignore the zero
      ↑   ↑   before the ·5)

  c 84 660 = 84 700 to 3 sf   **d** 0·040 31 = 0·0403 to 3 sf
      ↑   ↑

Exercise 32Ⓔ

In questions **1** to **8** write the numbers correct to three significant figures.

1 2·3462	**2** 0·814 38	**3** 26·241	**4** 35·55
5 112·74	**6** 210·82	**7** 0·8254	**8** 0·031 162

In questions **9** to **16** write the numbers correct to two significant figures.

9 5·894	**10** 1·232	**11** 0·5456	**12** 0·7163
13 0·1443	**14** 1·831	**15** 24·83	**16** 31·37

In questions **17** to **24** write the numbers correct to three
significant figures.

17 486·72	**18** 500·36	**19** 2·8888	**20** 3·1125
21 0·071 542	**22** 3·0405	**23** 2463·5	**24** 488 852

In questions **25** to **36** write the numbers to the degree of
accuracy indicated.

25 0·5126 (3 sf)	**26** 5·821 (2 sf)	**27** 65·89 (2 sf)
28 587·55 (3 sf)	**29** 0·581 (1 sf)	**30** 0·0713 (1 sf)
31 5·8354 (3 sf)	**32** 87·84 (2 sf)	**33** 2482 (2 sf)
34 52 666 (3 sf)	**35** 0·0058 (1 sf)	**36** 6568 (1 sf)

> sf is shorthand for
> significant figures.

5.7.4 Rounding to one, two or three decimal places

EXAMPLE

> Write these numbers correct to two decimal places (2 dp).
>   **a** 8·358   **b** 0·0328   **c** 74·355
>
> ....
>
> In each case look at the decimal place marked with an arrow to see
> if it is 'five or more'.
> Here you count figures after the decimal point.
>
>   **a** 8·358 = 8·34 to 2 dp   **b** 0·0328 = 0·03 to 2 dp   **c** 74·355 = 74·36 to 2 dp
>   ↑   ↑   ↑

Exercise 33Ⓔ

In questions **1** to **8** write the numbers correct to two decimal
places (2 dp).

1 5·381	**2** 11·0482	**3** 0·414	**4** 0·3666
5 8·015	**6** 87·044	**7** 9·0062	**8** 0·0724

> dp is shorthand for
> decimal places.

In questions **9** to **16** write the numbers correct to one decimal
place.

9 8·424	**10** 0·7413	**11** 0·382	**12** 0·095
13 6·083	**14** 19·53	**15** 8·111	**16** 7·071

In questions **17** to **28** write the numbers to the degree of
accuracy indicated.

17 8·155 (2 dp)	**18** 3·042 (1 dp)	**19** 0·5454 (3 dp)
20 0·005 55 (4 dp)	**21** 0·7071 (2 dp)	**22** 6·8271 (2 dp)
23 0·8413 (1 dp)	**24** 19·646 (2 dp)	**25** 0·071 35 (4 dp)
26 60·051 (1 dp)	**27** 7·30 (1 dp)	**28** 5·424 (2 dp)

29 Use a ruler to measure the dimensions of the rectangles.
  a Write the length and width in cm correct to one dp.
  b Work out the area of each rectangle and give the
    answer in cm^2 correct to one dp.

i    **ii**

Exercise 34Ⓔ

Write the answers to the degree of accuracy indicated.

1 0.153×3.74 (2 dp)   **2** $18.09 \div 5.24$ (3 sf)
3 184×2.342 (3 sf)   **4** $17.2 \div 0.89$ (1 dp)
5 $58 \div 261$ (2 sf)   **6** 88.8×44.4 (1 dp)
7 $(8.4 - 1.32) \times 7.5$ (2 sf) **8** $(121 + 3758) \div 211$ (3 sf)
9 $(1.24 - 1.144) \times 0.61$ (3 dp) **10** $1 \div 0.935$ (1 dp)
11 78.3524^2 (3 sf)   **12** $(18.25 - 6.941)^2$ (2 dp)
13 $9.245^2 - 65.2$ (1 dp)   **14** $(2 - 0.666) \div 0.028$ (3 sf)
15 8.43^3 (1 dp)   **16** $0.924^2 - 0.835^2$ (2 dp)

5.8 Estimating

5.8.1 Estimating answers

In some circumstances it is not realistic to work out the exact
answer to a problem. You can give an estimate for the
answer. For example, a builder does not know **exactly** how
many bricks a new garage will need. He may estimate that he
needs 2500 bricks and place an order for that number. In
practice he may need only 2237.
Estimate the answers to these calculations.

a $9.7 \times 3.1 \approx 10 \times 3$ About 30.
b $81.4 \times 98.2 \approx 80 \times 100$ About 8000.
c $19.2 \times 49.1 \approx 20 \times 50$ About 1000.
d $102.7 \div 19.6 \approx 100 \div 20$ About 5.

● Estimates can also help you check that your exact answer
  is correct.

Exercise 35 Ⓔ

Write each question and decide (by estimating) which answer is closest to the exact answer. Do not do the calculations exactly.

	Question	Answer A	Answer B	Answer C
1	$7{\cdot}79 \div 1{\cdot}9$	8	4	2
2	$27{\cdot}03 \div 5{\cdot}1$	5	0·5	9
3	$59{\cdot}78 \div 9{\cdot}8$	12	3	6
4	$58{\cdot}4 \times 102$	600	6000	2400
5	$6{\cdot}8 \times 11{\cdot}4$	19	280	80
6	$97 \times 1{\cdot}08$	100	50	1000
7	$972 \times 20{\cdot}2$	2000	20 000	9000
8	$7{\cdot}1 \times 103$	70	700	7000
9	$18{\cdot}9 \times 21$	400	60	200
10	$1{\cdot}078 \div 0{\cdot}98$	6	10	1
11	$1250{\cdot}5 \div 6{\cdot}1$	20	200	60
12	$20{\cdot}48 \div 3{\cdot}2$	6	12	3
13	$25{\cdot}11 \div 3{\cdot}1$	8	15	20
14	$216 \div 0{\cdot}9$	50	24	240
15	$19{\cdot}2 + 0{\cdot}41$	23	8	20
16	$207 + 18{\cdot}34$	25	230	1200
17	$68{\cdot}2 - 1{\cdot}38$	100	48	66
18	$7 - 0{\cdot}64$	6	1·5	0·5
19	$974 \times 0{\cdot}11$	9	100	500
20	$551{\cdot}1 \div 11$	7	50	5000
21	$207{\cdot}1 + 11{\cdot}65$	300	20	200
22	$664 \times 0{\cdot}51$	150	300	800
23	$(5{\cdot}6 - 0{\cdot}21) \times 39$	400	200	20
24	$\dfrac{17{\cdot}5 \times 42}{2{\cdot}5}$	300	500	100
25	$(906 + 4{\cdot}1) \times 0{\cdot}31$	600	300	30
26	$\dfrac{543 + 472}{18{\cdot}1 + 10{\cdot}9}$	60	30	100
27	$18{\cdot}9 \times 21{\cdot}4$	200	400	4000
28	$5{\cdot}14 \times 5{\cdot}99$	15	10	30
29	$811 \times 11{\cdot}72$	8000	4000	800
30	99×98	1 million	100 000	10 000
31	$1{\cdot}09 \times 29{\cdot}6$	20	30	60
32	$81\,413 \times 10{\cdot}96$	8 million	1 million	800 000
33	$601 \div 3{\cdot}92$	50	100	150

Exercise 36Ⓔ

1 For a wedding the caterers provide food at £39·75 per head. When Janet married John there were 207 guests at the wedding. Estimate the total cost of the food.

2 985 people share the cost of hiring an ice rink. About how much does each person pay if the total cost is £6017?

3 A home-made birthday card costs £3·95. Estimate the total cost of 107 cards.

4 The rent for a flat is £104 per week. Estimate the total spent on rent in one year.

In questions **5** and **6** there are six calculations and six answers. Write each calculation and insert the correct answer from the list given. Use estimation.

5 a $8·9 \times 10·1$ **b** $7·98 \div 1·9$ **c** $112 \times 3·2$
   d $11·6 + 47·2$ **e** $2·82 \div 9·4$ **f** $262 \div 100$

> Answers: 2·62, 58·8, 0·3, 89·89, 358·4, 4·2

6 a $49·5 \div 11$ **b** 21×22 **c** $9·1 \times 104$
   d $86 - 8·2$ **e** $2·4 \div 12$ **f** $651 \div 31$

> Answers: 21, 946·4, 0·2, 4·5, 462, 77·8

In questions **7** to **14** estimate which answer is closest to the actual answer.

7 The height of a double-decker bus

A	B	C
3 m	6 m	10 m

8 The height of the tallest player in the Olympic basketball competition

A	B	C
1·8 m	3·0 m	2·2 m

9 The mass of a £1 coin

A	B	C
1 g	10 g	100 g

10 The volume of your classroom

A	B	C
700 cubic feet	7000 cubic feet	70 000 cubic feet

11 The top speed of a Grand Prix racing car
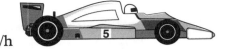

    A   B   C

  600 km/h   80 km/h   300 km/h

12 The number of times your heart beats in one day (24 h)

    A   B   C

  10 000   100 000   1 000 000

13 The thickness of one page in this book

    A   B   C

  0·01 cm   0·001 cm   0·0001 cm

14 The number of cars in a traffic jam 10 km long
on a 3-lane motorway. (Assume each car takes up
10 m of road.)

    A   B   C

  3000   30 000   300 000

15 A rectangular floor measures 28 metres by 43 metres. It
costs £6·95 per square metre to treat the floor for damp.
Estimate the cost of this job, showing all your working.

*16 A shopkeeper wants to buy
28 windows at £97   each
63 lights   at £4·95   each
32 shelves   at £19·99 each.
Do a rough calculation to find the total cost.

5.8.2 Checking answers

Here are five calculations followed by appropriate checks, using
inverse operations.

a $22·5 \div 5 = 4·5$ check $4·5 \times 5$

b $29·5 - 1·47 = 28·03$ check $28·03 + 1·47$

c $78·5 \times 20 = 1570$ check $1570 \div 20$

d $\sqrt{11} = 3·316\,62$ check $3·316\,62^2$

e $14·7 + 28·1 + 17·4 + 9·9$ check $9·9 + 17·4 + 28·1 + 14·7$ (add in reverse order)

Exercise 37Ⓔ

1 Do these and check using inverse operations.

    a $83·5 \times 20 = \square$ check $\square \div 20$

    b $104 - 13·2 = \square$ check $\square + 13·2$

    c $228·2 \div 7 = \square$ check $\square \times 7$

    d $\sqrt{28} = \square$ check $\square^2$

    e $11·5 + 2·7 + 9·8 + 20·7$ check $20·7 + 9·8 + 2·7 + 11·5$

2 a Will the answer to 64×0.8 be larger or smaller than 64?

   b Will the answer to $210 \div 0.7$ be larger or smaller than 210?

   c Will the answer to 17.4×0.9 be larger or smaller than 17.4?

3 Here are the answers found by six students. Some are correct but some are clearly impossible or highly unlikely. Decide which answers are 'OK' and which are 'impossible' or 'highly unlikely'.

   a The height of the three boxes on this trolley = 110 cm
   b Time taken to walk 1 mile to school = 21 minutes
   c The height of the classroom door = 220 cm
   d The number of schools in the UK = 500 000
   e The mean value of the numbers
      32, 35, 31, 36, 32 = 37.8
   f One per cent of the UK population = 60 000 people.

5.9 Solving numerical problems 2

Exercise 38 ©

1 Four dozen bags of grain weigh 2016 kg. How much does each bag weigh?

2 An office building has twelve floors and each floor has twenty windows.
A window cleaner charges 50p per window.
How much will he charge to clean all the windows in the building?

3 Answer 'always true' or 'sometimes false' for each statement.
   a $3 \times n = 3 + n$ **b** $n \times n = n^2$ **c** $m + n = n + m$
   d $3n - n = 3$ **e** $a - b = b - a$ **f** $n + n^2 = n^3$

4 A rectangular wheat field is 200 m by 400 m. One hectare is
10 000 m² and each hectare produces 3 tonnes of wheat.
   a What is the area of the field in hectares?
   b How much wheat is produced in this field?

5 A powerful computer is hired out at a rate of 50p per minute. How much will it cost to hire the computer from 06:30 to 18:00?

6 An old sailor keeps all of his savings in gold. Altogether the gold weighs ten pounds. The price of gold is $520 an ounce. How much is his gold worth?

1 pound = 16 ounces

7 This packet of sugar cubes costs 60p.

How much would you have to pay for this packet?

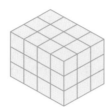

8 A wall measuring 3 m by 2 m is to be covered with square tiles of side 10 cm.
a How many tiles are needed?
b If the tiles cost £3·40 for ten how much will it cost?

9 Draw the next member of each sequence.

a

b

c

Exercise 39 **C**

1 The table shows the results of a test given to 60 students.

Mark	0	1	2	3	4	5
Number of students	1	4	10	12	15	18

a How many students scored fewer than 3 marks?
b Find the percentage of students who scored
  i 2 marks   **ii** 5 marks.

2 The thirteenth number in the sequence
1, 3, 9, 27, ... is 531 441.
  a What is the rule for the sequence?
  b What is the twelfth number?
  c What is the fourteenth number?

3 Six sacks of corn will feed 80 hens for 12 days.
Copy and complete these statements.

  a 18 sacks of corn will feed 80 hens for __ days.
  b 6 sacks of corn will feed 40 hens for __ days.
  c 60 sacks of corn will feed 40 hens for __ days.

4 Copy each sequence and write the next two numbers.
  a 2, 8, 20, —, —   **b** 1, 8, 27, —, —
  c 144, 121, 100, —, —   **d** $\dfrac{3}{4}, \dfrac{4}{5}, \dfrac{5}{6},$ —, —

5 What is the smaller angle between the hands of a clock at
four o'clock?

6 Write the coordinates of the points
which make up the '3.' You must
give the points in the correct
order starting at the
bottom.

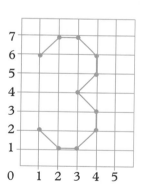

7 I think of a number. If I add 7 and then multiply the
result by 10 the answer is 90. What number was
I thinking of?

8 How many 31p stamps can be bought for £2 and how much
change will there be?

9 How many 10 mm pieces of string can be cut from a string
of length 1 metre?

10 Here are some function machines. Copy them and fill in the missing numbers.

a $30 \rightarrow \boxed{\times 3} \rightarrow \boxed{?}$ **b** $0\cdot1 \rightarrow \boxed{\times 10} \rightarrow \boxed{?}$ **c** $207 \rightarrow \boxed{\div 9} \rightarrow \boxed{?}$

d $\boxed{?} \rightarrow \boxed{+18} \rightarrow 62$ **e** $\boxed{?} \rightarrow \boxed{\div 7} \rightarrow 11$ **f** $\boxed{?} \rightarrow \boxed{\times 6} \rightarrow 666$

Exercise 40 C

1 A slimmer's calorie guide shows how many calories are contained in various foods.

Bread 1·2 calories per g
Cheese 2·5 calories per g
Meat 1·6 calories per g
Butter 6 calories per g

Calculate the number of calories in this meal:
50 g bread, 40 g cheese, 100 g meat, 15 g butter.

2 Write these as single numbers.
   a 8^2 **b** 1^4 **c** 10^2
   d 3×10^3 **e** 2^5 **f** 3^4

3 Here is a row of numbers

   1 2 3 4 5 6 7 8 9 10 11 12 13 14 15 16 17 18

   a Find **two** numbers next to each other which add up to 27.
   b Find **three** numbers next to each other which add up to 24.
   c Find **four** numbers next to each other which add up to 46.

4 Find area of the shaded shape.

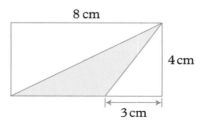

5 Write the number that is ten more than
   a 263   **b** 7447   **c** 74·5   **d** 295

6 Write the next number in each sequence.

a | 2 | 9 | 16 | 23 | ?

b | 2 | 4 | 8 | 16 | ?

c | 80 | 40 | 20 | 10 | ?

7 A group of four adults are planning a holiday in France. The box shows the ferry costs for the return journey.

> Adult £25
> Car £62

Travel around France is estimated at 2000 km and petrol costs 1·4 euros per litre. The car travels 10 km on one litre of petrol.
  a Calculate the total cost of the return journey on the ferry.
  b Calculate the number of litres of petrol to be used.
  c Calculate the total cost, in euros, of the petrol.

8 A journey by boat takes 2 hours 47 minutes. How long will it take at half the speed?

9 Ten posts are equally spaced in a straight line. It is 450 m from the first to the tenth post. What is the distance between successive posts?

10 Find the smallest whole number that is exactly divisible by all the numbers 1 to 5 inclusive.

Exercise 41 C

1 Seven fig rolls together weigh 560 g. A calorie guide shows that 10 g of fig roll contains 52 calories.
  a How much does one fig roll weigh?
  b How many calories are there in 1 g of fig roll?
  c How many calories are there in one fig roll?

2 Find each of these mystery numbers.
   a I am an odd number and a prime number. I am a factor of 14.
   b I am a two-digit multiple of 50.
   c I am one less than a prime number which is even.
   d I am odd, greater than one and a factor of both 20 and 30.

3 To the nearest whole number 5·84 is 6, 7·781 is 8 and 16·23 is 16.
   a Use these approximate values to obtain an approximate result for $\dfrac{5 \cdot 84 \times 16 \cdot 23}{7 \cdot 781}$.
   b Use the same approach to obtain approximate results for

   i $\dfrac{15 \cdot 72 \times 9 \cdot 78}{20 \cdot 24}$ **ii** $\dfrac{23 \cdot 85 \times 9 \cdot 892}{4 \cdot 867}$

4 King Richard is given three coins which look identical, but in fact one of them is an overweight fake.
Describe how he could discover the fake using an ordinary balance and only **one** weighing operation.

5 A light aircraft flies 375 km on 150 litres of fuel. How much fuel does it need for a journey of 500 km?

6 The headmaster of a school discovers an oil well in the school playground. As is the custom in such cases, he receives all the money from the oil. The oil comes out of the well at a rate of £15 for every minute of the day and night. How much does the headmaster receive in a 24-hour day?

7 A map uses a scale of 1 to 100 000. Calculate the actual length, in km, of a canal which is 5·4 cm long on the map.

8 Copy and complete this multiplication square.

×	6	3		
		15		35
			36	
2				
			32	56

9 I think of a number. If I subtract 4 and then divide the result by 4 the answer is 3. What number was I thinking of?

10 Copy the 9 points. Try to draw four straight lines which pass through all of the 9 points, without taking your pen from the paper and without going over any line twice.

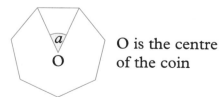

> The lines can go outside the pattern of dots.

Exercise 42 C

1 A car travels 35 m in 0·7 seconds. How far does it travel in
  a 0·1 s?   **b** 1 s?   **c** 2 minutes?

2 Twelve calculators cost £102. How many calculators can you buy for £76·50?

3 The outline of a 50p coin is shown.

O is the centre of the coin

Calculate the size of the angle, *a*, to the nearest $\frac{1}{10}$ of a degree.

4 The diagram shows the map of a farm which grows four different crops in the regions shown.

Each square represents one acre.
 a What is the total area of the farm?
 b What area is used for crop A?
 c What percentage of the farm is used for crop C?
 d What percentage of the farm is used for crop D?

5 A man smokes 40 cigarettes a day and each packet of 20 cigarettes costs £4·50. How much does he spend on cigarettes in a whole year of 365 days?

6 Adam worked 7 hours per day from Monday to Friday and 4 hours overtime on Saturday. The rate of pay from Monday to Friday is £6 per hour and the overtime rate is time and a half. How much did he earn during the week?

7 An examination is marked out of a total of 120 marks. How many marks did Alan get if he scores 65% of the marks?

8 A shopkeeper buys coffee at £3·65 per kg and sells it at 95p per 100 g. How much profit does he make per kg?

9 Five 2s can be used to make 25 as shown.

$$22 + 2 + \frac{2}{2} = 25$$

 a Use four 9s to make 100   **b** Use three 6s to make 7
 c Use three 5s to make 60   **d** Use five 5s to make 61
 e Use four 7s to make 1   **f** Use three 8s to make 11

10 Copy and find the missing digits.

a

	2	
+	5	4
		7

b

	1	7
+		6
	6	

c

	5		2
+	1	3	
		1	8

d

	4		4
+		5	
	8	2	4

e

	8	
−		4
	5	2

f

	8		2
−		5	
	2	3	2

Test yourself

 You may **not** use a calculator in questions **1** to **12**.

1 a Which **two** of these fractions are equivalent to $\frac{1}{4}$?

$\frac{2}{8}$ $\frac{5}{16}$ $\frac{6}{24}$ $\frac{11}{40}$

b Change $\frac{1}{4}$ to a decimal.

(AQA, 2003)

2 a i What fraction of this shape is shaded?
Write your fraction in its simplest form.

ii Copy this shape. Shade $\frac{1}{4}$ of it.

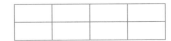

9 is the number that is halfway between 6 and 12.

6 ....9..... 12

b Work out the number that is halfway between
i   20 and 60
ii   100 000 and 200 000
iii 6·5 and 6·6
iv $\frac{1}{4}$ and $\frac{1}{2}$

(Edexcel, 2004)

3 a In a test Dipak scored 37 out of 50.
What is this as a percentage?

b Work out.
i $\frac{3}{4} - \frac{1}{3}$

ii $\frac{3}{8} \div 6$
Give your answer in its lowest terms.

(OCR, 2005)

4 Heather is revising fractions for her homework.
This is how she answers one of the questions.

$$\frac{1}{2} + \frac{1}{3} = \frac{2}{5}$$

Heather is wrong.
Show the correct way to work out $\frac{1}{2} + \frac{1}{3}$.

(AQA, 2004)

5 Ken and Susan share £20 in the ratio 1 : 3.
Work out how much money each person gets.

(Edexcel, 2003)

6 Kath knows a quick way to work out 15% of any amount of money.

To work out 15% of £160
10% of £160 = £16
So 5% of £160 = £8
So 15% of £160 = £24

Use Kath's method to work out 15% of £420.

(AQA, 2004)

7 Work out
  a $2655 \div 9$
  b 417×28
  c 40% of 150

(OCR, 2004)

8

| 60% of £40 | $\frac{2}{5}$ of £55 |

Which is the larger amount?
You **must** show your working.

(AQA, 2004)

9 A group of 23 people share some money equally and each person gets £476. What is the total amount of money they have shared?

All working must be shown.

(WJEC)

10 a Round 8640 to one significant figure.
  b A school buys 62 textbooks costing £10·25 each.
    i Write a calculation you could do in your head to **estimate** the total cost of the textbooks.
    ii Is your estimate of the total cost bigger or smaller than the actual cost? Explain how you know.

(OCR, 2004)

11 Hannah, Gemma and Jo use their calculators to work out the value of

$$\frac{28\cdot78}{4\cdot31 \times 0\cdot47}$$

Hannah gets 142·07, Gemma gets 14·207 and Jo gets 3·138.
Use approximations to show which one of them is correct.
You **must** show your working.

(AQA, 2004)

12 Estimate the value of

$$\frac{813 \times 19\cdot8}{97\cdot6}$$

(Edexcel, 2003)

13 Ben bought a car for £12 000.
Each year the value of the car depreciated by 10%.
Work out the value of the car two years after he bought it.

(Edexcel, 2003)

You may use a calculator in questions **14** to **22**.

14 A map has a scale of 1 : 5000.
  a The distance between two places on the map is 6 cm. What is the real life distance, in metres, between these two places?
  b The real life distance between two landmarks is 1·2 kilometres. What is the distance, in cm, between these landmarks on the map?

(WJEC)

15 Work out 23% of £64.

(Edexcel, 2003)

16 Alan, Brendan and Chloe shared £768 in the ratio 5 : 4 : 3. How much did each receive?

(OCR, 2005)

17 The diagram shows the measuring scale on a petrol tank.

$$0 \qquad \frac{1}{4} \qquad \frac{1}{2} \qquad \frac{3}{4} \qquad \text{Full}$$

a What fraction of the petrol tank is empty?
The petrol tank holds 28 litres when full.
A litre of petrol cost 74p.
b Work out the cost of the petrol which has to be added
to the tank so that it is full.
c What is the reading on each of these scales?

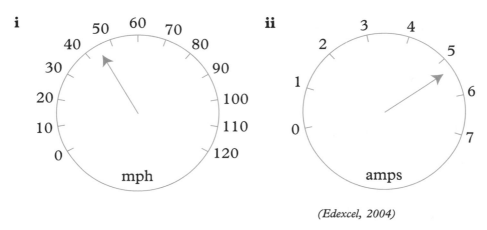

(Edexcel, 2004)

18 Sachin paid £250 for 36 sweatshirts.
He sold them all at £11·99 each.
Calculate this percentage profit.

(OCR, 2004)

19 This is a list of ingredients for making a pear and almond
crumble for 4 people.

Ingredients for **4** people
80 g plain flour
60 g ground almonds
90 g soft brown sugar
60 g butter
4 ripe pears

Work out the amount of each ingredient needed to make a
pear and almond crumble for **10** people.

(Edexcel, 2004)

20 In 1997 Quentin bought an antique table for £7500.
The next year the value of the table increased by 6%.
In each of the following 2 years the value of the table
increased by 8% of its value in the previous year.

What was the value of the table when Quentin sold it in
2000?

(OCR, 2003)

21 a Write these five fractions in order of size.
Start with the smallest fraction.

$$\frac{3}{4} \qquad \frac{1}{2} \qquad \frac{3}{8} \qquad \frac{2}{3} \qquad \frac{1}{6}$$

b Write these numbers in order of size.
Start with the smallest number.

$$65\% \qquad \frac{3}{4} \qquad 0{\cdot}72 \qquad \frac{2}{3} \qquad \frac{3}{5}$$

(Edexcel, 2004)

22 Simon repairs computers.
He charges

£56·80 for the first hour he works on a computer and
£42·50 for each extra hour's work.

Yesterday Simon repaired a computer and charged a total
of £269·30

a Work out how many hours Simon worked yesterday on
this computer.

Simon reduces his charges by 5% when he is paid promptly.
He was paid promptly for yesterday's work on the
computer.

b Work out how much he was paid.

(Edexcel, 2003)

6 Algebra 2

In this unit you will:
- learn how to use trial and improvement to find a solution
- revise using brackets in equations
- learn how to change the subject of a formula
- learn how to display and interpret inequalities
- revise straight-line graphs and learn about the equation of a straight line
- learn how to solve simultaneous equations
- learn how to plot and solve quadratic functions.

6.1 Trial and improvement

Some problems cannot be solved using straightforward equations. In such cases, the method of 'trial and improvement' is often a help.

EXAMPLE

Find the length and height of this rectangle using trial and improvement.

height

Its length is twice its height.
Its area is $162\,\text{cm}^2$

Area $= \text{length} \times \text{height}$
$= 2 \times \text{height} \times \text{height}$
$= 2 \times \text{height}^2$
So $162\,\text{cm}^2 = 2 \times \text{height}^2$

Try a number, for example:
  5    $2 \times 5^2 = 50$ Too small
 10    $2 \times 10^2 = 200$ Too large
  9    $2 \times 9^2 = 162$ ✓
So the height is 9 cm and the length is 18 cm.

Exercise 1 Ⓔ

1 Think of a rectangle having an area of $72\,\text{cm}^2$ whose base is twice its height.

Write the length of the base.

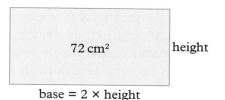

72 cm²   height

base = 2 × height

2 Find a rectangle of area $75\,\text{cm}^2$ so that its base is three times its height.

3 In each of these rectangles, the base is twice the height. The area is shown inside the rectangle. Find the base and the height of each rectangle.

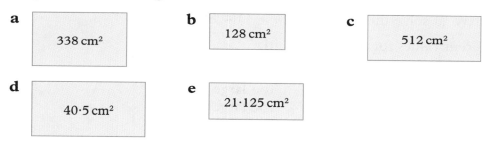

a   338 cm²

b   128 cm²

c   512 cm²

d   40·5 cm²

e   21·125 cm²

4 In this rectangle, the base is 1 cm more than the height and the area is $90\,\text{cm}^2$.

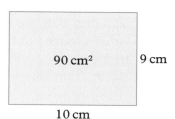

90 cm²   9 cm

10 cm

In each of the rectangles below, the base is 1 cm more than the height. Find each base and height.

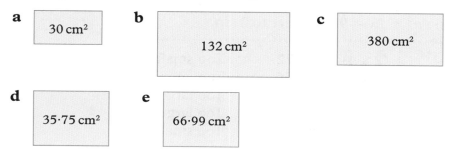

a   30 cm²

b   132 cm²

c   380 cm²

d   35·75 cm²

e   66·99 cm²

5 Once again the base of this rectangle is 1 cm more than the height. Try to find the base and the height of the rectangle.

75 cm²

6.1.1 Inexact answers

In some questions it is not possible to find an answer which is exactly correct. However, you can find answers which are nearer and nearer to the exact one, perhaps to the nearest 0·1 cm or even to the nearest 0·01 cm.

EXAMPLE

In this rectangle, the base is 1 cm more than the height h cm. The area is 80 cm^2. Find the height, h, correct to 1 dp.

height × length = area

Here you have to solve the equation $h(h + 1) = 80$.

a Try different values for h. Be systematic.

$h = 8$	:	$8(8 + 1)$	$= 72$	Too small
$h = 9$	:	$9(9 + 1)$	$= 90$	Too large
$h = 8·5$	:	$8·5(8·5 + 1)$	$= 80·75$	Too large
$h = 8·4$	:	$8·4(8·4 + 1)$	$= 78·96$	Too small

b You can see that the answer is between 8·4 and 8·5. You can also see that the value of $h = 8·5$ gave the value **closest** to 80. Take the answer as $h = 8·5$, correct to 1 dp.

Strictly speaking, to ensure that the answer **is** correct to 1 decimal place, you should try $h = 8·45$. This degree of complexity is unnecessary at this stage.

EXAMPLE

Solve the equation $x^3 + 2x = 90$, correct to 1 dp.

Try	$x = 4$	:	$4^3 + (2 \times 4)$	$= 72$	Too small
	$x = 5$	:	$5^3 + (2 \times 5)$	$= 135$	Too large
	$x = 4·5$	:	$4·5^3 + (2 \times 4·5)$	$= 100·125$	Too large
	$x = 4·4$	:	$4·4^3 + (2 \times 4·4)$	$= 93·984$	Too large
	$x = 4·3$	:	$4·3^3 + (2 \times 4·3)$	$= 88·107$	Too small

Now 88·107 is closer to 90 than 93·984.

The solution is $x = 4·3$, correct to 1 decimal place.

Exercise 2Ⓔ

1 In these two rectangles, the base is 1 cm more than the height.

a

Area = 100 cm² h

$h + 1$

b

Area = 65 cm² h

$h + 1$

Find the value of h for each one, correct to 1 decimal place.

2 Find a solution to each equation correct to 1 decimal place. Do not use the cube root key on your calculator. Show your working.
  a $x^3 = 40$ **b** $x^3 = 100$ **c** $x^3 = 300$

3 Find the cube roots of these numbers correct to 1 decimal place. Do not use the cube root key on your calculator. Show your working.
  a 4·7   **b** 28   **c** 225

4 Find the positive solutions to these equations, giving the answer correct to 1 decimal place.
  a $x(x - 3) = 11$ **b** $x(x - 2) = 7$
  c $x(x + 1) = 20$ **d** $x^3 + x = 40$

5 An engineer wants to make a solid metal cube of volume 526 cm³. Call the edge of the cube x and write an equation.
Find x, giving your answer correct to 1 dp.

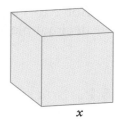

x

6 In this rectangle, the base is 2 cm more than the height. If the area is 20 cm², find h correct to 2 dp.

h

$h + 2$

7 Here you need a calculator with an x^y button or a $\wedge$ button.

Find x, correct to 1 decimal place.

  a $x^5 = 313$ **b** $5^x = 77$ **c** $x^x = 100$

Reminder:
To find 2^5 press
$\boxed{2}\ \boxed{\wedge}\ \boxed{5}\ \boxed{=}$

*8 The diagram represents a rectangular piece of paper ABCD which has been folded along EF so that C has moved to G.

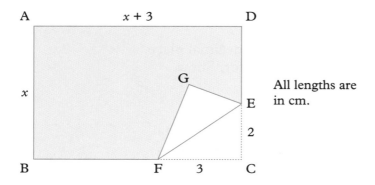

All lengths are in cm.

  a Calculate the area of △ECF.
  b Find an expression for the shaded area ABFGED in terms of x.

Given that the shaded area is 20 cm², show that $x(x + 3) = 26$.

Solve this equation, giving your answer correct to one decimal place.

6.2 Brackets and factors

6.2.1 Two brackets

Suppose you need to work out $(x + 3)(x + 2)$.
You can use the area of a rectangle to help.
Let $x + 3$ and $x + 2$ be the length and width of a rectangle.

Total area $= (x + 3)(x + 2)$
$\qquad\qquad = x^2 + 2x + 3x + 6$
$\qquad\qquad = x^2 + 5x + 6$

After a little practice, it is possible to do without the diagram.

Find   **a** $(x + 2)(x + 4)$ **b** $3(x + 1)(x + 2)$.

$$\begin{aligned}
\textbf{a} \quad (x + 2)(x + 4) &= x(x + 4) + 2(x + 4)\\
&= x^2 + 4x + 2x + 8\\
&= x^2 + 6x + 8\\
\textbf{b} \quad 3(x + 1)(x + 2) &= 3[x(x + 2) + 1(x + 2)]\\
&= 3[x^2 + 2x + x + 2]\\
&= 3x^2 + 9x + 6
\end{aligned}$$

> Leave the '3' outside a new bracket.

Exercise 3 Ⓔ

Remove the brackets and simplify.

1 $(x + 1)(x + 3)$
2 $(x + 3)(x + 2)$
3 $(y + 4)(y + 5)$
4 $(x + 3)(x + 4)$
5 $(x + 5)(x + 2)$
6 $(x + 8)(x + 2)$
7 $(a + 7)(a + 5)$
8 $(z + 9)(z + 2)$
9 $(x + 3)(x + 3)$
10 $(k + 11)(k + 11)$
11 $(2x + 1)(x + 3)$
12 $(3x + 4)(x + 2)$
13 $(2y + 3)(y + 1)$
14 $(7y + 1)(7y + 1)$
15 $(3x + 2)(3x + 2)$
16 $(5 + x)(4 + x)$
17 $2(x + 1)(x + 2)$
18 $3(x + 1)(2x + 3)$
19 $4(2y + 1)(3y + 2)$
20 $2(3x + 1)(x + 2)$

Be careful with an expression like $(x + 3)^2$. The answer is not $x^2 + 9$.

$$\begin{aligned}
(x + 3)^2 &= (x + 3)(x + 3)\\
&= x(x + 3) + 3(x + 3)\\
&= x^2 + 6x + 9
\end{aligned}$$

Exercise 4 Ⓔ

Remove the brackets and simplify.

1 $(x + 4)^2$
2 $(x + 2)^2$
3 $(x + 1)^2$
4 $(2x + 1)^2$
5 $(y + 5)^2$
6 $(3y + 1)^2$
7 $3(x + 2)^2$
8 $(3 + x)^2$
9 $(3x + 2)^2$
10 $2(x + 1)^2$

11 $(x + 1)^2 + (x + 2)^2$
12 $(x + 2)^2 + (x + 3)^2$
13 $(x + 2)^2 + (2x + 1)^2$
14 $(y + 3)^2 + (y + 4)^2$
15 $(x + 2)^2 + (2x + 3)^2$
16 $(x + 3)^2 + (x + 1)^2$

6.2.2 Factors

● You can factorise expressions by taking out common factors.

EXAMPLE

Factorise these expressions. **a** $12a - 15b$
  b $3x^2 - 2x$
  c $2xy + 6y^2$

....

a 3 goes into 12 and 15, so 3 is a **common factor**.
  $12a - 15b = 3(4a - 5b)$
b x is a common factor.
  $3x^2 - 2x = x(3x - 2)$
c 2 and y are common factors.
  $2xy + 6y^2 = 2y(x + 3y)$

EXAMPLE

Cancel the common factors in these expressions.

a $\dfrac{(x + 2)(x + 1)}{(x + 2)}$ **b** $\dfrac{3(x + 4)^2}{(x + 4)}$

....

a $\dfrac{\cancel{(x + 2)}\,(x + 1)}{\cancel{(x + 2)}} = x + 1$ **b** $\dfrac{3(x + 4)^{\cancel{2}}}{\cancel{(x + 4)}} = 3(x + 4)$

> Remember:
> $(x + 4)^2$ means
> $(x + 4) \times (x + 4)$.

Exercise 5Ⓔ

In questions **1** to **10** copy and complete the statements.

1 $6x + 4y = 2(3x + \Box)$ **2** $9x + 12y = 3(\Box + 4y)$
3 $10a + 4b = 2(5a + \Box)$ **4** $4x + 12y = 4(\Box + \Box)$
5 $10a + 15b = 5(\Box + \Box)$ **6** $18x - 24y = 6(3x - \Box)$
7 $8u - 28v = \Box(\Box - 7v)$ **8** $15s + 25t = \Box(3s + \Box)$
9 $24m + 40n = \Box(3m + \Box)$ **10** $27c - 72d = \Box(\Box - 8d)$

In questions **11** to **31** factorise the expressions.

11 $20a + 8b$ **12** $30x - 24y$ **13** $27c - 33d$
14 $35u + 49v$ **15** $12s - 32t$ **16** $40x - 16t$
17 $24x + 84y$ **18** $12x + 8y + 16z$ **19** $12a - 6b + 9c$
20 $10x - 20y + 25z$ **21** $20a - 12b - 28c$ **22** $48m + 8n - 24x$
23 $42x + 49y - 21z$ **24** $6x^2 + 15y^2$ **25** $20x^2 - 15y^2$
26 $7a^2 + 28b^2$ **27** $27a + 63b - 36c$ **28** $12x^2 + 24xy + 18y^2$
29 $64p - 72q - 40r$ **30** $36x - 60y + 96z$ **31** $9x + 6xy - 3x^2$

Factorise these expressions.

32 $x^2 - 5x$	**33** $2x^2 - 3x$	**34** $7x^2 + x$	**35** $y^2 + 4y$
36 $2x^2 + 8x$	**37** $4y^2 - 4y$	**38** $p^2 - 2p$	**39** $6a^2 + 2a$
40 $2ab - a$	**41** $3xy + 2x$	**42** $3t + 9t^2$	**43** $4 - 8x^2$
44 $5x - 10x^3$	**45** $4\pi r^2 + \pi rh$	**46** $\pi r^2 + 2\pi r$	

47 Simplify   **a** $\dfrac{(x + 3)(x - 1)}{(x + 3)}$ **b** $\dfrac{x(x + 2)}{x}$ **c** $\dfrac{2(x + 5)^2}{(x + 5)}$

6.3   Changing the subject of a formula

● You can rearrange any formula to have a different symbol before the equals sign. This is called 'changing the subject' of the formula.

You need to remember to do the same operation to both sides to keep the formula correct.

6.3.1   Simple formulae

EXAMPLE

Make x the subject in these formulae.
a $ax - p = t$
b $y(x + y) = v^2$

··

a $ax - p = t$
$\quad ax = t + p$ (Add p)
$\quad x = \dfrac{t + p}{a}$ (Divide by a)

b $y(x + y) = v^2$
$\quad yx + y^2 = v^2$
$\quad yx = v^2 - y^2$
$\quad x = \dfrac{v^2 - y^2}{y}$

Exercise 6Ⓔ

Make x the subject.

1 $x + b = e$	**2** $x - t = m$	**3** $x - f = a + b$
4 $x + h = A + B$	**5** $x + t = y + t$	**6** $a + x = b$
7 $k + x = m$	**8** $v + x = w + y$	**9** $ax = b$
10 $hx = m$	**11** $mx = a + b$	**12** $kx = c - d$
13 $vx = e + n$	**14** $3x = y + z$	**15** $xp = r$
16 $xm = h - m$	**17** $ax + t = a$	**18** $mx - e = k$
19 $ux - h = m$	**20** $ex + q = t$	**21** $kx - u^2 = v^2$
22 $gx + t^2 = s^2$	**23** $xa + k = m^2$	**24** $xm - v = m$
25 $a + bx = c$	**26** $t + sx = y$	**27** $y + cx = z$

28 $a + hx = 2a$ **29** $mx - b = b$ **30** $kx + ab = cd$

31 $a(x - b) = c$ **32** $c(x - d) = e$ **33** $m(x + m) = n^2$

34 $k(x - a) = t$ **35** $h(x - h) = k$ **36** $m(x + b) = n$

37 $a(x - a) = a^2$ **38** $c(a + x) = d$ **39** $m(b + x) = e$

> In questions **31** to **39** multiply out the brackets first.

6.3.2 Formulae involving fractions

EXAMPLE

Make x the subject in these formulae.

a $\dfrac{x}{a} = p$ **b** $\dfrac{m}{x} = t$ **c** $\dfrac{a^2}{m} = \dfrac{d}{x}$

....

a $\dfrac{x}{a} = p$ (multiply by a)   **b** $\dfrac{m}{x} = t$ **c** $\dfrac{a^2}{m} = \dfrac{d}{x}$

$x = ap$ $m = xt$ $xa^2 = dm$

$\dfrac{m}{t} = x$ $x = \dfrac{dm}{a^2}$

Exercise 7Ⓔ

Make x the subject.

1 $\dfrac{x}{t} = m$ **2** $\dfrac{x}{e} = n$ **3** $\dfrac{x}{p} = a$

4 $am = \dfrac{x}{t}$ **5** $bc = \dfrac{x}{a}$ **6** $e = \dfrac{x}{y^2}$

7 $\dfrac{x}{a} = (b + c)$ **8** $\dfrac{x}{t} = (c - d)$ **9** $\dfrac{x}{m} = s + t$

10 $\dfrac{x}{k} = h + i$ **11** $\dfrac{x}{b} = \dfrac{a}{c}$ **12** $\dfrac{x}{m} = \dfrac{z}{y}$

13 $\dfrac{x}{h} = \dfrac{c}{d}$ **14** $\dfrac{m}{n} = \dfrac{x}{e}$ **15** $\dfrac{b}{e} = \dfrac{x}{h}$

16 $\dfrac{x}{(a + b)} = c$ **17** $\dfrac{x}{(h + k)} = m$ **18** $\dfrac{x}{u} = \dfrac{m}{y}$

> When x is on the bottom, multiply both sides by x.

19 $\dfrac{x}{(h - k)} = t$ **20** $\dfrac{x}{(a + b)} = (z + t)$ **21** $t = \dfrac{e}{x}$

22 $a = \dfrac{e}{x}$ **23** $m = \dfrac{h}{x}$ **24** $\dfrac{a}{b} = \dfrac{c}{x}$

25 $\dfrac{u}{x} = \dfrac{c}{d}$ **26** $\dfrac{m}{x} = t^2$ **27** $\dfrac{h}{x} = \sin 20°$

28 $\dfrac{e}{x} = \cos 40°$ **29** $\dfrac{m}{x} = \tan 46°$ **30** $\dfrac{a^2}{b^2} = \dfrac{c^2}{x}$

6.3.3 Formulae with x^2- and negative x-terms

Make x the subject of these formulae.

a $t - x = a^2$ **b** $h - bx = m$ **c** $ax^2 + b = c$

. . . .

a $t - x = a^2$

$t = a^2 + x$

$t - a^2 = x$

or $x = t - a^2$

b $h - bx = m$

$h = m + bx$

$h - m = bx$

$\dfrac{h - m}{b} = x$

c $ax^2 + b = c$

$ax^2 = c - b$

$x^2 = \dfrac{c - b}{a}$

$x = \pm\sqrt{\dfrac{c - b}{a}}$

Exercise 8 Ⓔ

Make x the subject.

1 $cx^2 = h$
2 $bx^2 = f$
3 $x^2 t = m$
4 $x^2 y = (a + b)$

5 $mx^2 = (t + a)$
6 $x^2 - a = b$
7 $x^2 + c = t$
8 $x^2 + y = z$

9 $x^2 - a^2 = b^2$
10 $x^2 + t^2 = m^2$
11 $x^2 + n^2 = a^2$
12 $ax^2 = c$

13 $hx^2 = n$
14 $cx^2 = z + k$
15 $ax^2 + b = c$
16 $dx^2 - e = h$

17 $gx^2 - n = m$
18 $x^2 m + y = z$
19 $a + mx^2 = f$
20 $a^2 + x^2 = b^2$

21 $a - x = y$
22 $h - x = m$
23 $z - x = q$
24 $v = b - x$

25 $m = k - x$
26 $h - cx = d$
27 $y - mx = c$
28 $k - ex = h$

29 $a^2 - bx = d$
30 $m^2 - tx = n^2$
31 $v^2 - ax = w$
32 $y - x = y^2$

33 $k - t^2 x = m$
34 $e = b - cx$
35 $z = h - gx$
36 $a + b = c - dx$

37 $y^2 = v^2 - kx$
38 $h = d - fx$
39 $a(b - x) = c$
40 $h(m - x) = n$

Exercise 9 Ⓔ

1 A formula for calculating velocity is $v = u + at$.

    a Rearrange the formula to express a in terms of v, u and t.

    b Calculate a when $v = 20$, $u = 4$, $t = 8$.

> Make a the subject.

2 $P = \dfrac{mk}{y}$

    a Express k in terms of P, m and y.

    b Express y in terms of P, m and k.

3 A formula for calculating repair bills, R, is $R = \dfrac{n-d}{p}$.

a Express n in terms of R, p and d.
b Calculate n when $R = 400$, $p = 3$ and $d = 55$.

Exercise 10Ⓔ

This exercise is different as it contains a mixture of questions from the last four exercises.

Make the letter in brackets the subject.

1 $ax - d = h$ $[x]$

2 $zy + k = m$ $[y]$

3 $d(y + e) = f$ $[y]$

4 $m(a + k) = d$ $[k]$

5 $a + bm = c$ $[m]$

6 $ae^2 = b$ $[e]$

7 $yt^2 = z$ $[t]$

8 $x^2 - c = e$ $[x]$

9 $my - n = b$ $[y]$

10 $a(z + a) = b$ $[z]$

11 $\dfrac{a}{x} = d$ $[x]$

12 $\dfrac{k}{m} = t$ $[k]$

13 $\dfrac{u}{m} = n$ $[u]$

14 $\dfrac{y}{x} = d$ $[x]$

15 $\dfrac{a}{m} = t$ $[m]$

16 $\dfrac{d}{g} = n$ $[g]$

17 $\dfrac{t}{k} = (a + b)$ $[t]$

18 $y = \dfrac{v}{e}$ $[e]$

19 $c = \dfrac{m}{y}$ $[y]$

20 $\dfrac{a}{m} = b$ $[a]$

21 $g(m + a) = b$ $[m]$

22 $h(h + g) = x^2$ $[g]$

23 $y - t = z$ $[t]$

24 $me^2 = c$ $[e]$

25 $a(y + x) = t$ $[x]$

26 $uv - t^2 = y^2$ $[v]$

27 $k^2 + t = c$ $[k]$

28 $k - w = m$ $[w]$

29 $b - an = c$ $[n]$

30 $m(a + y) = c$ $[y]$

31 $pq - x = ab$ $[x]$

32 $a^2 - bk = t$ $[k]$

33 $v^2z = w$ $[z]$

34 $c = t - u$ $[u]$

35 $xc + t = 2t$ $[c]$

36 $m(n + w) = k$ $[w]$

37 $v - mx = t$ $[m]$

38 $c = a(y + b)$ $[y]$

39 $m(a - c) = e$ $[c]$

40 $ba = ce$ $[a]$

41 $\dfrac{a}{p} = q$ $[p]$

42 $\dfrac{a}{n^2} = e$ $[n]$

6.4 Inequalities

6.4.1 Symbols

- There are four inequality symbols.
  $x < 4$ means 'x is less than 4'
  $y > 7$ means 'y is greater than 7'
  $z \leqslant 10$ means 'z is less than or equal to 10'
  $t \geqslant -3$ means 't is greater than or equal to -3'

- If there are two symbols in one statement look at each part separately.

For example, if n is an **integer** and $3 < n \leqslant 7$, n has to be greater than 3 but at the same time it has to be less than or equal to 7.
So n could be 4, 5, 6 or 7 only.

> An integer is a whole number.

EXAMPLE

Show on a number line the range of values of x stated.
a $x > 1$ **b** $x \leqslant -2$ **c** $1 \leqslant x < 4$

a $x > 1$ Use an open circle to show
                            that 1 is not included.

b $x \leqslant -2$

                            Use a filled in circle to
c $1 \leqslant x < 4$ show that -2 is included.

Exercise 11Ⓔ

1 Write each statement with either $>$ or $<$ in the box.
  a $3 \ \square \ 7$ **b** $0 \ \square \ -2$ **c** $3 \cdot 1 \ \square \ 3 \cdot 01$
  d $-3 \ \square \ -5$ **e** $100 \, \text{m} \ \square \ 1 \, \text{m}$ **f** $1 \, \text{kg} \ \square \ 1 \, \text{lb}$

2 Write the inequality shown. Use x for the variable.

  a   2   **b**   5   **c**   100

  d -2 2   **e** -6 **f**   3   8

3 Draw number lines to show these inequalities.
  a $x \geqslant 7$ **b** $x < 2 \cdot 5$ **c** $1 < x < 7$ **d** $0 \leqslant x \leqslant 4$ **e** $-1 < x \leqslant 5$

4 Write an inequality for each statement.
  a You must be at least 16 to get married.
    [Use A for age.]
  b Vitamin J1 is not recommended for people over 70 or for children 3 years or under.
  c To braise a rabbit the oven temperature should be between 150 °C and 175 °C.
    [Use T for temperature.]
  d Applicants for training as paratroopers must be at least 1·75 m tall.
    [Use h for height.]

5 Answer 'true' or 'false'.
  a n is an integer and $1 < n \leqslant 4$, so n can be 2, 3 or 4 only.
  b x is an integer and $2 \leqslant x < 5$, so x can be 2, 3 or 4 only.
  c p is an integer and $p \geqslant 10$, so p can be 10, 11, 12, 13 …

6.4.2 Solving inequalities

Follow the same procedure that you use for solving equations except that when you multiply or divide by a **negative** number the inequality is **reversed**.

For example, $4 > -2$ but after multiplying by -2 you get $-8 < 4$

It is best to avoid dividing by a negative number.

> You are aiming for just an x on one side of the inequality.

EXAMPLE

a $2x - 1 > 5$
  $2x > 5 + 1$ (add 1)
  $x > \dfrac{6}{2}$ (divide by 2)
  $x > 3$

b $x + 1 \leqslant 3x$
  $1 \leqslant 3x - x$ (subtract x from both sides)
  $1 \leqslant 2x$
  $\dfrac{1}{2} \leqslant x$ (divide by 2)

Exercise 12Ⓔ

Solve these inequalities.

1 $x - 3 > 10$ **2** $x + 1 < 0$ **3** $5 > x - 7$
4 $2x + 1 \leqslant 6$ **5** $3x - 4 > 5$ **6** $10 \leqslant 2x - 6$
7 $5x < x + 1$ **8** $2x \geqslant x - 3$ **9** $4 + x < -4$
10 $3x + 1 < 2x + 5$ **11** $2(x + 1) > x - 7$ **12** $7 < 15 - x$
13 $9 > 12 - x$ **14** $4 - 2x \leqslant 2$ **15** $3(x - 1) < 2(1 - x)$
16 $7 - 3x < 0$ **17** $\dfrac{x}{3} < -1$ **18** $\dfrac{2x}{5} > 3$
19 $2x > 0$ **20** $\dfrac{x}{4} < 0$

*21 The height of the picture has to be greater than the width.
Find the range of possible values of x.

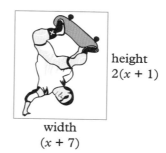

height
$2(x + 1)$

width
$(x + 7)$

In questions **22** to **27**, solve the two inequalities separately.

*22 $10 \leqslant 2x \leqslant x + 9$ ***23** $x < 3x + 2 < 2x + 6$
*24 $10 \leqslant 2x - 1 \leqslant x + 5$ ***25** $3 < 3x - 1 < 2x + 7$
*26 $x - 10 < 2(x - 1) < x$ ***27** $4x + 1 < 8x < 3(x + 2)$

Exercise 13Ⓔ

1 The area of the rectangle must be greater than the area of the triangle.
Find the range of possible values of x.

4

$x + 1$

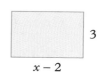

3

$x - 2$

For questions **2** to **8**, list the solutions which satisfy the given condition.

2 $3a + 1 < 20$; a is a positive integer.

3 $b - 1 \geqslant 6$; b is a prime number less than 20.

4 $1 < z < 50$; z is a square number.

5 $2x > -10$; x is a negative integer.

6 $x + 1 < 2x < x + 13$; x is an integer.

7 $0 \leqslant 2z - 3 \leqslant z + 8$; z is a prime number.

8 $\frac{a}{2} + 10 > a$; a is a positive even number.

9 Given that $4x > 1$ and $\frac{x}{3} \leqslant 1\frac{1}{3}$, list the possible integer values of x.

10 State the smallest integer n for which $4n > 19$.

11 Given that $-4 \leqslant a \leqslant 3$ and $-5 \leqslant b \leqslant 4$, find
a the largest possible value of a^2
b the smallest possible value of ab
c the largest possible value of ab
d the value of b if $b^2 = 25$.

12 For any shape of triangle ABC, copy and complete the statement

'AB + BC $\square$ AC'

by writing $<$, $>$ or $=$ inside the box.

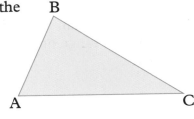

13 Find a simple fraction r such that $\frac{1}{3} < r < \frac{2}{3}$.

14 Find the largest prime number p such that $p^2 < 400$.

15 Find the integer n such that $n < \sqrt{300} < n + 1$.

16 If $f(x) = 2x - 1$ and $g(x) = 10 - x$ for what values of x is $f(x) > g(x)$?

*17 If $2^r > 100$, what is the smallest integer value of r?

*18 Given $\left(\frac{1}{3}\right)^x < \frac{1}{200}$, what is the smallest integer value of x?

*19 Find the smallest integer value of x which satisfies $x^x > 10\,000$.

*20 What integer values of x satisfy $100 < 5^x < 10\,000$?

6.5 Straight-line graphs

6.5.1 Horizontal and vertical lines

- The points P, Q, R and S have coordinates $(4, 4)$, $(4, 3)$, $(4, 2)$ and $(4, 1)$. They all lie on a straight line. Since the x-coordinate of all the points is 4, you say the **equation** of the line is $x = 4$.

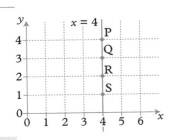

- All vertical lines have equations $x = $ a number

- The points A, B, C and D have coordinates $(1, 3)$, $(2, 3)$, $(3, 3)$ and $(4, 3)$. They all lie on a straight line. Since the y-coordinate of all the points is 3, you say the **equation** of the line is $y = 3$.

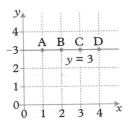

- All horizontal lines have equations $y = $ a number

Exercise 14 Ⓒ

1 Write the equations for the lines marked A, B and C.

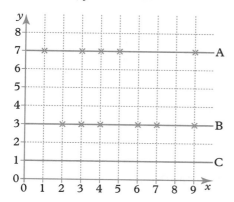

2 Write the equations for the lines marked P, Q and R.

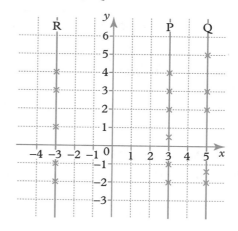

3 a Draw axes with values of x and y as in question **1**.

  b Draw the lines $y = 3$ and $x = 2$. At what point do they cross?

  c Draw the lines $y = 1$ and $x = 5$. At what point do they cross?

  d Draw the lines $x = 8$ and $y = 6$. At what point do they cross?

4 In the diagram, E and N lie on the line with equation $y = 1$. B and K lie on the line $x = 5$. Find the equation of the line passing through these points.

a A and D	**e** L and E
b A, B and I	**f** D, K and G
c M and P	**g** C, M, L and H
d I and H	**h** P and F

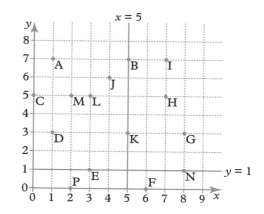

6.5.2 Relating x and y

● The sloping line passes through these points $(1, 1)$, $(2, 2)$, $(3, 3)$, $(4, 4)$, $(5, 5)$.

For each point, the y-coordinate is equal to the x-coordinate.

The equation of the line is $y = x$ (or $x = y$).

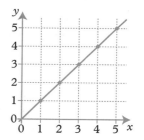

- This line passes through

  $(0, 1), (1, 2), (2, 3), (3, 4), (4, 5).$

  For each point the y-coordinate is one more than the x-coordinate. The equation of the line is $y = x + 1$.

  You can also say that the x-coordinate is always one less than the y-coordinate. You can then write the equation of the line as $x = y - 1$.
  (Most mathematicians use the equation beginning '$y =$'.)

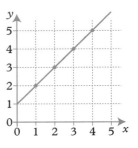

6.5.3 Drawing graphs

- Think about the graph $y = 2x - 3$. There are two operations.

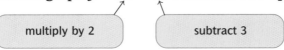

| multiply by 2 | subtract 3 |

You multiply x before you subtract 3.
A function machine shows this.

$$x \longrightarrow \boxed{\times 2} \longrightarrow \boxed{-3} \longrightarrow y$$

So when $x = 2$, $y = 2 \times 2 - 3 = 1$. On the graph you plot $(2, 1)$.
  when $x = 3$, $y = 2 \times 3 - 3 = 3$. On the graph you plot $(3, 3)$.
  when $x = 5$, $y = 2 \times 5 - 3 = 7$. On the graph you plot $(5, 7)$.

Here is the graph of $y = 2x - 3$.

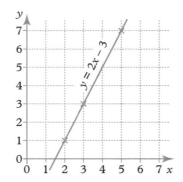

Remember:
a Start both axes at 0 and take care with scales around 0.
b Label the axes 'x' and 'y'.
c Label the graph with its equation.

- Think about the graph of $y = 3(x + 5)$.
  You do the operation in brackets first.
  The function machine is

$$x \longrightarrow \boxed{+5} \longrightarrow \boxed{\times 3} \longrightarrow y$$

When $x = 1$, $y = (1 + 5) \times 3 = 18$. You plot $(1, 18)$.
  $x = 2$, $y = (2 + 5) \times 3 = 21$. You plot $(2, 21)$.
  $x = 6$, $y = (6 + 5) \times 3 = 33$. You plot $(6, 33)$.

Exercise 15 **C**

For each question, copy and complete the table using the function machine. Then draw the graph using the scales given.

1 $y = 2x + 3$ for x from 0 to 6. $\left(\begin{array}{l} x: 1 \text{ cm} = 1 \text{ unit} \\ y: 1 \text{ cm} = 1 \text{ unit} \end{array} \right)$

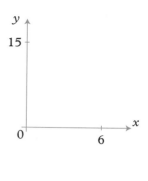

$x \rightarrow \boxed{\times 2} \rightarrow \boxed{+3} \rightarrow y$

x	0	1	2	3	4	5	6
y	3				11		
coordinates					(4,11)		

2 $y = x + 3$ for x from 0 to 7. $\left(\begin{array}{l} x: 1 \text{ cm} = 1 \text{ unit} \\ y: 1 \text{ cm} = 1 \text{ unit} \end{array} \right)$

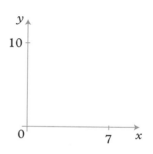

$x \rightarrow \boxed{+3} \rightarrow y$

x	0	1	2	3	4	5	6	7
y			5					
coordinates			(2,5)					

3 $y = 2x$ for x from 0 to 5. $\left(\begin{array}{l} x: 1 \text{ cm} = 1 \text{ unit} \\ y: 1 \text{ cm} = 1 \text{ unit} \end{array} \right)$

$x \rightarrow \boxed{\times 2} \rightarrow y$

x	0	1	2	3	4	5
y		2				
coordinates		(1,2)				

4 $y = 2x + 4$ for x from 0 to 5. $\left(\begin{array}{l} x: 1 \text{ cm} = 1 \text{ unit} \\ y: 1 \text{ cm} = 1 \text{ unit} \end{array} \right)$

$x \rightarrow \boxed{\times 2} \rightarrow \boxed{+4} \rightarrow y$

x	0	1	2	3	4	5
y			8			

5 $y = 2x - 1$ for x from 0 to 5. $\left(\begin{array}{l} x: 1 \text{ cm} = 1 \text{ unit} \\ y: 1 \text{ cm} = 1 \text{ unit} \end{array} \right)$

$x \rightarrow \boxed{\times 2} \rightarrow \boxed{-1} \rightarrow y$

x	0	1	2	3	4	5
y	−1					

6 $y = 3x + 1$ for x from 0 to 3. $\left(\begin{array}{l} x: 1 \text{ cm} = 1 \text{ unit} \\ y: 1 \text{ cm} = 1 \text{ unit} \end{array}\right)$

7 $y = 2(x + 1)$ for x from 0 to 5.

$x \rightarrow \boxed{+1} \rightarrow \boxed{\times 2} \rightarrow y$

8 $y = 3(x + 2)$ for x from 0 to 4. $\left(\begin{array}{l} x: 1 \text{ cm} = 1 \text{ unit} \\ y: 1 \text{ cm} = 1 \text{ unit} \end{array}\right)$

$x \rightarrow \boxed{+2} \rightarrow \boxed{\times 3} \rightarrow y$

9 $y = \dfrac{x}{2}$ for x from 0 to 7. $\left(\begin{array}{l} x: 1 \text{ cm} = 1 \text{ unit} \\ y: 1 \text{ cm} = 1 \text{ unit} \end{array}\right)$

$x \rightarrow \boxed{\div 2} \rightarrow y$

x	0	1	2	3	4	5	6	7
y						$2\frac{1}{2}$		

10 $y = 5 + x$ for x from 0 to 5.

11 $y = 6 - x$ for x from 0 to 6.

12 $y = 2x - 3$ for x from 0 to 5.

Exercise 16 C

In this exercise the x-values may be positive and negative.
For each question make a table and then draw the graph.

1 $y = 2x + 1$ for x from -1 to 4.

$x \rightarrow \boxed{\times 2} \rightarrow \boxed{+1} \rightarrow y$

x	-1	0	1	2	3	4
y	-1	1				

2 $y = 3x + 2$ for x from -2 to 3.

x	-2	-1	0	1	2	3
y	-4		2			

3 $y = x - 3$ for x from -2 to 5.

4 $y = x + 6$ for x from -3 to 3.

5 $y = 2x - 1$ for x from -3 to 3.

6 $y = 4x + 1$ for x from -2 to 2.

*7 $y = 5 - x$ for x from 0 to 5. $\left(\begin{array}{l} x:\ 1\ \text{cm} = 1\ \text{unit} \\ y:\ 1\ \text{cm} = 1\ \text{unit} \end{array}\right)$

$x \longrightarrow \boxed{\begin{array}{c}\text{subtract}\\\text{from 5}\end{array}} \longrightarrow y$

*8 $y = 10 - 2x$ for x from 0 to 6. $\left(\begin{array}{l} x:\ 1\ \text{cm} = 1\ \text{unit} \\ y:\ 1\ \text{cm} = 1\ \text{unit} \end{array}\right)$

$x \longrightarrow \boxed{\times 2} \longrightarrow \boxed{\begin{array}{c}\text{subtract}\\\text{from 10}\end{array}} \longrightarrow y$

*9 $y = 4 - 2x$ for x from 0 to 3.   ***10** $y = 2(6 - x)$ for x from 0 to 6.

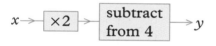   

*11 $y = 2(x + 1)$ for x from -2 to 4.   **12** $y = 3(x - 1)$ for x from -2 to 2.

*13 **a** Draw the graph of $y = 2x + 1$ for x from 0 to 4.
  b Draw the graph of $y = 8$.
  c Write the coordinates of the point where the two lines meet.

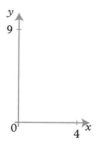

*14 **a** Draw the graph of $y = x$ for x from 0 to 6.
  b Draw the graph of $y = 1$.
  c Draw the graph of $x = 5$.
  d Shade in the triangle formed by the three lines.

*15 Draw the graph of $x + y = 6$ for x from 0 to 6.

*16 **a** Draw, on the same page, the graphs of $x + y = 8$ and $y = x$, for values of x from 0 to 8.
  b Write the coordinates of the point where the two lines meet.

6.6 Equations of lines in the form $y = mx + c$

6.6.1 Gradient

- The gradient of a straight line is a measure of how steep it is.

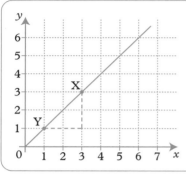

Gradient of line XY $= \dfrac{3-1}{3-1} = \dfrac{2}{2} = 1$

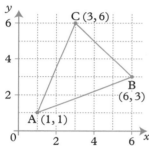

Gradient of line AB $= \dfrac{3-1}{6-1} = \dfrac{2}{5}$

Gradient of line AC $= \dfrac{6-1}{3-1} = \dfrac{5}{2}$

Gradient of line BC $= \dfrac{6-3}{3-6} = -1$

- A line which slopes upwards to the right has a **positive** gradient.
- A line which slopes upwards to the left has a **negative** gradient.
- Gradient $= \dfrac{\text{difference in } y\text{-coordinates}}{\text{difference in } x\text{-coordinates}}$

Think of N for negative gradient.

Exercise 17Ⓔ

1 Find the gradients of AB, BC, AC.

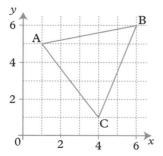

2 Find the gradients of PQ, PR, QR.

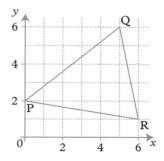

3 Find the gradients of the lines joining these pairs of points.

   a $(5, 2) \rightarrow (7, 8)$ **b** $(-1, 3) \rightarrow (1, 6)$

   c $(1, 4) \rightarrow (5, 6)$ **d** $(4, 10) \rightarrow (0, 20)$

4 Find the value of a if the line joining the points $(3a, 4)$ and $(a, 2)$ has a gradient of 1.

5 **a** Write the gradient of the line joining the points $(6, n)$ and $(3, 4)$.

   b Find the value of n if the line is parallel to the x-axis.

6.6.2 The form $y = mx + c$

Here are two straight lines.

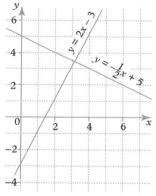

For $y = 2x - 3$, the gradient is 2 and the y-intercept is -3.

For $y = -\dfrac{1}{2}x + 5$, the gradient is $-\dfrac{1}{2}$ and the y-intercept is 5.

These two lines illustrate a general rule.

● When the equation of a straight line is in the form $y = mx + c$, the gradient of the line is m and the intercept on the y-axis is c.

> An intercept is where a line crosses an axis.

EXAMPLE

Draw the line $y = 2x + 3$ on a sketch graph.

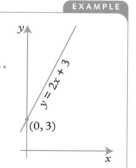

The line $y = 2x + 3$ has a gradient of 2 and cuts the y-axis at $(0, 3)$.

> The word 'sketch' means that you do not plot a series of points but simply show the position and slope of the line.

Exercise 18Ⓔ

1 Write the equations of these lines.

a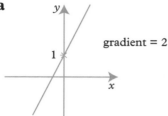
gradient = 2

b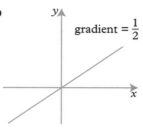
gradient = $\frac{1}{2}$

c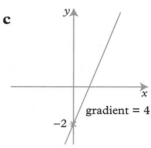
gradient = 4

d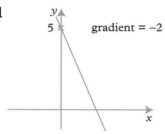
gradient = −2

e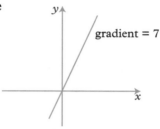
gradient = 7

f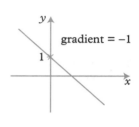
gradient = −1

g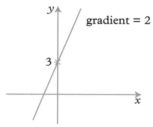
gradient = 2

h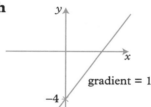
gradient = 1

i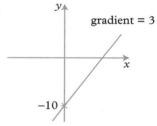
gradient = 3

2 Here are the equations of four lines.

$$y = 3x - 7 \qquad y = 2x - 7 \qquad y = -2x \qquad y = 3x$$

> Parallel lines have the same gradient.

Which two lines are parallel?

3 Write the equation of the line with
    a gradient 4 and y-intercept −3
    b gradient −3 and y-intercept 5
    c gradient $\frac{1}{3}$ and y-intercept −2.

4 Write the equation of any line which is parallel to the line $y = 5x - 11$.

Draw the line $x + 2y - 6 = 0$ on a sketch graph.

First rearrange the equation to make y the subject.
$$x + 2y - 6 = 0$$
$$2y = -x + 6$$
$$y = -\frac{1}{2}x + 3$$

The line has a gradient of $-\frac{1}{2}$ and cuts the y-axis at $(0, 3)$.

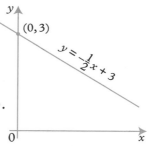

Exercise 19Ⓔ

In questions **1** to **20**, write
a the gradient of the line
b the intercept on the y-axis.
Then draw a small sketch graph of each line.

1 $y = x + 3$ **2** $y = x - 2$ **3** $y = 2x + 1$

4 $y = 2x - 5$ **5** $y = 3x + 4$ **6** $y = \frac{1}{2}x + 6$

7 $y = 3x - 2$ **8** $y = 2x$ **9** $y = \frac{1}{4}x - 4$

10 $y = -x + 3$ **11** $y = 6 - 2x$ **12** $y = 2 - x$

13 $y + 2x = 3$ **14** $3x + y + 4 = 0$ **15** $2y - x = 6$

16 $3y + x - 9 = 0$ **17** $4x - y = 5$ **18** $3x - 2y = 8$

19 $10x - y = 0$ **20** $y - 4 = 0$

*21 Find the equations of the lines A and B.

*22 Find the equations of the lines C and D.

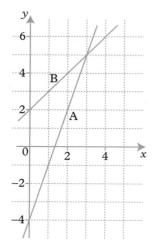

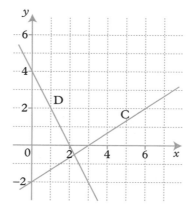

6.7 Simultaneous equations

6.7.1 Graphical solution

EXAMPLE

Louise and Philip are two children.
Louise is 5 years older than Philip.
The sum of their ages is 12 years.
How old is each child?

Let Louise be x years old and Philip be y years old.

So you can say $\quad x + y = 12 \quad$ (sum = 12)
and $\qquad\qquad\quad x - y = 5 \quad$ (difference = 5)

Draw on the same page the graphs of $\quad x + y = 12$
and $\quad x - y = 5.$

$x + y = 12$ goes through $(0, 12), (2, 10), (6, 6), (12, 0).$
$x - y = 5$ goes through $(5, 0), (7, 2), (10, 5).$

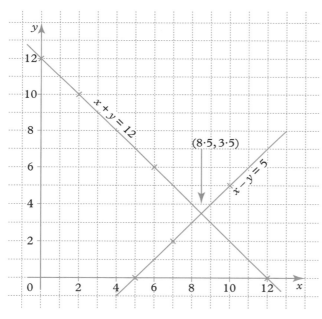

The point $(8 \cdot 5, 3 \cdot 5)$ lies on both lines at the same time.
It is where the lines cross.

So $x = 8 \cdot 5$, $y = 3 \cdot 5$ are the solutions of the **simultaneous** equations $x + y = 12$, $x - y = 5.$

So Louise is $8\frac{1}{2}$ years old and Philip is $3\frac{1}{2}$ years old.

There are lots of values of x and y that fit the equation $x + y = 12$, and lots of values that fit the equation $x - y = 5$, **but** only the values $x = 8 \cdot 5$ and $y = 3 \cdot 5$ fit **both** equations.

Exercise 20Ⓔ

1 Use the graphs to solve the simultaneous equations.

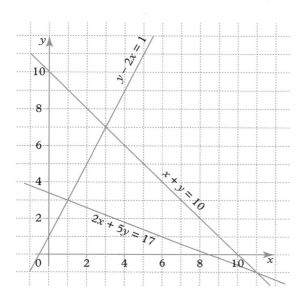

a $x + y = 10$ **b** $2x + 5y = 17$ **c** $x + y = 10$
     $y - 2x = 1$ $y - 2x = 1$ $2x + 5y = 17$

In questions **2** to **6**, solve the simultaneous equations by drawing graphs. Use a scale of 1 cm to 1 unit on both axes.

2 $x + y = 6$
   $2x + y = 8$
   Draw axes with x and y
   from 0 to 8.

3 $x + 2y = 8$
   $3x + y = 9$
   Draw axes with x and y
   from 0 to 9.

4 $x + 3y = 6$
    $x - y = 2$
   Draw axes with
   x from 0 to 8
   and y from -2 to 4.

5 $5x + y = 10$
    $x - y = -4$
   Draw axes with
   x from -4 to 4
   and y from 0 to 10.

6 $a + 2b = 11$
   $2a + b = 13$
   Here the unknowns are a and b. Draw the a-axis across the
   page from 0 to 13 and the b-axis up the page, also from
   0 to 13.

7 There are four lines drawn here.

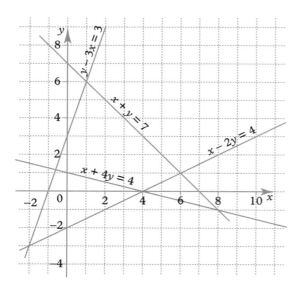

Write the solutions to these pairs of equations.

a $x - 2y = 4$
 $x + 4y = 4$

b $x + y = 7$
 $y - 3x = 3$

c $y - 3x = 3$
 $x - 2y = 4$

d $x + 4y = 4$
 $x + y = 7$

e $x + 4y = 4$
 $y - 3x = 3$ (For **e** give x and y correct to 1 dp.)

6.7.2 Algebraic solution

You can also solve simultaneous equations without drawing graphs. There are two methods: substitution and elimination.

You can choose for yourself which one to use in any question.

1 Substitution method

This method is used when one equation contains a single 'x' or 'y', as in equation [2] of this example.

> EXAMPLE
>
> Solve the simultaneous equations
> $$3x - 2y = 0 \qquad \dots [1]$$
> $$2x + y = 7 \qquad \dots [2]$$
> ....
>
> From [2] $2x + y = 7$
> $$y = 7 - 2x$$
>
> | Substituting for y in [1] | Substituting for x in [2] |
> |---|---|
> | $3x - 2(7 - 2x) = 0$ | $2 \times 2 + y = 7$ |
> | $3x - 14 + 4x = 0$ | $y = 3$ |
> | $7x = 14$ | |
> | $x = 2$ | |
>
> The solutions are $x = 2$, $y = 3$.
>
> These values of x and y are the only pair which
> simultaneously satisfy **both** equations.

> Number the
> equations, then
> you can refer to
> them.

> Substitute
> $(7 - 2x)$ for y in
> equation [1] to find
> the x-value.

> Check your answer
> by substituting the
> solutions in one of
> the original
> equations.

Exercise 21Ⓔ

Use the substitution method to solve these simultaneous equations.

1 $2x + y = 5$
$\quad\; x + 3y = 5$

2 $x + 2y = 8$
$\quad\; 2x + 3y = 14$

3 $3x + y = 10$
$\quad\; x - y = 2$

4 $2x + y = -3$
$\quad\; x - y = -3$

5 $4x + y = 14$
$\quad\; x + 5y = 13$

6 $x + 2y = 1$
$\quad\; 2x + 3y = 4$

7 $2x + y = 5$
$\quad\; 3x - 2y = 4$

8 $2x + y = 13$
$\quad\; 5x - 4y = 13$

9 $7x + 2y = 19$
$\quad\; x - y = 4$

10 $b - a = -5$
$\quad\;\; a + b = -1$

11 $a + 4b = 6$
$\quad\;\; 8b - a = -3$

12 $a + b = 4$
$\quad\;\; 2a + b = 5$

13 $3m = 2n - 6\frac{1}{2}$
$\quad\;\; 4m + n = 6$

14 $2w + 3x - 13 = 0$
$\quad\;\; x + 5w - 13 = 0$

15 $x + 2(y - 6) = 0$
$\quad\;\; 3x + 4y = 30$

16 $2x = 4 + z$
$\quad\;\; 6x - 5z = 18$

*17 $3m - n = 5$
$\quad\;\; 2m + 5n = 9$

*18 $5c - d - 11 = 0$
$\quad\;\; 4d + 3c = -5$

It is useful, at this point, to revise the operations of addition
and subtraction with negative numbers.

> EXAMPLE
>
> Simplify
> **a** $-7 + -4 = -7 - 4 = -11$ **b** $-3x + (-4x) = -3x - 4x = -7x$
> **c** $4y - (-3y) = 4y + 3y = 7y$ **d** $3a + (-3a) = 3a - 3a = 0$

2 Elimination method

Use this method when the first method is unsuitable (some people prefer to use it for every question).

EXAMPLE

Solve the simultaneous equations. $2x + 3y = 5$ ... [1]
$5x - 2y = -16$ ... [2]

[1] ×5
[2] ×2
[3] − [4]

$$10x + 15y = 25 \quad \ldots [3]$$
$$10x - 4y = -32 \ldots [4]$$
$$15y - (-4y) = 25 - (-32)$$
$$19y = 57$$
$$y = 3$$

Substitute for y in [1]

$$2x + 3 \times 3 = 5$$
$$2x = 5 - 9 = -4$$
$$x = -2$$

The solutions are $x = -2$, $y = 3$.

Remember to check your solutions by substituting the x- and y-values in equation [2].

Exercise 22Ⓔ

Use the elimination method to solve these simultaneous equations.

1 $2x + 5y = 24$
$4x + 3y = 20$

2 $5x + 2y = 13$
$2x + 6y = 26$

3 $3x + y = 11$
$9x + 2y = 28$

4 $x + 2y = 17$
$8x + 3y = 45$

5 $3x + 2y = 19$
$x + 8y = 21$

6 $2a + 3b = 9$
$4a + b = 13$

7 $2x + 3y = 11$
$3x + 4y = 15$

8 $3x + 8y = 27$
$4x + 3y = 13$

9 $2x + 7y = 17$
$5x + 3y = -1$

10 $5x + 3y = 23$
$2x + 4y = 12$

11 $7x + 5y = 32$
$3x + 4y = 23$

12 $3x + 2y = 4$
$4x + 5y = 10$

13 $3x + 2y = 11$
$2x - y = -3$

14 $3x + 2y = 7$
$2x - 3y = -4$

15 $x - 2y = -4$
$3x + y = 9$

16 $5x - 7y = 27$
$3x - 4y = 16$

17 $3x - 2y = 7$
$4x + y = 13$

18 $x - y = -1$
$2x - y = 0$

19 $y - x = -1$
$3x - y = 5$

20 $x - 3y = -5$
$2y + 3x + 4 = 0$

6.7.3 Problems solved by simultaneous equations

Exercise 23Ⓔ

Solve each problem by writing a pair of simultaneous equations.

1 Find two numbers with a sum of 15 and a difference of 4. (Let the numbers be x and y.)

2 Twice one number added to three times another gives 21. Find the numbers, if the difference between them is 3.

3 Twice one number plus the other number add up to 12. The sum of the two numbers is 7. Find the numbers.

4 Double the larger number plus three times the smaller number makes 31. The difference between the numbers is 3. Find the numbers.

5 Here is a puzzle from a newspaper. The ? and ★ stand for numbers which you have to find. The totals for the rows and columns are given.

Write two equations involving ? and ★ and solve them to find the values of ? and ★.

?	★	?	★	36
?	★	★	?	36
★	?	★	★	33
?	★	?	★	36
39	33	36	33	

6 The line, with equation $y + ax = c$ passes through the points $(1, 5)$ and $(3, 1)$. Find a and c.

> For the point $(1, 5)$ put $x = 1$ and $y = 5$ into $y + ax = c$

7 The line $y = mx + c$ passes through $(2, 5)$ and $(4, 13)$. Find m and c.

6.8 Graphs of quadratic functions

A quadratic function has an x^2-term and usually an x-term and a number term. For example $y = x^2 + 2x$, $y = x^2 + x + 3$ and $y = x^2 - 7$ are all quadratic functions.

6.8.1 Table of values

You are most likely to make a mistake in this topic in making the **table of values**. It helps you to get it right if you put a row in the table for each term in the equation and input the correct values.

EXAMPLE

Work out a table of values for the graph of $y = x^2 + 1$ for x from -3 to $+3$.

When $x = -3$, $x^2 = (-3)^2 = 9$.

x	-3	-2	-1	0	1	2	3
x^2	9	4	1	0	1	4	9
1	1	1	1	1	1	1	1
y	10	5	2	1	2	5	10

Add these two numbers to get y.

EXAMPLE

Work out a table of values for the graph of $y = x^2 + 2x$ for values of x from -3 to $+3$.

Be careful with negative numbers: when $x = -3$, $2x = 2 \times (-3)$ $= -6$.

x	-3	-2	-1	0	1	2	3
x^2	9	4	1	0	1	4	9
$2x$	-6	-4	-2	0	2	4	6
y	3	0	-1	0	3	8	15

Add these two numbers to get y.

Exercise 24Ⓔ

Copy and complete the tables of values. Add in as many rows as there are terms in each equation.

1 $y = x^2 + 4$

x	-3	-2	-1	0	1	2	3
x^2							
4	4	4	4	4			
y							

2 $y = x^2 - 7$

x	-3	-2	-1	0	1	2	3
x^2	9						
-7	-7	-7					
y							

3 $y = x^2 + 3x$

x	−4	−3	−2	−1	0	1	2
x^2	16	9	4				
$3x$	−12	−9	−6				
y	4	0	−2				

4 $y = x^2 + 5x$

x	−4	−3	−2	−1	0	1	2
x^2	16						
$5x$	−20						
y	−4						

5 $y = x^2 - 2x$

x	−3	−2	−1	0	1	2	3
x^2							9
$-2x$	6						−6
y							3

6 $y = x^2 - 4x$

x	−2	−1	0	1	2	3	4
x^2							
$-4x$	6						
y							

7 $y = x^2 + 2x + 3$

x	−3	−2	−1	0	1	2	3
x^2	9	4					
$2x$	−6	−4					
3	3	3	3	3			
y	6	3					

8 $y = x^2 + 3x - 2$

x	−4	−3	−2	−1	0	1	2
x^2							
$3x$							
-2	−2	−2					
y							

9 $y = x^2 + 4x - 5$

x	−3	−2	−1	0	1	2	3
x^2							
$+4x$							
-5							
y							

10 $y = x^2 - 2x + 6$

x	−3	−2	−1	0	1	2	3
x^2							
$-2x$							
6							
y							

In questions **11** to **18** draw and complete a table of values for the values of x given.

11 $y = x^2 + 4x$; x from −3 to +3.

12 $y = x^2 - 6x$; x from −4 to +2.

13 $y = x^2 + 8$; x from −3 to +3.

14 $y = x^2 + 3x + 1$; x from −4 to +2.

15 $y = x^2 - 5x + 3$; x from −3 to +3.

16 $y = x^2 - 3x - 5$; x from −2 to +4.

17 $y = x^2 - 3$; x from −5 to +1.

18 $y = 2x^2 + 1$; x from −3 to +3.

6.8.2 Curved graphs

Draw the graph of $y = x^2 + x - 2$ for values of x from -3 to $+3$.

a

x	-3	-2	-1	0	1	2	3
x^2	9	4	1	0	1	4	9
$+x$	-3	-2	-1	0	1	2	3
-2	-2	-2	-2	-2	-2	-2	-2
y	4	0	-2	-2	0	4	10

b Plot the x- and y-values from the table.

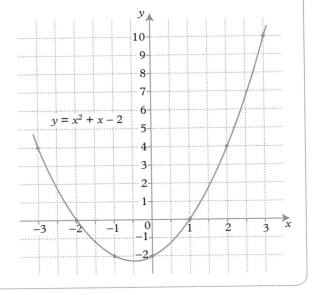

$y = x^2 + x - 2$

Exercise 25Ⓔ

For each question make a table of values and then draw the graph. Suggested scales: 2 cm to 1 unit on the x-axis and 1 cm to 1 unit on the y-axis.

1 $y = x^2 + 2$; x from -3 to $+3$.

x	-3	-2	-1	0	1	2	3
x^2	9	4	1	0	1		
$+2$	2	2	2				
y	11	6	3				

2 $y = x^2 + 5$; x from -3 to $+3$.

3 $y = x^2 - 4$; x from -3 to $+3$.

4 $y = x^2 - 8$; x from -3 to $+3$.

5 $y = x^2 + 2x$; x from -4 to $+2$.

x	−4	−3	−2	−1	0	1	2
x^2	16	9					4
+ 2x	−8	−6					4
y	8	3					8

6 $y = x^2 + 4x$; x from -5 to $+1$.

7 $y = x^2 + 4x - 1$; x from -2 to $+4$.

8 $y = x^2 + 2x - 5$; x from -4 to $+2$.

9 $y = x^2 + 3x + 1$; x from -4 to $+2$.

6.8.3 Graphical solution of equations

You can find an approximate solution from an accurately drawn graph for a wide range of equations, many of which are impossible to solve exactly by other methods.

EXAMPLE

Draw the graphs of $y = x^2 - 2x$ and $y = x + 1$.
Hence find approximate solutions to the equation $x^2 - 2x = x + 1$.

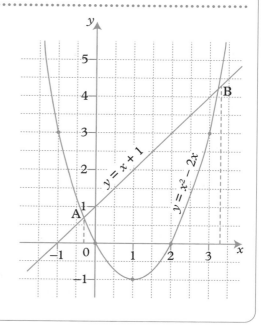

The solutions to the equation are the x- values at the two points of intersection.
At A $x \approx -0.3$ ⎞ These are the
At B $x \approx 3.3$ ⎠ approximate solutions.

Check the solutions by substituting 3·3 and -0.3 for x in the two equations.
You get

$3.3^2 - 2 \times 3.3 = 4.29$
$3.3 + 1 = 4.3$

and

$(-0.3)^2 - 2(-0.3) = 0.69$
$-0.3 + 1 = 0.7$

Exercise 26Ⓔ

1 The diagram shows the graphs of
$y = x^2 - 2x - 3$, $y = 3$ and $y = -2$.
Use the graphs to find approximate
solutions to these equations.
a $x^2 - 2x - 3 = 3$
b $x^2 - 2x - 3 = -2$
a $x^2 - 2x - 3 = 0$

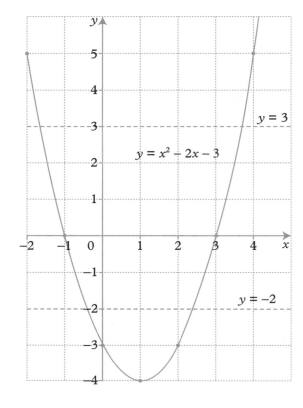

In questions **2** to **6** use a scale of 2 cm to 1 unit for x and 1 cm
to 1 unit for y.

2 Draw the graph of the function $y = x^2 - 2x$
for $-1 \leqslant x \leqslant 4$. Hence find approximate solutions
of the equations
a $x^2 - 2x = 1$ **b** $x^2 - 2x = 0$

> This means use
> x-values from -1
> up to 4.

3 Draw the graph of the function $y = x^2 - 3x + 5$ for
$-1 \leqslant x \leqslant 5$.
Hence find approximate solutions of the equations
a $x^2 - 3x + 5 = 5$ **b** $x^2 - 3x + 5 = 8$

4 Draw the graph of $y = x^2 - 2x + 2$ for $-2 \leqslant x \leqslant 4$.
By drawing other graphs, solve the equations
a $x^2 - 2x + 2 = 8$ **b** $x^2 - 2x + 2 = 3$

5 Draw the graph of $y = x^2 - 7x$ for $0 \leqslant x \leqslant 7$.
a Use the graph to find approximate solutions of the
equation $x^2 - 7x = -3$.
b Explain why the equation $x^2 - 7x = -14$ does not
have a solution.

6 **a** Draw axes with x from -6 to 4 and y from -20 to 10. Use a scale of 1 cm to 1 unit for x and 2 cm to 5 units for y.

b Draw the graphs of these equations.
$y = x^2 + 2x - 15$; $y = x$; $y = -5$;
$y = 0$; $y = -19$.

c Hence solve the equations
  i $x^2 + 2x - 15 = -5$
  ii $x^2 + 2x - 15 = 0$
  iii $x^2 + 2x - 15 = -19$
  iv $x^2 + 2x - 15 = x$

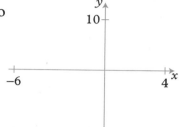

7 **a** Draw axes with x from -4 to 4 and y from -6 to 10. Use a scale of 1 cm to 1 unit for both axes.

b Draw the graphs of these equations.
$y = x^2 - 6$; $y = 4 - x$; $y = -3$; $y = 4$;
$y = 0$; $y = x$.

c Hence solve the equations
  i $x^2 - 6 = 4$
  ii $x^2 - 6 = 0$
  iii $x^2 - 6 = -3$
  iv $x^2 - 6 = x$
  v $x^2 - 6 = 4 - x$

Test yourself

1 Factorise $3y + y^2$

(Edexcel, 2003)

2 a Solve the inequality
$2n - 1 < n + 3$
  b List the solutions that are positive integers.

(Edexcel, 2003)

3 a Use the formula $y = mx + c$ to find the value of y
when $m = 2$, $x = -3$ and $c = 11$.
  b Rearrange the formula $y = mx + c$ to make x the subject.

(OCR, 2003)

4 a Write the integer values of n for which $-3 \leqslant 3n < 12$.
  b Solve the inequality $4x + 3 \geqslant 23$.

(OCR, 2004)

5 The diagram shows the graphs of the equations
$x + y = 1$ and $y = x - 5$

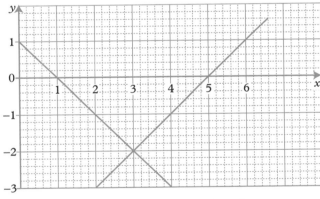

Use the diagram to solve the simultaneous equations
$x + y = 1$
$y = x - 5$

(Edexcel, 2003)

6 Solve these simultaneous equations by an algebraic
(not graphical) method.
You must show all your working.

$2x + 3y = 17$
$4x - y = 6$

(WJEC)

7 a Simplify

　　i $p^4 \times p^3$ **ii** $\dfrac{12t^5}{3t^2}$

b Solve
　　$3x + 19 > 4$

c Rearrange the formula to make w the subject.
　　$s = \dfrac{w + y}{2}$

(OCR, Spec.)

8 A is the point $(0, 3)$ and B is the point $(3, 9)$.

a Calculate the gradient of the line AB.

b Write the equation of the line AB.

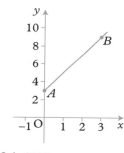

(AQA, 2003)

9 ABCD is a rectangle.
A is the point $(0, 1)$.
C is the point $(0, 6)$.

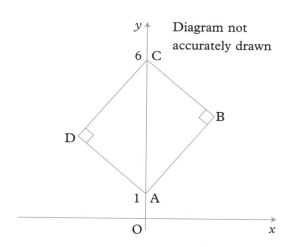

Diagram not
accurately drawn

The equation of the straight line through A and B is
$y = 2x + 1$.

Find the equation of the straight line through D and C.

(Edexcel, 2004)

10 A straight line has equation $y = \frac{1}{2}x + 1$

The point P lies on the straight line.
P has a y-coordinate of 5.

a Find the x-coordinate of P.

b Write the equation of a different straight line that is parallel to $y = \frac{1}{2}x + 1$.

c Rearrange $y = ax + c$ to make x the subject.

(Edexcel, 2003)

11 Use the method of trial and improvement to find a solution, to 1 decimal place, of the equation

$x^3 + x = 100$

Show all your trials in a copy of the table. A first trial has been completed for you.

Trial x	$x^3 + x$	Too high/too low
5	130	*too high*

(AQA, 2003)

12 Use trial and improvement to solve the equation

$x^3 + 2x = 60$

Show all your trials and their outcomes.
Give your answer correct to 1 decimal place.

(OCR, 2004)

13 a Factorise

    i $10x + 5$ **ii** $x^2 + 7x$

b Simplify

    $a^3b^4 \times \dfrac{a^2}{b}$

c Rearrange the formula $v = u + at$ to make a the subject.

d Solve

    i $\dfrac{x}{4} = 12$ **ii** $4(x - 2) = 2x + 5$

(OCR, 2004)

14 A cuboid has a square base of side x cm.
The height of the cuboid is 1 cm more than the
length x cm.
The volume of the cuboid is 230 cm³.

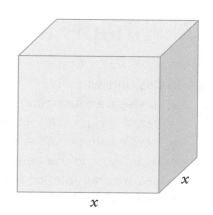

a Show that $x^3 + x^2 = 230$

The equation $x^3 + x^2 = 230$ has a solution
between $x = 5$ and $x = 6$.

b Use a trial and improvement method to find
this solution.
Give your answer correct to 1 decimal place.
You must show **all** your working.

(Edexcel, 2003)

15 a Copy and complete the table of values for

$y = x^2 - 3x - 1$

x	−1	0	1	2	3	4
y	3	−1			−1	3

b Draw the graph of $y = x^2 - 3x - 1$.
c Use your graph to solve the equation

$x^2 - 3x - 1 = 0$

(OCR, 2004)

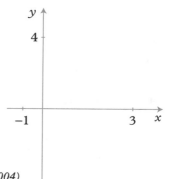

16 a Copy and complete the table of values for the curve

$y = x^2 - 10$

x	−4	−3	−2	−1	0	1	2	3	4
y	6	−1				−9	−6	−1	

b Draw the graph of $y = x^2 - 10$.
c Use your graph to find the value of x for which
$x^2 - 10 = 0$.

(OCR, 2005)

7 Shape and space 2

In this unit you will:
- revise transformations of shapes
- learn about circles, including their area and perimeter
- revise volume and surface area calculations
- learn how to use bearings
- learn about interior and exterior angles of polygons
- learn about loci
- learn how to use Pythagoras' theorem.

7.1 Transforming shapes

7.1.1 Reflection

- When you **reflect** an object in a line, all points map to equidistant points on the opposite side of the line.

- In a reflection the object and its image are **congruent** because they are the same size and shape.

EXAMPLE

a A′B′C′D′ is the image of ABCD after reflection in the broken line (the mirror line).

b △2 is the image of △1 after reflection in the diagonal mirror line.

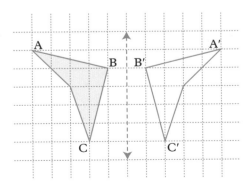

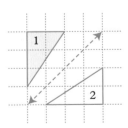

Point A is reflected to A′.

Exercise 1 C

On square grid paper draw the object and its image after reflection in the broken line.

1

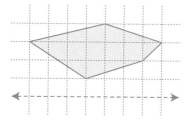

2

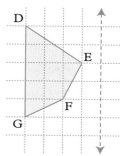

3

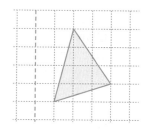

4

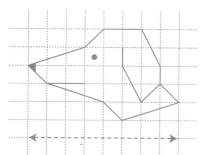

5

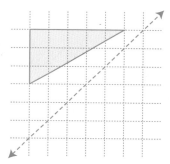

6

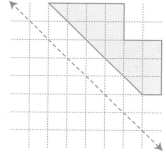

7 Copy the diagram.
   a Reflect △1 in the line AB.
      Label the image △2.
   b Reflect △2 in the line CD.
      Label the image △3.

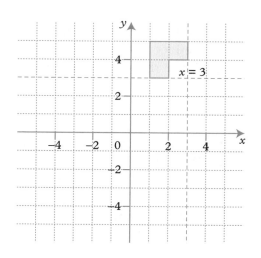

Exercise 2 C

1 Copy the diagram.
   Draw the image of the shape after
   reflection in
   a the x-axis; label it 1
   b the y-axis; label it 2
   c the line $x = 3$; label it 3.

For questions **2** to **5** draw a pair of axes so that both x and y can take values from -7 to $+7$.

2 a Plot and label P(7, 5), Q(7, 2), R(5, 2).
 b Draw the lines $y = -1$, $x = 1$ and $y = x$.
   Use dotted lines.
 c Draw the image of $\triangle$PQR after reflection in
   i the line $y = -1$; label it $\triangle$1
   ii the line $x = 1$; label it $\triangle$2
   iii the line $y = x$; label it $\triangle$3.
 d Write the coordinates of the image of point P in each case.

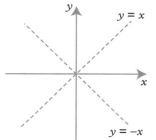

3 a Plot and label L(7, -7), M(7, -1), N(5, -1).
 b Draw the lines $y = x$ and $y = -x$.
   Use dotted lines.
 c Draw the image of $\triangle$LMN after reflection in
   i the x-axis; label it $\triangle$1
   ii the line $y = x$; label it $\triangle$2
   iii the line $y = -x$; label it $\triangle$3.
 d Write the coordinates of the image of point L in each case.

4 a Draw the line $x + y = 7$. (It passes through (0, 7) and (7, 0).)
 b Draw $\triangle$1 at $(-3, -1)$, $(-1, -1)$, $(-1, -4)$.
 c Reflect $\triangle$1 in the y-axis on to $\triangle$2.
 d Reflect $\triangle$2 in the x-axis on to $\triangle$3.
 e Reflect $\triangle$3 in the line $x + y = 7$ on to $\triangle$4.
 f Reflect $\triangle$4 in the y-axis on to $\triangle$5.
 g Write the coordinates of $\triangle$5.

5 a Draw the lines $y = 2$, $x = -1$ and $y = x$.
 b Draw $\triangle$1 at $(1, -3)$, $(-3, -3)$, $(-3, -5)$.
 c Reflect $\triangle$1 in the line $y = x$ on to $\triangle$2.
 d Reflect $\triangle$2 in the line $y = 2$ on to $\triangle$3.
 e Reflect $\triangle$3 in the line $x = -1$ on to $\triangle$4.
 f Reflect $\triangle$4 in the line $y = x$ on to $\triangle$5.
 g Write the coordinates of $\triangle$5.

6 Find the equation of the mirror line
for these reflections.
  a △1 on to △2
  b △1 on to △3
  c △1 on to △4
  d △1 on to △5.

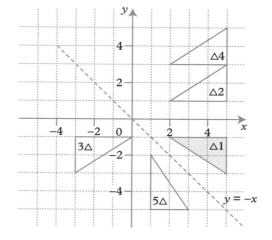

7.1.2 Rotation

● You need three things to describe a **rotation**:
  1 the centre
  2 the angle
  3 the direction (for example, clockwise).

△A′B′C′ is the image of △ABC after a
90° clockwise rotation about centre O.

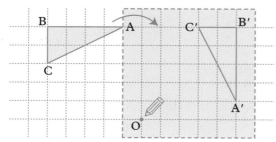

Draw △ABC on tracing paper
and then put the tip of your pencil
on O. Turn the tracing paper 90°
clockwise about O. The tracing
paper now shows the position of
△A′B′C′.

Exercise 3 C

Draw the object and its image after the rotation given.
Take O as the centre of rotation in each case.

1

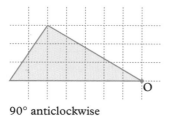

90° anticlockwise

2

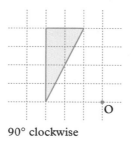

90° clockwise

3

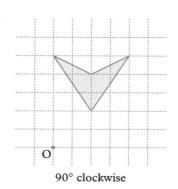

90° clockwise

4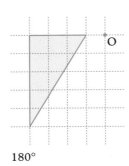

180°

5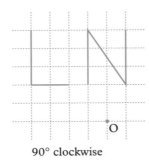

90° clockwise

6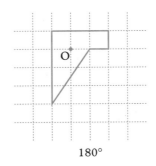

180°

7 The shape below has been rotated about several different centres to form this pattern.

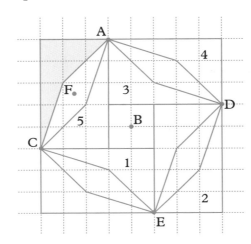

Describe the rotation that takes the shaded shape on to shape 1, shape 2, shape 3, shape 4 and shape 5. For each one, give the centre (A, B, C, D, E or F), the angle and the direction of the rotation (for example, 'centre C, 90°, clockwise').

Exercise 4 ©

1 Copy the diagram.

  a Rotate △ABC 90° clockwise about (0, 0). Label it △1.

  b Rotate △DEF 180° clockwise about (0, 0). Label it △2.

  c Rotate △GHI 90° clockwise about (0, 0). Label it △3.

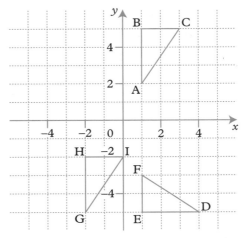

For questions **2** and **3** draw a pair of axes with values
of x and y from -7 to $+7$.

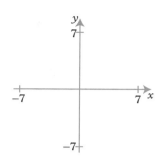

2 a Plot $\triangle 1$ at $(2, 3)$, $(6, 3)$, $(3, 6)$.
  b Rotate $\triangle 1$ 90° clockwise about $(2, 1)$ on to $\triangle 2$.
  c Rotate $\triangle 2$ 180° about $(0, 0)$ on to $\triangle 3$.
  d Rotate $\triangle 3$ 90° anticlockwise about $(1, 1)$ on to $\triangle 4$.
  e Write the coordinates of $\triangle 4$.

3 a Plot $\triangle 1$ at $(4, 4)$, $(6, 6)$, $(2, 6)$.
  b Rotate $\triangle 1$ 90° anticlockwise about $(6, 0)$ on to $\triangle 2$.
  c Rotate $\triangle 2$ 90° anticlockwise about $(-3, -4)$ on to $\triangle 3$.
  d Rotate $\triangle 3$ 90° clockwise about $(-3, 2)$ on to $\triangle 4$.
  e Write the coordinates of $\triangle 4$.

7.1.3   Finding the centre of a rotation

Exercise 5 ©

In questions **1** to **3** copy the diagram exactly on square
grid paper and then use tracing paper to find the centre
of the rotation that takes the shaded shape on to the
unshaded shape. Mark the centre of rotation with a
cross.

1

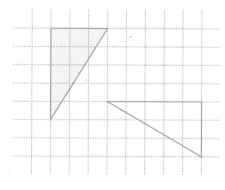

2

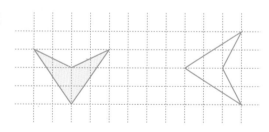

3

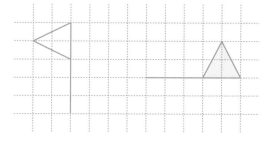

4 Copy the diagram.

Find the coordinates
of the centres of these
rotations.

a $\triangle 1 \rightarrow \triangle 2$
b $\triangle 1 \rightarrow \triangle 3$
c $\triangle 1 \rightarrow \triangle 4$
d $\triangle 1 \rightarrow \triangle 5$

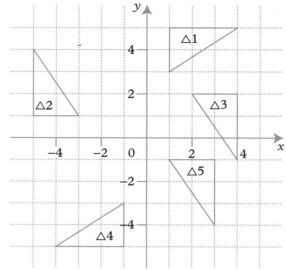

For questions **5** and **6** draw a pair of axes with values of
x and y from -7 to $+7$.

5 a Plot and label these triangles.
$\triangle 1$: $(3, 4)$, $(7, 4)$, $(3, 7)$
$\triangle 2$: $(3, 2)$, $(6, 2)$, $(3, -2)$
$\triangle 3$: $(-7, -4)$, $(-3, -4)$, $(-3, -7)$
$\triangle 4$: $(-2, 1)$, $(-5, 1)$, $(-2, 5)$
$\triangle 5$: $(2, -3)$, $(5, -3)$, $(2, -7)$

b Find the coordinates of the centres of these
rotations.
i $\triangle 1 \rightarrow \triangle 2$ **ii** $\triangle 1 \rightarrow \triangle 3$
iii $\triangle 1 \rightarrow \triangle 4$ **iv** $\triangle 1 \rightarrow \triangle 5$

6 a Plot and label these triangles.
$\triangle 1$: $(-4, -3)$, $(-4, -7)$, $(-6, -7)$
$\triangle 2$: $(-3, 4)$, $(-7, 4)$, $(-7, 6)$
$\triangle 3$: $(-2, 1)$, $(2, 1)$, $(2, -1)$
$\triangle 4$: $(0, 7)$, $(4, 7)$, $(4, 5)$
$\triangle 5$: $(2, -3)$, $(4, -3)$, $(2, -7)$

b Find the coordinates of the centres of these
rotations.
i $\triangle 1 \rightarrow \triangle 2$ **ii** $\triangle 1 \rightarrow \triangle 3$
iii $\triangle 1 \rightarrow \triangle 4$ **iv** $\triangle 1 \rightarrow \triangle 5$

7.1.4 Enlargement

A

B

C

Photo A has been enlarged to give photos B and C. Notice that the shape of the face is exactly the same in all the pictures

Photo A measures 22 mm by 27 mm
Photo B measures 44 mm by 54 mm
Photo C measures 66 mm by 81 mm

From A to B both the width and the height have been multiplied by 2.
You say B is an enlargement of A with a **scale factor** of 2.
Similarly C is an enlargement of A with a scale factor of 3.

Also C is an enlargement of B with a scale factor of $1\frac{1}{2}$.

You can find the scale factor of an enlargement by dividing corresponding lengths on two pictures.

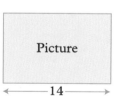

In this enlargement the scale factor is $\frac{21}{14}$ (= 1·5).

Exercise 6 C

1 This picture is to be enlarged to fit exactly in a frame.
Which of these frames will the picture fit?
Write 'yes' or 'no'.
  a 100 mm by 70 mm
  b 110 mm by 70 mm
  c 150 mm by 105 mm
  d 500 mm by 300 mm.

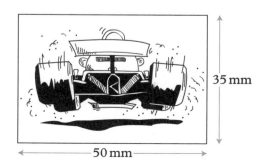

2 This picture is to be enlarged so that it fits exactly into the frame. Find the length x.

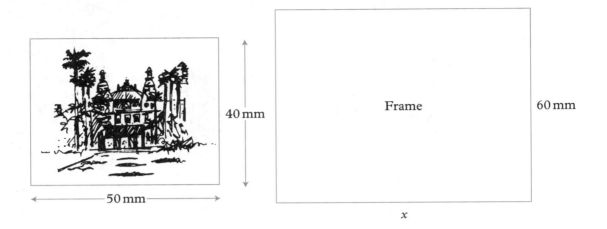

3 a This picture is enlarged to fit into the frame on the right.

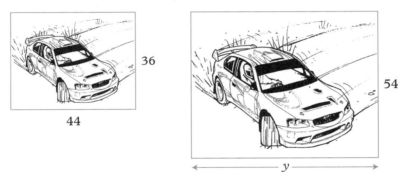

Calculate y.

b The picture is reduced to fit into this frame.

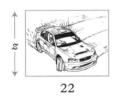

22

Calculate z.

4 Janine has started to draw a two times enlargement of a house using the squares.

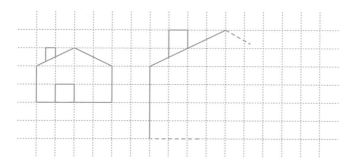

Draw the complete enlargement (use square grid paper).

5 Draw a three times enlargement of this figure on square grid paper.

6 Draw a two times enlargement of this shape on square grid paper.
Measure the angles a and b on the shape and its image.
Write the correct version of each sentence.

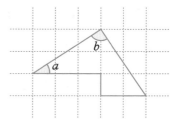

 a 'In an enlargement, the angles in a shape are changed/unchanged.'
 b 'In an enlargement, the object and the image are congruent/not congruent.'

7 This diagram shows an arrowhead and its enlargement. Notice that lines drawn through corresponding points (A, A′ or B, B′) all go through one point O.
This point is called the centre of enlargement.
Copy and complete
OA′ = __ × OA
OB′ = __ × OB.

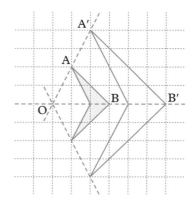

8 Copy this shape and its enlargement onto square grid paper. Draw construction lines to find the centre of enlargement.

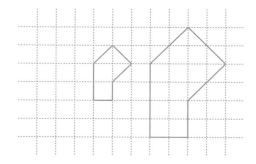

9 Copy the diagram, leaving space on the left for construction lines.
  a Mark the centre of enlargement for
    i △1⟶△2
    ii △1⟶△3
    iii △2⟶△3
  b Write the scale factor for the enlargement △2⟶△3.

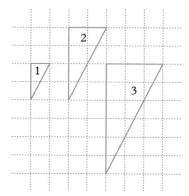

7.1.5 Centre of enlargement

● For a mathematical description of an enlargement you need two things:
  1 the scale factor   **2** the centre of enlargement.

The triangle ABC is enlarged on to triangle A′ B′ C′ with a scale factor of 3 and centre O.

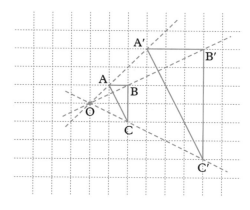

All lengths are measured from the **centre of enlargement**.

Note: OA′ = 3 × OA;   OB′ = 3 × OB;   OC′ = 3 × OC.

Exercise 7 C

Copy each diagram and draw an enlargement using the centre
O and the scale factor given.

1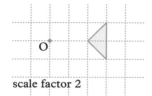
scale factor 2

2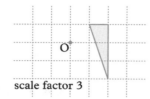
scale factor 3

3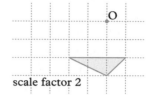
scale factor 2

4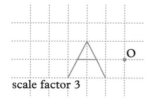
scale factor 3

5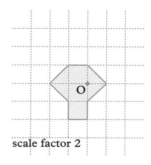
scale factor 2

6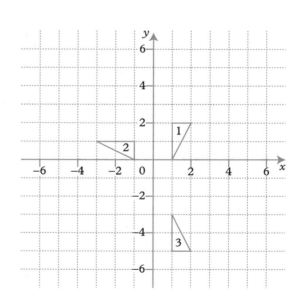
scale factor 2

7 a Copy the diagram.
   b Draw the image of △1 after
         enlargement with scale factor 3,
         centre $(0, 0)$. Label the image △4.
   c Draw the image of △2 after
         enlargement with scale factor 2,
         centre $(-1, 3)$. Label the
         image △5.
   d Draw the image of △3 after
         enlargement with scale factor 2,
         centre $(-1, -5)$. Label the
         image △6.
   e Write the coordinates of the
         'pointed ends' of △4, △5 and △6.
         (The 'pointed end' is the vertex of
         the triangle with the smallest
         angle.)
   f The smallest angle in △1 is 27°.
         How much larger is the smallest angle in the
         enlargement of △1 with scale factor 3?
         (Part **b**.)

For questions **8** and **9** draw a pair of axes with values from
-7 to $+7$.

8 a Plot and label the triangles
    △1: $(5, 5)$, $(5, 7)$, $(4, 7)$
    △2: $(-6, -5)$, $(-3, -5)$, $(-3, -4)$
    △3: $(1, -4)$, $(1, -6)$, $(2, -6)$
  b Draw the image of △1 after enlargement with scale
      factor 2, centre $(7, 7)$. Label the image △4.
  c Draw the image of △2 after enlargement with scale
      factor 3, centre $(-6, -7)$. Label the image △5.
  d Draw the image of △3 after enlargement with scale
      factor 2, centre $(-1, -5)$. Label the image △6.
  e Write the coordinates of the 'pointed ends' of △4, △5 and △6.

9 a Plot and label the triangles
    △1: $(5, 3)$, $(5, 6)$, $(4, 6)$
    △2: $(4, -3)$, $(1, -3)$, $(1, -2)$
    △3: $(-4, -7)$, $(-7, -7)$, $(-7, -6)$
  b Draw the image of △1 after enlargement with scale
      factor 2, centre $(7, 7)$. Label the image △4.
  c Draw the image of △2 after enlargement with scale
      factor 3, centre $(5, -4)$. Label the image △5.
  d Draw the image of △3 after enlargement with scale
      factor 4, centre $(-7, -7)$. Label the image △6.
  e Write the coordinates of the 'pointed ends' of △4, △5 and △6.

7.1.6 Enlargements with fractional scale factors (reductions)

- Even though a shape is smaller, mathematicians still call it an
  enlargement with a fractional scale factor.

The unshaded shape is the image of the shaded shape after an
enlargement with scale factor $\frac{1}{2}$, centre O.

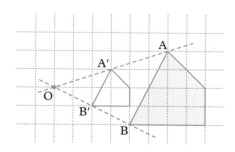

Note that $OA' = \frac{1}{2} \times OA$

$OB' = \frac{1}{2} \times OB$

Exercise 8Ⓔ

Copy each diagram and draw an enlargement using the
centre O and the scale factor given.

1

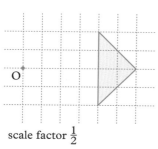

scale factor $\frac{1}{2}$

2

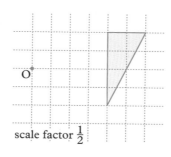

scale factor $\frac{1}{2}$

3

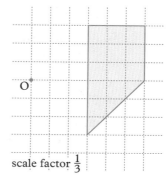

scale factor $\frac{1}{3}$

4 a Plot and label the triangles

△1: $(7, 6), (1, 6), (1, 3)$
△2: $(7, -1), (7, -7), (3, -7)$
△3: $(-5, 7), (-5, 1), (-7, 1)$

b Draw △4, the image of △1 after an enlargement with
scale factor $\frac{1}{3}$, centre $(-2, 0)$.

c Draw △5, the image of △2 after an enlargement with
scale factor $\frac{1}{2}$, centre $(-5, -7)$.

d Draw △6, the image of △3 after an enlargement with
scale factor $\frac{1}{2}$, centre $(-7, -5)$.

7.1.7 Translation

> ● A **translation** is simply a 'shift', for example '3 units to the right'
> or '4 units down'. There is no turning or reflection and the
> object stays the same size.

a △1 is mapped on to △2 by the
translation with vector $\begin{pmatrix} 4 \\ 2 \end{pmatrix}$

b △2 is mapped on to △3 by the
translation with vector $\begin{pmatrix} 2 \\ -3 \end{pmatrix}$

c △3 is mapped on to △2 by the
translation with vector $\begin{pmatrix} -2 \\ 3 \end{pmatrix}$

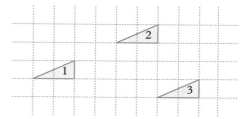

> Look at one vertex
> on the shape.

● In a vector the top number gives the number of units across (positive to the right) and the bottom number gives the number of units up/down (positive upwards).

So $\begin{pmatrix} 4 \\ 2 \end{pmatrix}$ is 4 across → $\begin{pmatrix} -2 \\ 3 \end{pmatrix}$ is 2 across ← $\begin{pmatrix} 3 \\ 0 \end{pmatrix}$ is 3 to the right
2 up ↑ 3 up ↑

Exercise 9Ⓔ

1 Look at the diagram.
Write the vector for each of these translations.

  a △1 → △2   **b** △1 → △3
  c △1 → △4   **d** △1 → △5
  e △1 → △6   **f** △6 → △5
  g △1 → △8   **h** △2 → △3
  i △2 → △4   **j** △2 → △5
  k △2 → △6   **l** △2 → △8
  m △3 → △5   **n** △8 → △2

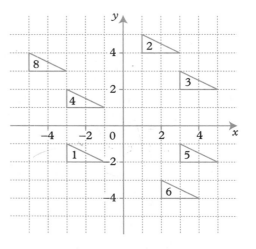

7.1.8 Successive transformations

Exercise 10Ⓔ

1 Copy the diagram.
  a Rotate △1 180° about (4, 2).
    Label the image △2.
  b Reflect △2 in the line $y = 2$.
    Label the image △3.
  c Describe the **single** transformation which maps △1 on to △3.

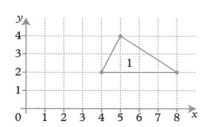

2 Copy the diagram.
  a Reflect △A in the line $y = x$.
    Label the image △B.
  b Reflect △B in the x-axis.
    Label the image △C.
  c Describe fully the **single** transformation which maps △A on to △C.

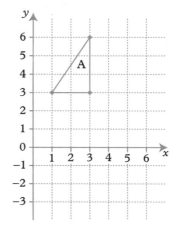

3 ABCD is transformed onto A'B'C'D' by a reflection followed by a translation parallel to the x-axis.

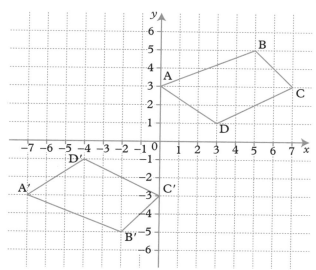

  a Describe these two transformations as fully as possible.
  b Do you get the same image if you do the translation before the reflection?

4 Draw axes for both x and y between -8 and $+8$. Plot the points $(1, 1)$, $(3, 1)$, $(3, 2)$, $(2, 2)$, $(2, 4)$ and $(1, 4)$ and join up to make an 'L' shape.
This shape is transformed onto the points $(-2, -2)$, $(-2, -6)$, $(-4, -6)$, $(-4, -4)$, $(-8, -4)$, $(-8, -2)$, by **two** transformations: an enlargement with centre $(0, 0)$ followed by a reflection.
Describe each of these transformations as fully as possible.

7.2 Circle calculations

7.2.1 Parts of a circle

- All points on a circle are an equal distance from the centre.

- A line from the centre to any point on the circle is a radius.

- A line across the circle through the centre is a diameter.

- The circumference is the perimeter of the circle.

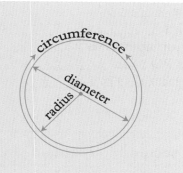

Exercise 11 C

1 Write the radius and the diameter of each circle.

a

6 cm

b

18 cm

c

10 cm

d

2 m

2 Find the radius and diameter by measuring.

3 Find the radius and diameter by measuring.

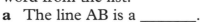

4 The diagrams show a regular pentagon ABCDE and a regular hexagon ABCDEF.
What angles would there be at the centre of a regular octagon (8 sides)?

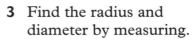

5 Copy and complete each sentence by choosing the correct word from the list.
 a The line AB is a _____.
 b The line CD is a _____.
 c The angle OPC is _____.
 d The shaded region above AB is a _____.
 e The curved line PQ is an _____.
 f The shaded region OPQ is a _____.

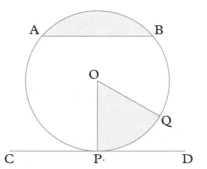

| arc | chord | sector | segment | tangent | ninety degrees |

7.2.2   Circumference of a circle

● The circumference of a circle is given by the formula
$$C = \pi d$$

> Learn this important formula.

EXAMPLE

Find the circumference.

a

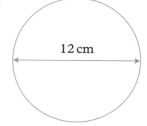

12 cm

b

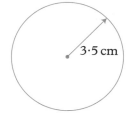

3·5 cm

> π is a special number. Its value is 3·142 approximately.

> Use the π button on your calculator.

a $C = \pi d$
   $C = \pi \times 12$ cm
   $C = 37{\cdot}7$ cm (to 3 sf)

b $C = \pi \times 7$ cm
   $C = 22{\cdot}0$ cm (to 3 sf)

> In part **b** you need the **diameter**. The diameter is twice the radius, so d = 7 cm.

Sometimes it is convenient to give the answer in a form involving π. In this example you could write '$C = 12\pi$ cm' and '$C = 7\pi$ cm'.

Exercise 12 Ⓔ

Find the circumference of the circles in questions **1** to **12**. Use the π button on a calculator or take $\pi = 3{\cdot}142$. Give the answers correct to 3 significant figures.

1

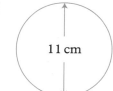

11 cm

2

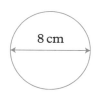

8 cm

3

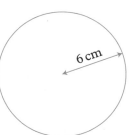

6 cm

4

5 cm

5

5 cm

6

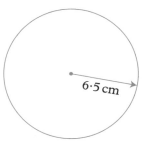

6·5 cm

7

1 cm

8

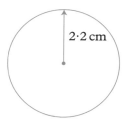

2·2 cm

9 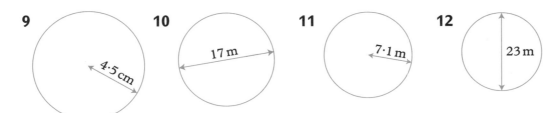 4·5 cm

10 17 m

11 7·1 m

12 23 m

13 A 10p coin has a diameter of 2·4 cm and a 5p coin has a diameter of 1·6 cm.

How much longer, to the nearest mm, is the circumference of the 10p coin?

14 A circular pond has a diameter of 2·7 m. Calculate the length of the perimeter of the pond.

15 A running track has two semicircular ends of radius 34 m and two straights of 93·2 m as shown.

Calculate the total distance around the track to the nearest metre.

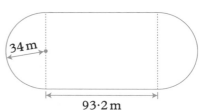
34 m

93·2 m

*16 A fly, perched on the tip of the minute hand of a grandfather clock, is 14·4 cm from the centre of the clock face. How far does the fly move between 12:00 and 13:00?

*17 A penny-farthing bicycle is shown. In a journey the front wheel rotates completely 156 times. How far does the bicycle travel?

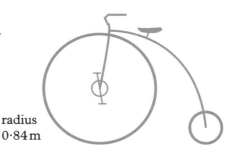

radius 0·84 m

*18 How many complete revolutions does a cycle wheel of diameter 60 cm make in travelling 400 m?

*19 Circle A has radius 5 cm and circle B has diameter 13 cm. Find the circumference of each circle, leaving π in your answer.

*20 The diagram shows a framework for a target, consisting of 2 circles of wire and 6 straight pieces of wire. The radius of the outer circle is 30 cm and the radius of the inner circle is 15 cm.

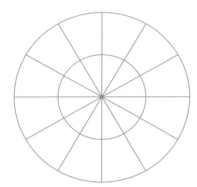

Calculate the total length of wire needed for the whole framework.

7.2.3   Area of a circle

● The area of a circle of radius r is given by the formula
$$A = \pi r^2$$

> Learn this important formula.

> EXAMPLE

Find the area of this circle.

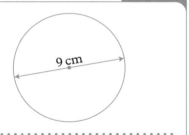

9 cm

....

In this circle $r = 4{\cdot}5$ cm
Area of circle $= \pi \times 4{\cdot}5^2$
$\qquad\qquad\quad = 63{\cdot}6$ cm^2 (to 3 sf)
Remember the formula is $\pi(r^2)$ **not** $(\pi r)^2$.
On a calculator, work out the answer like this:

| 4·5 | × | 4·5 | × | π | = |

or, on a modern calculator like this:

| 4·5 | x^2 | × | π | = |

> Note that on many calculators you need to press SHIFT before the π button.

Exercise 13Ⓔ

In questions **1** to **8** find the area of the circle. Use the π button on a calculator or use π = 3·14. Give the answers correct to three significant figures.

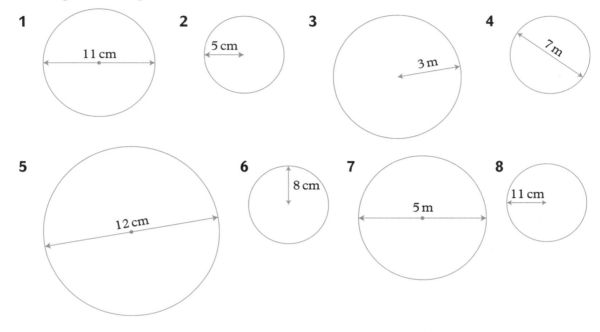

1 11 cm

2 5 cm

3 3 m

4 7 m

5 12 cm

6 8 cm

7 5 m

8 11 cm

9 A spinner of radius 7·5 cm is divided into six equal sectors.
Calculate the area of each sector.

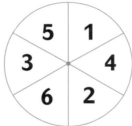

10 A circular swimming pool of diameter 12·6 m is to be covered by a plastic sheet to keep out leaves and insects. Work out the area of the pool.

11 A circle of radius 5 cm is inscribed inside a square as shown. Find the area shaded.

5 cm

12 A large circular lawn is sprayed with weedkiller. Each square metre of grass requires 2 g of weedkiller. How much weedkiller is needed for a lawn of radius 27 m?

13 Discs of radius 4 cm are cut from a rectangular plastic sheet of length 84 cm and width 24 cm.

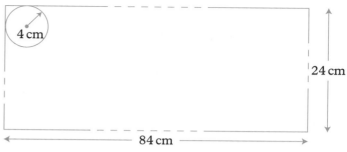

How many complete discs can be cut out? Find
  a the total area of the discs cut
  b the area of the sheet wasted.

14 A circular pond of radius 6 m is surrounded by a path of width 1 m. Find the area of the path.

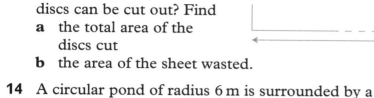

In questions **15** and **16** find the shaded area. Lengths are in cm.

15

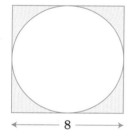

16

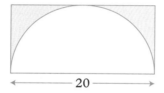

7.2.4 More complicated shapes

EXAMPLE

For this shape find
a the perimeter
b the area.

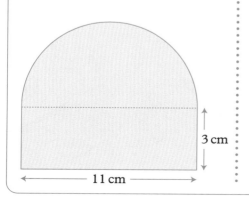

a Perimeter $= \left(\dfrac{\pi \times 11}{2}\right) + 11 + 3 + 3$

$\qquad = 34 \cdot 3$ cm (3 sf)

b Area $= \left(\dfrac{\pi \times 5 \cdot 5^2}{2}\right) + (11 \times 3)$

$\qquad = 80 \cdot 5$ cm^2 (3 sf)

Exercise 14Ⓔ

Use the π button on a calculator or take π = 3·14. Give the answers correct to 3 sf. For each shape find the perimeter.

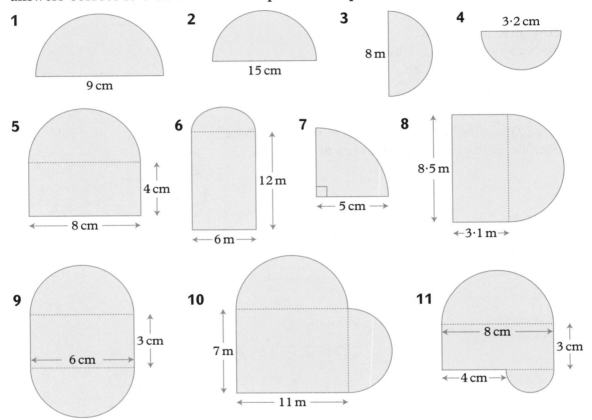

7.3 Volume

7.3.1 Cuboids

> ● The amount of space which an object occupies is called its **volume**.

A cube with edges 1 cm long has a volume of one cubic centimetre, which you write as 1 cm³.

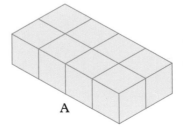

● A **cuboid** is a solid object whose faces are all rectangles. Shapes A and B are cuboids.

Do not confuse a cuboid with a **cube**, where all the edges are the same length.

A cube is a special kind of cuboid.

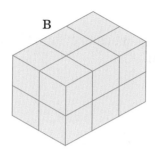

●   The volume of a cuboid is given by
Volume = length × width × height

For cuboid A, volume = $4 \times 2 \times 1 = 8 \, \text{cm}^3$

For cuboid B, volume = $2 \times 3 \times 2 = 12 \, \text{cm}^3$

●   **Faces, edges and vertices**
The **faces** of the cuboid are the flat surfaces on the shape.
There are 6 faces on a cuboid.

The **edges** of the cuboid are the lines that make up the shape.
There are 12 edges on a cubiod.

The **vertices** of the cuboid are where the edges meet at a point.
There are 8 vertices on a cuboid.

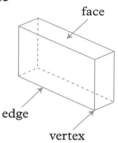

EXAMPLE

a   Find the volume of this cuboid.

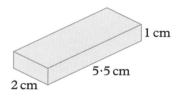

b   Find the volume of a cube of side 5 cm.

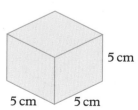

- -

a   Volume = length × width × height
         $= 5.5 \quad \times \quad 2 \quad \times \quad 1$
         $= 11 \, \text{cm}^3$

b   Volume = $5 \times 5 \times 5$
         $= 125 \, \text{cm}^3$

Exercise 15Ⓔ

In questions **1** to **6** work out the volume of each cuboid. All lengths are in cm.

1

2

3

4

5

6

7 This large cube is cut into lots of identical small cubes as shown.
Calculate the volume of each small cube.

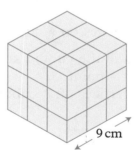

For questions **8** to **13** draw the solid on isometric paper and write the volume of the object. All the objects are made from centimetre cubes.

8   **9**   **10**

11   **12**   **13**

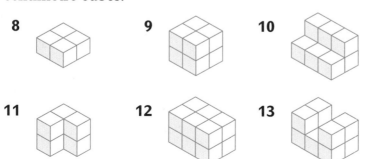

14 The diagram shows an empty swimming pool. Water is pumped into the pool at a rate of $1\,m^3$ per minute. How long will it take to fill the pool?

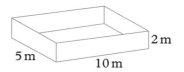

15 The diagram shows the net for a cube. Calculate the volume of the cube.

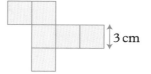

16 Find the missing side x in each case. All lengths are in cm.

a

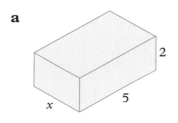

Volume = $30\,cm^3$

b

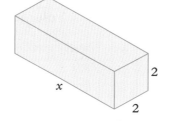

Volume = $16\,cm^3$

c

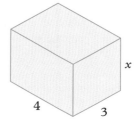

Volume = $24\,cm^3$

17 A cylindrical tin of volume $120\,cm^3$ is filled with sand from a rectangular box. How many times can the tin be filled if the dimensions of the box are 50 cm by 40 cm by 20 cm?

18 The diagram shows a cube of volume $1\,m^3$. How many cubic centimetres (cm^3) are there in $1\,m^3$?

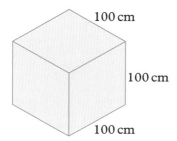

*19 Mr Gibson decided to build a garage and began by calculating the number of bricks required. The garage was to be 6 m by 4 m and 2·5 m in height. Each brick measures 22 cm by 10 cm by 7 cm. Mr Gibson estimated that he would need about 40 000 bricks. Is this a reasonable estimate?

7.3.2 Prisms

- A prism is an object with a uniform cross-section.

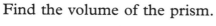

Area = A

Volume = $A \times l$

- A cuboid is a prism whose cross-section is a rectangle.

Volume = $l \times w \times h$

EXAMPLE

Find the volume of the prism.

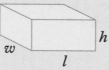

Area of end
= 12 cm²

Volume = $A \times l$
         = 12 × 8
         = 96 cm³

Exercise 16(E)

Find the volume of each prism in questions **1** to **9**. Give your answers in the correct units.

1 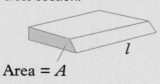 Area of end
= 15 cm²

10 cm

2 Area of end
= 5 m²

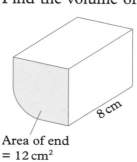

12 m

3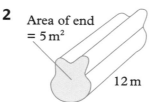

12 cm

10 cm

← 8 cm →

4 10 cm

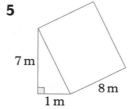

10 cm

3 cm

5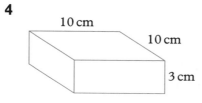

7 m

1 m

8 m

6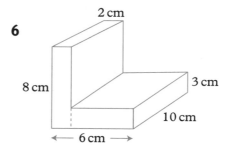

2 cm

8 cm

3 cm

6 cm

10 cm

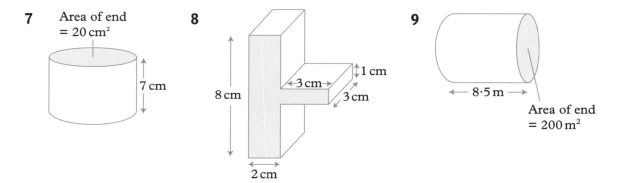

7 Area of end
= 20 cm²
7 cm

8 8 cm, 3 cm, 1 cm, 3 cm, 2 cm

9 8·5 m
Area of end
= 200 m²

10 A cylindrical bar has a cross-sectional area of 12 cm² and a length of two **metres**. Calculate the volume of the bar in cm³.

11 The diagram represents a building.
   a Calculate the area of the shaded end.
   b Calculate the volume of the building.

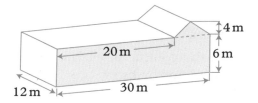

20 m, 4 m, 6 m, 12 m, 30 m

Find the volume of each prism in questions **12** to **14**. Give your answers in the correct units.

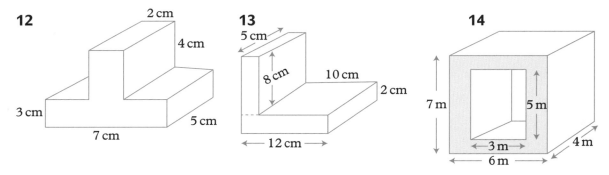

12 2 cm, 4 cm, 3 cm, 7 cm, 5 cm

13 5 cm, 8 cm, 10 cm, 2 cm, 12 cm

14 7 m, 5 m, 3 m, 6 m, 4 m

7.3.3 Cylinders

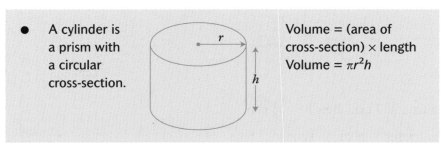

- A cylinder is a prism with a circular cross-section.

  r, h

  Volume = (area of cross-section) × length
  Volume = $\pi r^2 h$

EXAMPLE

Find the volumes of these cylinders.

a

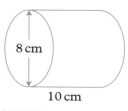

8 cm

10 cm

b

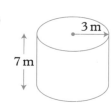

3 m

7 m

···

a Diameter = 8 cm
   Radius   = 4 cm
   Volume = $\pi \times 4^2 \times 10$
        = 503 cm^3 (3 sf)

b Volume = $\pi \times 3^2 \times 7$
        = 198 m^3 (3 sf)

Exercise 17Ⓔ

Find the volume of each cylinder. Use the π button on
a calculator or use $\pi = 3\cdot14$. Give the answers correct
to 3 sf.

1

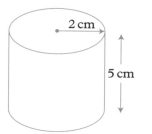

2 cm

5 cm

2

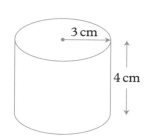

3 cm

4 cm

3

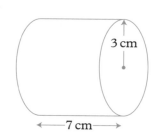

3 cm

7 cm

4

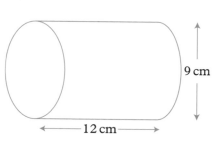

9 cm

12 cm

5

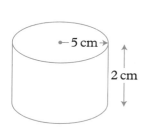

5 cm

2 cm

6

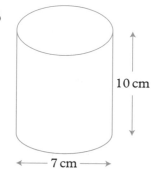

10 cm

7 cm

7 A cylinder with radius = 7 cm, height = 5 cm.

8 A cylinder with diameter = 8 m, height = 3·5 m.

9 Find the capacity in litres of this oil drum.

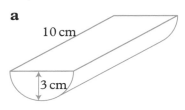

1000 cm³ = 1 litre

10 Cylinders are cut along the axis of symmetry to form these objects. Find the volume of each object.

a

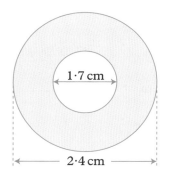

10 cm

3 cm

b

2 cm

8 cm

*11 A rectangular block has dimensions 20 cm × 7 cm × 7 cm. Find the volume of the largest solid cylinder which can be cut from this block.

*12 A washer is made by cutting a circle of metal from the centre of a metal disc. This washer is 2 mm thick.

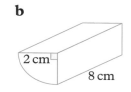

1·7 cm

2·4 cm

2 mm

a Find the area of the flat surface of the washer.
b Calculate the volume of the washer.

*13 A cylindrical water tank has internal diameter 40 cm and height 50 cm and a cylindrical mug has internal diameter 8 cm and height 10 cm. If the tank is initially full, how many mugs can be filled from the tank?

*14 The diagram shows the semicircular cross-section of a water trough of radius 21 cm. The length of the trough is 2 metres.
What is the volume of the trough? Don't forget to give the units for your answer.

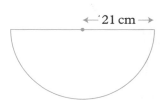

21 cm

7.4 Surface area

● The surface area of a shape is the sum of the area of all its faces.

7.4.1 Cylinders and cuboids

If you unwrap the label around a tin of baked beans, you will get a rectangle whose length is equal to the circumference of the tin ($2\pi r$).

So you can see that the **curved surface area** = $2\pi rh$.

The total surface area = curved surface area +
$$\text{area of two end circles.}$$
$$= 2\pi rh + 2\pi r^2$$

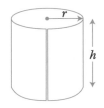

EXAMPLE

Calculate the **total** surface area of a solid cylinder of radius 4 cm and height 7 cm.

···

$$\text{Curved surface area} = 2\pi rh$$
$$= 2\times\pi\times4\times7$$
$$= 56\pi \text{ cm}^2$$
$$\text{Area of two ends} = 2\times\pi r^2$$
$$= 2\times\pi\times4^2$$
$$= 32\pi \text{ cm}^2$$
$$\text{Total surface area} = (56\pi + 32\pi)\text{ cm}^2$$
$$= 276 \text{ cm}^2 \text{ (3 sf)}$$

Remember: curved surface area of a cylinder = $2\pi rh$

Exercise 18Ⓔ

1 Calculate the **curved** surface area of each cylinder. All lengths are in cm.

a

$r = 4$
$h = 5$

b

$r = 3$
$h = 7$

c

$r = 2\cdot5$
$h = 6$

d

$r = 1\cdot2$
$h = 10$

2 Work out the total surface area of these solid cylinders.

a

$r = 3$
$h = 5$

b

$r = 5$
$h = 8$

c

$r = 2$
$h = 4·5$

3 A tin of tomatoes has radius 3·7 cm and
height 10·2 cm.
Calculate the area of the paper label
wrapped around the tin.

4 a How many faces has a cube?
  b Work out the total surface area of this cube.

4 cm

5 Calculate the total surface area of each solid cuboid.
All lengths are in cm.

a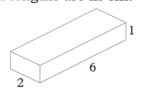

1
6
2

b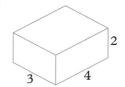

2
3   4

c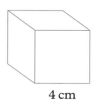

7
3   2

6 The solid object shown is made from 27
small cubes each 1 cm by 1 cm by 1 cm. The
small cubes are glued together and then the
outside is painted red.
Calculate
  a the number of cubes with one face painted
  b the number of cubes with two faces painted
  c the number of cubes with three faces painted
  d the number of cubes with no faces painted.
(Check that the answers to **a**, **b**, **c** and **d** add up
to the correct number.)

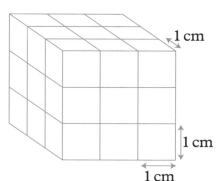

1 cm

1 cm

1 cm

7 This object is made from four 1 cm cubes.
Find the total surface area of the object.

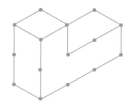

7.5 Bearings and scale drawing

7.5.1 Bearings

Bearings are used where there are no
roads to guide the way. Ships, aircraft
and mountaineers use bearings to
work out where they are.

● Bearings are measured **clockwise** from north.

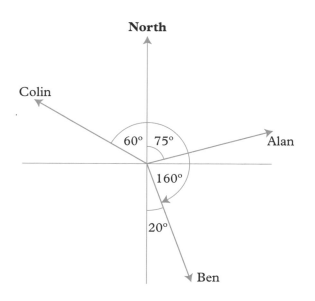

Alan is walking on a bearing of 075°.
Ben is walking on a bearing of 160°.
Colin is walking on a bearing of 300°.

> Notice that you
> put 3 digits in a
> bearing – the
> direction 75° east
> of north is 075°.

Exercise 19 C

The diagrams show the directions in which several people are travelling. Work out the bearing for each person.

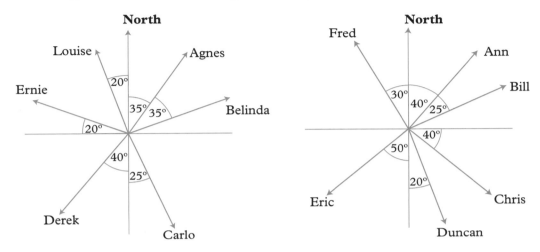

Exercise 20 C

The map of North America shows six radar tracking stations, A, B, C, D, E, F.

1 From A, measure the bearing of
  a F
  b B
  c C.

2 From C, measure the bearing of
  a E
  b B
  c D.

3 From F, measure the bearing of
  a D
  b A.

4 From B, measure the bearing of
  a A
  b E
  c C.

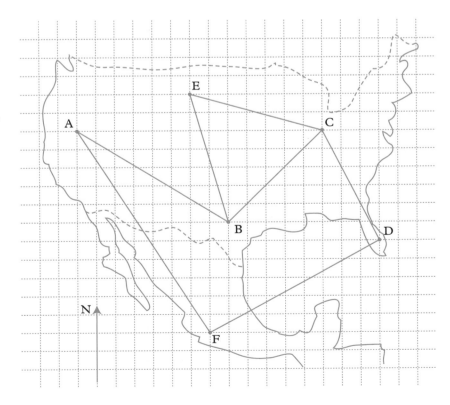

7.5.2 Scale drawing

You can solve problems involving compass directions using a
scale drawing.

A ship sails 10 km north, then changes course to south-east, and sails
a further 8 km.
How far is the ship from its starting point?

Use a scale of 1 cm to 1 km.
  a Mark a starting point S and draw a vertical line. This is
     the north line.
  b Mark a point A, 10 cm from S.
  c Draw a line at 45° to SA, as shown, and mark a point
     F, 8 cm from A.
  d Measure the distance SF.

$$SF = 7{\cdot}1 \text{ cm}$$

The ship is 7·1 km from its starting point.
(An answer between 7·0 km and 7·2 km would be
acceptable.)

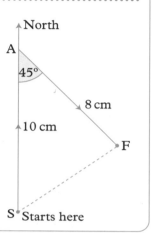

Exercise 21 Ⓒ

Use a scale of 1 cm to represent 1 km.
1 A ship sails 8 km due east and then a further
     6 km due south. Find the distance of the ship
     from its starting point.

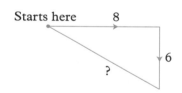

2 A ship sails 7 km due west and then a further 4 km north-east.
     Find the distance of the ship from its starting point.

3 A ship sails 8 km due north and then a further 7 km on a
     bearing 080°, as in the diagram (which is not drawn to scale).
     How far is the ship now from its starting point?

4 A ship sails 9 km on a bearing 090° and
     then a further 6 km on a bearing 050°,
     as shown in the diagram. How far is the
     ship now from its starting point?

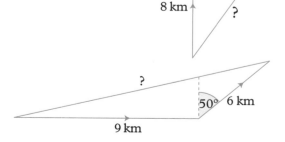

5   A ship sails 6 km on a bearing 160° and then a
further 10 km on a bearing 240°, as shown.
   a   How far is the ship from its starting point?
   b   On what bearing must the ship sail so that it
   returns to its starting point?

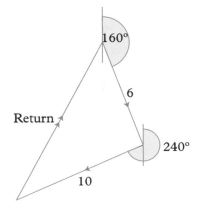

Exercise 22 Ⓒ

Draw the points P and Q in the middle of a clean page
of square grid paper. Mark the points A, B, C, D and E
accurately, using the information given.

1   A is on a bearing of 040° from P and
015° from Q.

2   B is on a bearing of 076° from P and
067° from Q.

3   C is on a bearing of 114° from P and
127° from Q.

4   D is on a bearing of 325° from P and 308° from Q.

5   E is on a bearing of 180° from P and 208° from Q.

7.6   Angles in polygons

7.6.1   Polygons

●   A polygon is a flat shape with 3 or more edges.

A quadrilateral is a polygon with 4 edges.

The table shows the names of the more
common polygons.

Name	Number of edges
Quadrilateral	4
Pentagon	5
Hexagon	6
Octagon	8

● A regular polygon has edges of equal length and all its angles are equal.

Here is a regular hexagon.

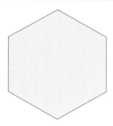

Exercise 23(E)

1 **a** Draw a sketch of a pentagon.
   b On your sketch draw all the diagonals of the pentagon. (There should be five altogether.)

2 **a** Draw a hexagon and draw all the diagonals.
   b How many are there?

3 What is the name for a regular polygon with
   a three sides   **b** four sides?

4 **a** Draw a pentagon with two right angles.
   b Draw a pentagon with one pair of parallel sides.

7.6.2   Sum of the interior angles of a polygon

● You have already seen that the sum of the angles in a triangle is 180° and the sum of the angles in a quadrilateral is 360°.

● You can divide a pentagon into 3 triangles and a hexagon into 4 triangles. The sum of the angles in each shape is equal to the sum of the angles in the triangles inside.

Pentagon, 5 sides

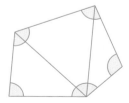

There are 3 triangles.
Sum of the interior angles = $3 \times 180°$
$= 540°$

Hexagon, 6 sides

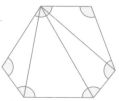

There are 4 triangles.
Sum of the interior angles = $4 \times 180°$
$= 720°$

In a **regular** pentagon all the angles are the same.
As there are five equal angles, each angle = 540° ÷ 5
= 108°

Exercise 24Ⓔ

1 Find the angles marked with letters.

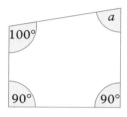 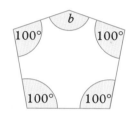 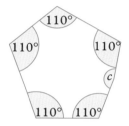 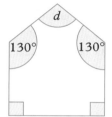

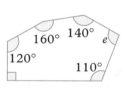  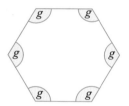

2 The diagram shows a regular pentagon.
  a Find the angle DCE.
  b Show that EC is parallel to AB.

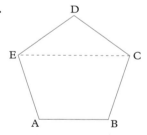

3 Find the angles marked with letters.

a   **b**   **c**

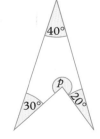 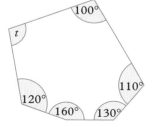 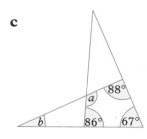

7.6.3 Exterior angles of a polygon

● The exterior angle of a polygon is the angle between an extended side and the adjacent side of the polygon.

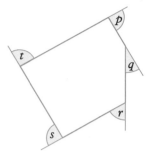

If you put all the exterior angles together you can see that the sum of the angles is 360°.
This is true for any polygon.

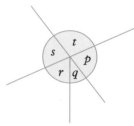

● The sum of the exterior angles of a polygon = 360°.

● In a **regular** polygon all exterior angles are equal.

● For a **regular** polygon with n sides, each exterior angle $= \dfrac{360°}{n}$.

So, for a regular pentagon, each exterior angle $= \dfrac{360°}{5} = 72°$

And, for a regular hexagon, each exterior angle $= \dfrac{360°}{6} = 60°$

EXAMPLE

The diagram shows a regular octagon (8 sides).
a Calculate the size of each exterior angle (marked e).
b Calculate the size of each interior angle (marked i).
....

a There are 8 exterior angles and the sum of these angles in 360°.
So angle $e = \dfrac{360°}{8} = 45°$

b $e + i = 180°$ (angles on a straight line)
So $i = 135°$

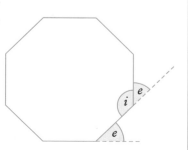

Exercise 25Ⓔ

1 Copy and complete: 'The sum of the exterior angles of any polygon is_____'.

2 Here is a regular pentagon with its exterior angles marked. How big is each exterior angle?

3 Look at the polygon in the diagram.
  a Calculate each exterior angle.
  b Check that the total of the exterior angles is 360°.

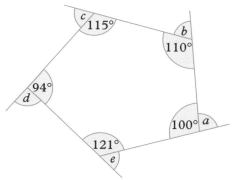

4 The diagram shows a regular decagon.
  a Calculate the angle a.
  b Calculate the interior angle of a regular decagon.

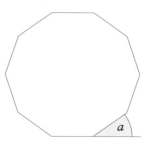

5 Find the exterior angle of a regular polygon with
  a 9 sides   **b** 18 sides   **c** 60 sides.

6 Find the angles marked with letters.

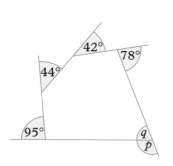

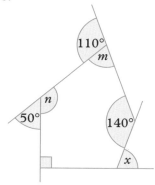

7 A regular dodecagon has 12 sides.
  a Calculate the size of each exterior angle, e.
  b Use your answer to find the size of each interior angle, i.

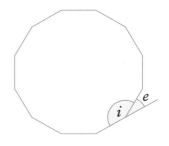

8 Each exterior angle of a regular polygon is 15°. How many sides has the polygon?

9 Each exterior angle of a regular polygon is 18°. How many sides has the polygon?

10 Each interior angle of a regular polygon is 140°. How many sides has the polygon?

7.7 Locus

● The word **locus** describes the position of points which obey a certain rule. The locus can be the path traced out by a moving point.

7.7.1 Three important loci

Circle
The locus of points which are equidistant from a fixed point O is drawn here. It is a circle with centre O.

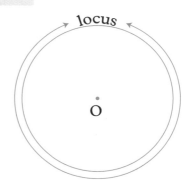

Perpendicular bisector
The locus of points which are equidistant from two fixed points A and B is drawn here.

The locus is the perpendicular bisector of the line AB. Use compasses to draw arcs, as shown, or use a ruler and a protractor.

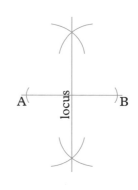

Angle bisector

The locus of points which are equidistant
from two fixed lines AB and AC is drawn here.

The locus is the line which bisects the
angle BAC. Use compasses to draw
arcs or use a protractor to construct
the locus.

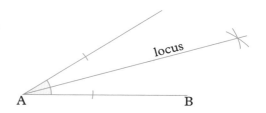

Exercise 26Ⓔ

1 Draw the locus of a point P which moves so that it is always
3 cm from a fixed point X.

•X

2 Mark two points P and Q which are 10 cm apart.
Draw the locus of points which are equidistant from
P and Q.

3 Draw two lines AB and AC of length 8 cm, where
∠BAC = 40°.
Draw the locus of points which are equidistant from AB
and AC.

4 A sphere rolls along a surface from A to B. Copy the diagram
and sketch the locus of the centre of the sphere as it moves
from A to B in each case.

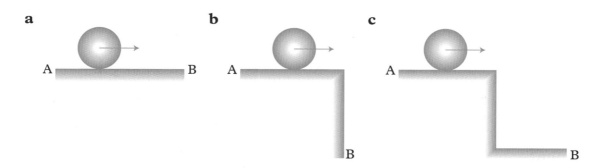

a   **b**   **c**

5 A rectangular slab ABCD is rotated around corner
B from position 1 to position 2.
Draw a diagram, on square grid paper, to show
a the locus of corner A
b the locus of corner C.

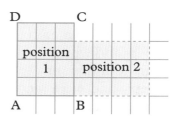

6 The diagram shows a section of coastline with a lighthouse L and coastguard C.
A sinking ship sends a distress signal.
The ship appears to be up to 40 km from L and up to 20 km from C.
Copy the diagram and show the region in which the ship could be.

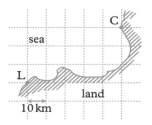

7 a Draw the triangle LMN full size.
 b Draw the locus of the points which are
  i equidistant from L and N
  ii equidistant from LN and LM
  iii 4 cm from M.
(Draw the three loci in different colours.)

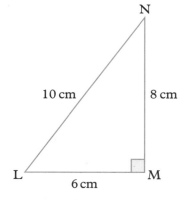

*8 Draw a line AB of length 6 cm. Draw the locus of a point P so that angle ABP = 90°.

*9 Mr Gibson's garden has a fence on two sides and trees at two corners.
He wants to build a sand pit so that it is
 a equidistant from the 2 fences
 b equidistant from the 2 trees.
Make a scale drawing (1 cm = 1 m) and mark where Mr Gibson can put the sand pit.

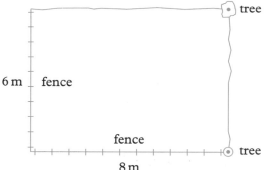

7.8 Pythagoras' theorem

Pythagoras (569–500 BC) was one of the first great mathematical names in Greek antiquity. He settled in southern Italy and formed a mysterious brotherhood with his students who were bound by an oath not to reveal the secrets of numbers and who exercised great influence. They laid the foundations of arithmetic through geometry and were among the first mathematicians to develop the idea of proof.

7.8.1 The theorem

● In a right-angled triangle the square on the hypotenuse is equal to the sum of the squares on the other two sides.

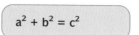

$$a^2 + b^2 = c^2$$

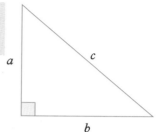

7.8.2 Finding the longest side (the hypotenuse)

EXAMPLE

a Find the length x.

b Find the length y.

$$x^2 = 3^2 + 5^2$$
$$x^2 = 9 + 25$$
$$x^2 = 34$$
$$x = \sqrt{34}$$
$$x = 5{\cdot}83 \text{ (2 dp)}$$

The side on its own in the equation is the hypotenuse.

$$y^2 = 5^2 + 3{\cdot}5^2$$
$$y^2 = 25 + 12{\cdot}25$$
$$y^2 = 37{\cdot}25$$
$$y = \sqrt{37{\cdot}25}$$
$$y = 6{\cdot}10 \text{ (2 dp)}$$

Exercise 27 Ⓔ

Give your answers correct to 2 dp, where necessary. The units are cm unless you are told otherwise.

1 Find x.

a
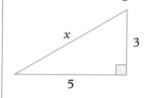

b

c

d
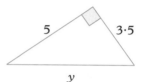

e
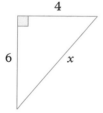

f
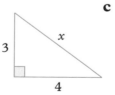

g
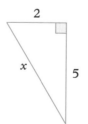

h
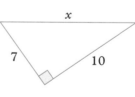

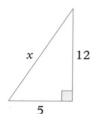

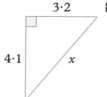

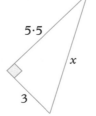

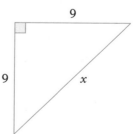

2 Find the length of a diagonal of a square of side 8 cm.

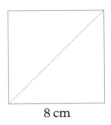

8 cm

3 **a** Find the length of a diagonal of a rectangle of length 7 cm and width 4·5 cm.

 b Find the length of a diagonal of a rectangle of length 10 cm and width 7·2 cm.

4 A ladder leans against a vertical wall. The foot of the ladder is 2 m from the bottom of the wall and the ladder reaches 6 m up the wall.
How long is the ladder?

5 Shruti walks 100 m north and then 55 m east. How far is she now from her starting point?

7.8.3 Finding a shorter side

EXAMPLE

a Find the length x.

2 | 3

x

b Find the length y.

10

4 | y

$$x^2 + 2^2 = 3^2$$
$$x^2 + 4 = 9$$
$$x^2 \quad = 5$$
$$x \quad = \sqrt{5}$$
$$x \quad = 2\cdot24 \ (2 \, \text{dp})$$

The longest side is on its own in the equation.

$$y^2 + 4^2 = 10^2$$
$$y^2 + 16 = 100$$
$$y^2 \quad = 84$$
$$y \quad = \sqrt{84}$$
$$y \quad = 9\cdot17 \ (2 \, \text{dp})$$

Exercise 28Ⓔ

1 Find y.

a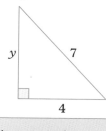

In part **a** write
$y^2 + 4^2 = 7^2$

b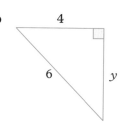

c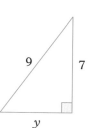

d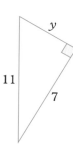

e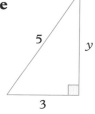

f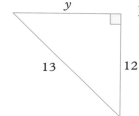

g

h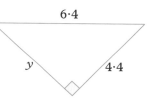

2 A 4 m ladder rests against a vertical wall with
its foot 2 m from the wall.
How far up the wall does the ladder reach?

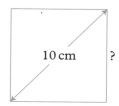

3 The diagonals of a square are of length 10 cm.
How long is each side of the square?

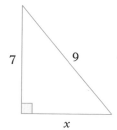

Exercise 29Ⓔ (Mixed questions)

In questions **1** to **8**, find x. All the lengths are in cm.

1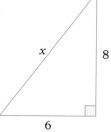

2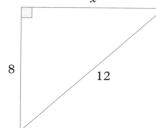

3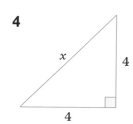

4

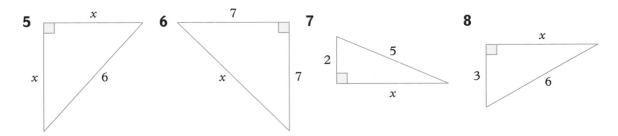

5 **6** **7** **8**

9 Find the length of a diagonal of a rectangle of length 9 cm and width 4 cm.

10 An isosceles triangle has sides 10 cm, 10 cm and 4 cm.
Find the height of the triangle.

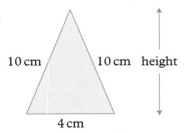

11 A ladder rests against a vertical wall with its foot 1·5 m from the wall. The ladder reaches 4 m up the wall. How long is the ladder?

12 A ship sails 20 km due north and then 35 km due east. How far is it from its starting point?

13 The square and the rectangle have the same length diagonal. Find x.

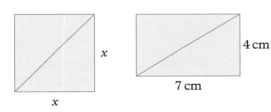

Exercise 30Ⓔ

1 a Find the height of the triangle, h.
 b Find the area of the triangle ABC.

2 A thin wire of length 18 cm is bent in the shape shown.
Calculate the length from A to B.

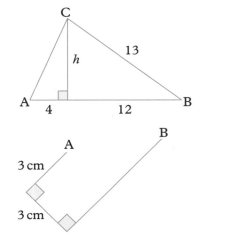

3 A paint tin is a cylinder of radius 12 cm and height 22 cm. Leonardo, the painter, drops his stirring stick into the tin and it disappears.
Work out the maximum length of the stick.

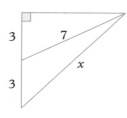

4 In the diagram A is $(1, 2)$ and B is $(6, 4)$

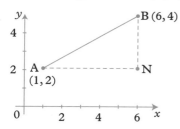

Work out the length AB. (First find the lengths of AN and BN.)

5 On square grid paper plot $P(1, 3)$, $Q(6, 0)$, $R(6, 6)$. Find the lengths of the sides of triangle PQR. Is the triangle isosceles?

Questions **6** to **12** are more difficult. You cannot find x immediately.

In questions **6** to **11** find x.

6

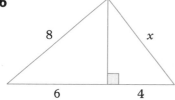

7

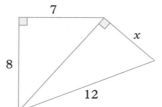

8

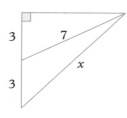

9

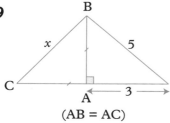

(AB = AC)

10

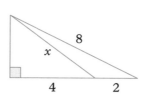

11

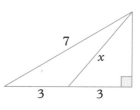

*12 The diagram shows a rectangular block.
Calculate   **a** AC   **b** AY

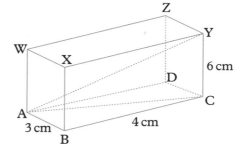

7.9 Areas and volumes of similar shapes

The two rectangles shown are similar.
The ratio of their corresponding sides is k.

area of ABCD $= ab$
area of WXYZ $= ka \times kb = k^2 ab$

so, $\dfrac{\text{area of WXYZ}}{\text{area ABCD}} = \dfrac{k^2 ab}{ab} = k^2$

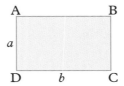

This illustrates an important general rule for all similar shapes:

- If two figures are similar and the ratio of corresponding sides is k, then the ratio of their areas is k^2.

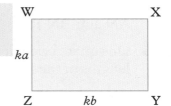

Note: k may be called the **linear scale factor** and k^2 may be called the **area scale factor**.

We can also show that:

- If two three-dimensional shapes are similar and the ratio of corresponding sides is k, then the ratio of their volumes is k^3.

EXAMPLE

Triangles A and B are similar.
Find the area of triangle B.

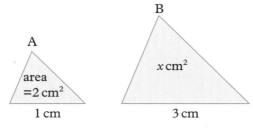

ratio of corresponding sides $= \dfrac{3}{1} = 3$

so, ratio of areas $= 3^2 = 9$

area of B $= 9 \times$ area of A
$= 9 \times 2$
$= 18 \text{ cm}^2$

EXAMPLE

Two similar cylinders have heights of 3 cm and 6 cm respectively. If the volume of the smaller cylinder is 30 cm³, find the volume of the larger cylinder.

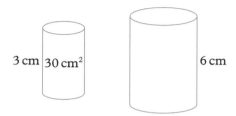

3 cm 30 cm²   6 cm

ratio of heights (k) = $\dfrac{6}{3}$ (linear scale factor)

= 2

so, ratio of volumes (k^3) = 2^3 (volume scale factor)

= 8

and volume of larger cylinder = 8 × 30

= 240 cm³

Exercise 31Ⓔ

In questions **1** to **4**, the number written inside a figure represents the area of the shape in cm². Numbers on the outside give linear dimensions in cm. In each case the shapes are similar. In questions **1** to **4**, find the unknown area A.

Remember:
If linear scale
factor = k, area
scale factor = k^2.

1

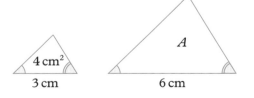

4 cm² A

3 cm   6 cm

2

6 cm

2 cm   3 cm² A

3

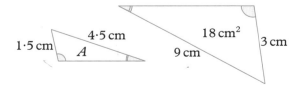

1·5 cm   4·5 cm   18 cm²   3 cm

A 9 cm

4

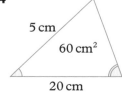

5 cm   60 cm²   2·5 A

20 cm   10

In questions **5** to **8**, the objects are similar and a number written inside a figure represents the volume of the object in cm^3.

Numbers on the outside give linear dimensions in cm. Find the unknown volume V.

> Remember:
> If linear scale factor = k, volume scale factor = k^3.

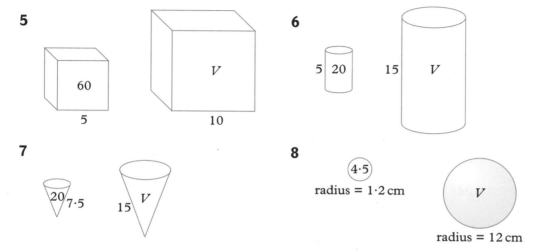

5

60

5

V

5

10

6

5 | 20

15 | V

7

20 | 7·5

15 | V

8

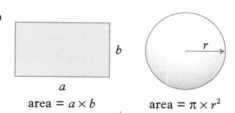

4·5

radius = 1·2 cm

V

radius = 12 cm

7.10 Dimensions of formulae

A formula for an area has two lengths multiplied together.

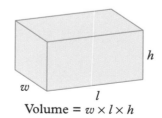

b

a

area = $a \times b$

r

area = $\pi \times r^2$

Numbers like 2, 5 or π are not lengths and have no dimensions.

A formula for a volume has three lengths multiplied together.

h

w

l

Volume = $w \times l \times h$

A formula with one length only represents a length.

Exercise 32Ⓔ

1 Letters a, b, c are lengths.

   a $s = 5bc$. Is this a formula for length, area or volume?

b $m = 3a + b$. Is this a formula for length or area?

2 Here are several expressions in which each letter is a length. Decide whether each expression is a length, an area or a volume.

a	$2r$	**b**	a^2	**c**	$4ab$	**d**	πd
e	πr^2	**f**	$3a + 2b$	**g**	$7b^2 + a^2$	**h**	abc
i	x^2y	**j**	$11a^2 + ab$	**k**	$5(a + b)$	**l**	$\pi(r^2 + t^2)$
m	$7 \times e \times f$	**n**	$6lmm$	**o**	$8cd$	**p**	$10y$

3 Copy and complete the table by putting a tick in the correct column.

expression	length	area	volume	none of these
$7r$				
$3r^2$				
$\pi r^2 h$				
abc^2				
$\pi(a^2 + c^2)$				

7.11 Three-dimensional coordinates

In three-dimensional space, we need three coordinates to describe how the position of a point relates to the common origin of three axes. In this diagram, O is the origin for the three axes. For A, $x = 5$, $y = 4$, $z = 2$, so A is written $(5, 4, 2)$.

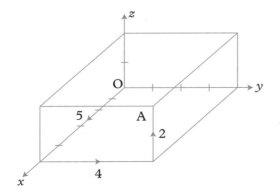

Exercise 33(E)

1 Write the coordinates of the points A, B, C, D.

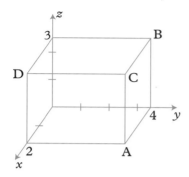

2 a Write the coordinates of B, C, Q, R.

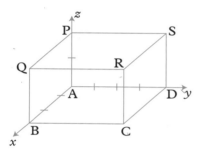

Each division on each axis represents 1 unit. So D has coordinates (0, 4, 0).

b Write the coordinates of the mid-points of
i AD   **ii** DS   **iii** DC
c Write the coordinates of the centre of the face
i ABCD   **ii** PQRS   **iii** RSDC
d Write the coordinates of the centre of the box.

3 a Write the coordinates of C, R, B, P, Q.

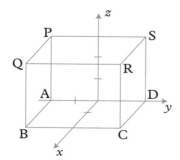

b Write the coordinates of the mid-points of
i QB   **ii** PQ

Test yourself

1 a i Find the area of the
     shaded shape.
  ii Find the perimeter of the
     shaded shape.

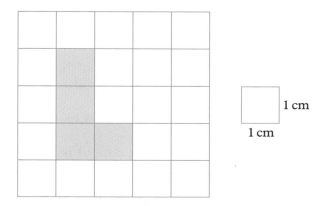

1 cm

1 cm

Here is a solid prism made from centimetre cubes.

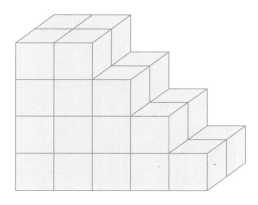

1 cm³

b Find the volume of the solid prism.

(Edexcel, 2004)

2 The diagram shows a wedge in the shape of
a triangular prism.

The cross-section of the prism is shown as a
shaded triangle.

The area of the triangle is 15 cm².
The length of the prism is 10 cm.

Work out the volume of the prism.

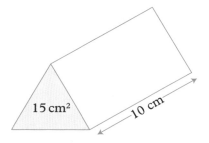

15 cm²   10 cm

Diagram not accurately drawn

(Edexcel, 2004)

3 a

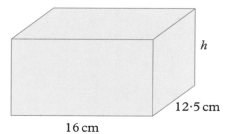

Diagram not accurately drawn

The base of a cuboid has length 16 cm and width 12·5 cm.
The volume of the cuboid is 1880 cm³.
Find the height, *h*, of the cuboid.

b The volume of another cuboid is 36 cm³.
The length, width and height of the cuboid are all **different whole** numbers.

Given one set of possible values of the length, width and height.

(OCR, 2005)

4 A cylinder has a radius of 5 cm.
a Calculate the circumference of a circular end of the cylinder.

b The cylinder has a volume of 250 cm³.
Calculate the height of the cylinder.

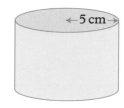

(AQA, 2004)

5 James plans a game.
He hides objects at X, Y and Z and marks the positions on a plan.

The scale of this plan is 1 cm to 100 m.
a Measure the bearing of Z from Y.
b **Calculate** the distance a competitor must walk from Y to Z.
Give your answer to the nearest metre. You **must** show all your working.

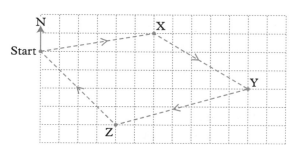

(AQA, 2004)

6 The diagram shows the position of Elaine's house, H, and her position, X, on a walk. The scale of the diagram is 1 cm represents 2 km.
   a Measure and write the bearing and distance, in km, of X from H.

   Elaine then walks to a position, Y, which is 15 km from H and on a bearing of 260° from H.
   b Copy the diagram and mark the position of Y.

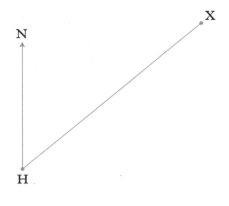

(OCR, Spec)

7 An ice hockey puck is in the shape of a cylinder with a radius of 3·8 cm, and a thickness of 2·5 cm.

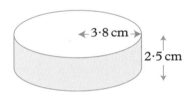   Diagram not accurately drawn

It is made out of rubber with a density of 1·5 grams per cm^3.

Work out the mass of the ice hockey puck.
Give your answer correct to 3 significant figures.

(Edexcel, 2004)

8 Shape A is rotated 90° anticlockwise, centre (0, 1), to shape B.
Shape B is rotated 90° anticlockwise, centre (0, 1), to shape C.
Shape C is rotated 90° anticlockwise, centre (0, 1), to shape D.
   a Copy the diagram and mark the position of shape D.
   b Describe the single transformation that takes shape C to shape A.

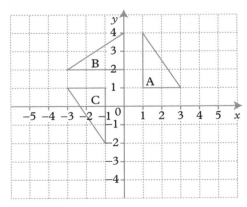

(Edexcel, 2003)

9 Describe fully a single transformation
that would map the shaded shape on to
a shape A
b shape B
c shape C.

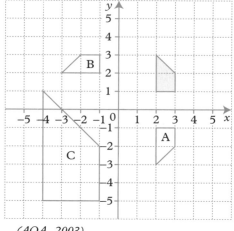

(AQA, 2003)

10 Copy the diagram.
a Reflect triangle A in the line $x = 4$.
Label the image P.
b Translate triangle A by 4 squares to
the left and 3 squares down.
Label the image Q.
c Triangle B is an enlargement of
triangle A.
Write down the scale factor of the
enlargement.
d Describe fully the single
transformation that maps triangle
A onto triangle C.

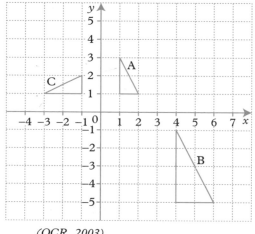

(OCR, 2003)

11 A circle has a radius of 6·1 cm.
Work out the area of the circle.

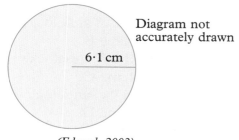

Diagram not
accurately drawn

6·1 cm

(Edexcel, 2003)

12 A circular pond has a radius of 2·2 m.
a Calculate the circumference of the pond.
b Calculate the area of the pond.

(AQA, 2003)

13 **In this question take π as 3·14 or use the π button on your calculator**
A piece of wire is bent to form a circle of radius 8 cm.
a What is the area of the circle?
b The same piece of wire is then bent to form a square.
Calculate the length of each side of the square.

(WJEC)

14 This design is made with three semicircles, each of diameter 8 cm.
Find the perimeter of the design.

(OCR, 2005)

15 P and Q are two points marked on the grid.
Copy the diagram.
Construct accurately the locus of all points which are equidistant from P and Q.

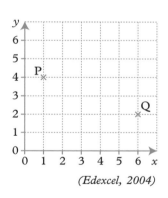

(Edexcel, 2004)

16 A point moves around the outside of an equilateral triangle.
It is always 2 cm from the nearest point on the perimeter of the triangle.
Copy the diagram and construct the locus of the point.

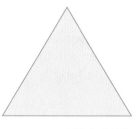

(AQA, 2003)

17 The diagram shows a regular octagon.

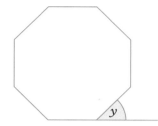

Diagram not accurately drawn

Calculate the size of the exterior angle of the regular octagon, marked y on the diagram.

(AQA, 2004)

18 a The exterior angle of a regular polygon is 30°.
How many sides does the polygon have?

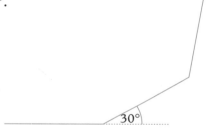

b AB is one side of another polygon.
Calculate the length of AB.

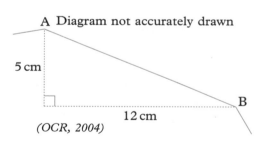

A Diagram not accurately drawn

5 cm

12 cm

B

(OCR, 2004)

19 The diagram shows a plot of land.
The measurements are in metres.

a The plot is to be turfed to make a lawn.
What will be the area of the lawn?

b i Find the length AB.
ii A fence is to be put around the perimeter of the plot.
What will be the length of the fence?

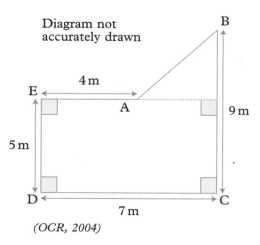

Diagram not accurately drawn

(OCR, 2004)

20 PQRS is a quadrilateral.
Angles RQS and QSP are right angles.
PS = 4 cm, QR = 12 cm and RS = 13 cm.

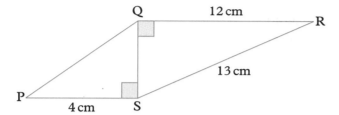

Show that the length of PQ is $\sqrt{41}$.

(AQA, 2004)

8 Data handling 2

In this unit you will:
- learn how to interpret a scatter diagram
- learn how to create and use two-way tables
- learn about collecting and interpreting data
- practise techniques for statistics coursework.

8.1 Scatter diagrams

Sometimes it is interesting to discover if there is a relationship (or **correlation**) between two sets of data.

Teacher's note: See page 375 for 'Scatter graphs on a computer'.

Examples

- Do tall people weigh more than short people?

- If you spend longer revising for a test, will you get a higher mark?

- Do people who have credit cards have more debts than other people?

- Do tall parents have tall children?

- If there is more rain, will there be less sunshine?

- Does the number of Olympic gold medals won by British athletes affect the rate of inflation?

If there is a relationship, it will be easy to spot if your data are plotted on a scatter diagram.

- A scatter diagram is a graph in which one set of data is plotted on the horizontal axis and the other set on the vertical axis.

EXAMPLE

Each month the average outdoor temperature was recorded together with the amount of gas (in therms) used to heat a house. The results are plotted on this scatter diagram.

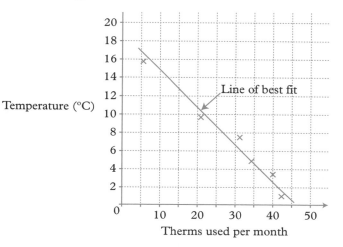

There is a high degree of **correlation** between these figures. Gas companies do use weather forecasts to predict future gas consumption over the whole country.

A **line of best fit** has been drawn 'by eye'. You can estimate that if the outdoor temperature was 12 °C then about 17 therms of gas would be used.

Note: You can only predict within the range of values given. If you extend the line below zero the line of best fit predicts that about 60 therms would be used when the temperature is −4 °C.
But −4 °C is well outside the range of the values plotted so the prediction is not **valid**.

Perhaps at −4 °C a lot of people might stay in bed and the gas consumption would not increase by much. The point is you don't know!

1   The line in the example has a negative gradient and so there is negative correlation.
2   If the line of best fit has a positive gradient there is positive correlation.
3   Some data when plotted on a scatter diagram does not appear to fit any line at all. In this case there is no correlation.

Remember: 'N' for negative.

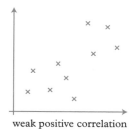

weak positive correlation

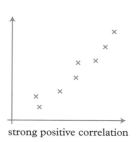

strong positive correlation

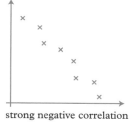

strong negative correlation

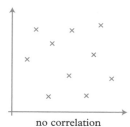

no correlation

Exercise 1(E)

1 For this question you need to make some measurements of people in your class.

a Measure everyone's height and 'armspan' to the nearest cm.

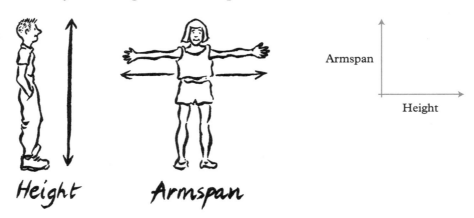

Height   Armspan

Plot the measurements on a scatter graph.
Is there any correlation?

b Now measure everyone's 'head circumference' just above the eyes.
Plot head circumference and height on a scatter graph.
Is there any correlation?

c Decide as a class which other measurements (for example, pulse rate) you can (fairly easily) take and plot these to see if any correlation exists.

d Which pair of measurements gave the best correlation?

2 Plot the points in each table on a scatter graph, with t across the page and z up the page. Draw axes with values from 0 to 20. Describe the correlation, if any, between the values of t and z (for example, 'strong positive', 'weak negative' etc.).

a

t	8	17	5	13	19	7	20	5	11	14
z	9	16	7	13	18	10	19	8	11	15

b

t	4	9	13	16	17	6	7	18	10
z	5	3	11	18	6	11	18	12	16

c

t	12	2	17	8	3	20	9	5	14	19
z	6	13	8	15	18	2	12	9	12	6

3 Describe the correlation, if any, in these scatter graphs.

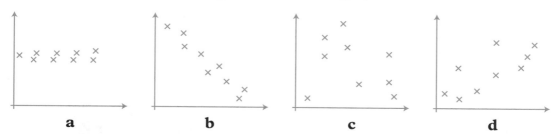

a	**b**	**c**	**d**

4 Plot the points in the table on a scatter graph, with s across the page and h up the page. Draw axes with values from 0 to 20.

s	3	13	20	1	9	15	10	17
h	6	13	20	6	12	16	12	17

   a Draw a line of best fit.
   b What value would you expect for h when s is 6?

5 The marks of 7 students in the two papers of a physics examination are shown in the table.

Paper 1	20	32	40	60	71	80	91
Paper 2	15	25	40	50	64	75	84

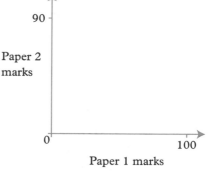

   a Plot the marks on a scatter diagram, using a scale of 1 cm to 10 marks, and draw a line of best fit.
   b A student scored a mark of 50 on Paper 1. What would you expect her to get on Paper 2?

6 The table shows
   i   the engine size in litres of various cars
   ii the distance travelled in km on one litre of petrol.

Engine	0·8	1·6	2·6	1·0	2·1	1·3	1·8
Distance	13	10·2	5·4	12	7·8	11·2	8·5

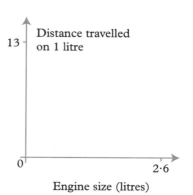

   a Plot the figures on a scatter graph using a scale of 5 cm to 1 litre across the page and 1 cm to 1 km up the page. Draw a line of best fit.
   b A car has a 2·3 litre engine. How far would you expect it to go on one litre of petrol?

7 The data show the latitude of 10 cities in the northern hemisphere and the average high temperatures.

City	Latitude (degrees)	Mean high temperature (°F)
Bogota	5	66
Casablanca	34	72
Dublin	53	56
Hong Kong	22	77
Istanbul	41	64
St Petersburg	60	46
Manila	15	89
Mumbai	19	87
Oslo	60	50
Paris	49	59

a Draw a scatter diagram and draw a line of best fit. Plot latitude across the page with a scale of 2 cm to 10°. Plot temperature up the page from 40 °F to 90 °F with a scale of 2 cm to 10 °F.

b Which city lies well off the line?
Do you know what factor might cause this apparent discrepancy?

c The latitude of Shanghai is 31° N. What do you think its mean high temperature is?

8 What sort of pattern would you expect if you took readings of these variables and drew a scatter diagram?
  a cars on roads; accident rate
  b sales of perfume; advertising costs
  c birth rate; rate of inflation
  d petrol consumption of car; price of petrol
  e outside temperature; sales of ice cream.

8.2 Two-way tables

A school collects data about the reading ability of six-year-olds. Children are given a short passage and they 'pass' if they can read over three-quarters of the words. The results are given in this two-way table.

	Boys	Girls	Total
Can read	464	682	
Cannot read	317	388	
Total			

Useful information can be found from the table. Begin by working out the totals for each row and column.

Find also the 'grand total' by adding together the totals in **either** the rows **or** columns.

	Boys	Girls	Total
Can read	464	682	1146
Cannot read	317	388	705
Total	781	1070	1851

↑
'Grand total'

Out of 781 boys, 464 can read.

Percentage of boys who can read $= \dfrac{464}{781} \times 100\%$

$= 59 \cdot 4\%$ (1 dp)

Similarly, the percentage of girls who can read

$$= \dfrac{682}{1070} \times 100\%$$

$$= 63 \cdot 7\% \text{ (1 dp)}$$

So a slightly higher percentage of the girls can read.

Exercise 2 ⓒ

Give percentages correct to one decimal place.

1 Here are ten shapes.

Copy and complete this two-way table to show the different shapes.

	Shaded	Unshaded
Squares		
Triangles		

2 The students in a class were asked to name their favourite sport. Here are the results.

Boy	Football	Boy	Swimming	Boy	Football
Girl	Hockey	Boy	Football	Girl	Football
Girl	Football	Girl	Hockey	Girl	Swimming
Boy	Hockey	Girl	Swimming	Boy	Football
Boy	Football	Girl	Hockey	Boy	Hockey

  a Record the results in a two-way table.

  b How many boys were in the class?

  c What percentage of the boys chose hockey?

3 This incomplete two-way table shows details of seven year-olds who can/cannot ride a bicycle without stabilisers.

	Girls	Boys	Total
Can cycle	95		215
Cannot cycle		82	
Total			476

  a Copy the table and work out the missing numbers.
  b What percentage of the girls can cycle?
  c What percentage of the boys can cycle?

4 Mrs Kotecha collected these data to help assess the risk of various drivers who apply for car insurance with her company. She needs to know if men drivers are more or less likely than women drivers to be involved in motor accidents.

	Men	Women	Total
Had accident	75	88	
Had no accident	507	820	
Total			

  a Copy the table and work out the totals for the rows and columns.
  b What percentage of the men had accidents?
  c What percentage of the women had accidents?
  d What conclusions, if any, can you draw?

8.3 Using a computer

8.3.1 Using a spreadsheet on a computer

This section is written for use with Microsoft Excel. Other spreadsheet programs work in a similar way.

Select Microsoft Excel from the desk top.

A spreadsheet appears on your screen as a grid with rows numbered 1, 2, 3, 4, ... and the columns lettered A, B, C, D, ...
The result should be a window like this one.

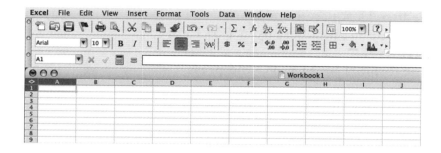

Cell   The spaces on the spreadsheet are called cells. Individual cells are referred to as A1, B3, F9, like grid references. Cells may contain **labels**, **values** or **formulae**. The current cell has a black border.

Label   Any words, headings or messages used to help the layout and organisation of the spreadsheet.

Value   A number placed in a cell. It may be used as input to a calculation.

Tasks 1, 2 and 3 are written for you to help you become familiar with how the main functions of a spreadsheet program work. Afterwards there are sections on different topics where spreadsheets can be used.

Task 1   To generate the whole numbers from 1 to 10 in column A.
  a In cell A1 type '1' and press *Return*. This will automatically take you to the cell below. (Note that you must use the *Return* button and not the arrow keys to move down the column.)
  b In cell A2 type the formula '= A1 + 1' and press *Return*. (Note that the = sign is needed before any formula.)
  c Now copy the formula in A2 down column A as far as A10. Click on A2 again and put the arrow in the bottom right corner of cell A2 (a + sign will appear) and drag down to A10.

Task 2   To generate the odd numbers in column B.
  a In B1 type '1' (press *Return*).
  b In B2 type the formula '= B1 + 2' (press *Return*).
  c Click in B2 and copy the formula down column B as far as B10.

Task 3 To generate the first 15 square numbers.

    a As before generate the numbers from 1 to 15 in cells A1 to A15.

    b In B1 put the formula '= A1 $\star$ A1' and press *Return*.

    c Click in B1 and copy the formula down as far as B15.

8.3.2 Pie charts and bar charts using a spreadsheet on a computer

EXAMPLE

Display the data about the activities in one day.

Enter the headings: *Sleep* in A1, *School* in B1 etc.
(Use the *Tab* key to move across the page.)

Enter the data: 8 in A2, 7 in B2 etc.

	A	B	C	D	E	F	G	H	I
1	Sleep	School	TV	Eating	Homework	Other			
2	8	7	2	1	2	5			
3									
4									
5									
6									
7									

Now highlight all the cells from A1 to F2. (Click on A1 and drag across to F2.)

Click on the ![chart icon] chart wizard on the toolbar.

Select 'pie' and then choose one of the examples displayed. Follow the on-screen prompts.

Alternatively, for a bar chart, select 'charts' after clicking on the chart wizard. Proceed as above.

You will be able to display your charts with various '3D' effects, possibly in colour. This approach is recommended when you are presenting data that you have collected as part of an investigation.

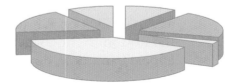

8.3.3   Scatter graphs on a computer

Plot a scatter graph showing the marks of
10 students in Maths and Science.

Enter the headings: *Maths* in A1, *Science* in B1.
Enter the data as shown.

Now highlight all the cells from A2 to B11.
(Click on A1 and drag across and down
to B11.)

	A	B
1	Maths	Science
2	23	30
3	45	41
4	73	67
5	35	74
6	67	77
7	44	50
8	32	41
9	66	55
10	84	70
11	36	32

Click on the [chart wizard icon] chart wizard on the
toolbar.

Select XY (Scatter) and select the picture
which looks like a scatter graph.

Follow the on-screen prompts.

On 'Titles' enter:
Chart title: Maths/Science results
Value (X) axis: Maths
Value (Y) axis: Science

Experiment with 'Axes', 'Gridlines', 'Legend' and 'Data Labels'.

Task   Enter the data on a spreadsheet and print a scatter graph.
What does each scatter graph show?

1

Height (cm)	Armspan (cm)
162	160
155	151
158	157
142	144
146	148
165	163
171	167
148	150
150	147

2

Temperature (°C)	Sales
23	7
18	14
7	23
20	9
4	30
12	19
15	15
18	15
10	20

8.4 Collecting data

8.4.1 The data handling cycle

1 Specifying the problem and planning	Start with a problem or a question you wish to answer, for example • Do most children go to the secondary school of their choice? • Do adults watch more television than children?
2 Collecting data	• You have to decide what data (information) you need to collect to answer the question. • You may decide to collect the data yourself. When you collect data by a survey or by experiment the data are called **primary data**. • If you look up published data that someone else has collected in a book or on the Internet the data are called **secondary data**.
3 Processing and presenting the data	• You usually have to **process** the data by, for example, calculating averages or frequencies so that you can **represent** the data in a pictorial form such as a pie chart or a frequency diagram.
4 Interpreting and discussing results	• Finally you need to **interpret** the results and draw conclusions so that you can answer your original question. • You need to look for patterns in the data and for any possible exceptions. • Sometimes the results you find suggest that you need to modify the way the survey was conducted.

8.4.2 Collecting data

Simple data collection sheet

A **tally chart** is often a good way to get started. Here is a tally chart for the responses to a question about the amount of homework set to students in a school.

Response	Tally	Frequency
Not enough	⌶⌶⌶ ||||	9
About right	⌶⌶⌶ ⌶⌶⌶ ⌶⌶⌶ ⌶⌶⌶ ||	22
Too much	⌶⌶⌶ ⌶⌶⌶ |||	13
Don't know	|||	3

> You could represent these data in a bar chart or pie chart.

● Here is another question which you can answer using statistical methods.

'Do different newspapers use words of different lengths or sentences of different lengths?'

You could conduct an experiment by choosing a similar page from different newspapers.

Number of words in a sentence	1–5	6–10	11–15	16–20	21+
Guardian					
Mirror					
Mail					

Questionnaires

● Surveys are made to find the popularity of various TV programmes. Advertisers are prepared to pay a large sum for a 30-second advertisement in a programme with an audience of 10 million people.

● Supermarkets use questionnaires to discover what things are most important to their customers. They might want to find out how people feel about ease of car parking, price of food, quality of food, length of time queueing to pay, etc.

- Most surveys are conducted using questionnaires. It is very important to design the questionnaire well so that:
  - **a** People will cooperate and will answer the questions honestly.
  - **b** The questions are not biased.
  - **c** The answers to the questions can be analysed and presented for ease of understanding.

Checklist

A Provide an introduction to the sheet so that the person you are asking knows the purpose of the questionnaire.

'Proposed new traffic lights'

B Make the questions easy to understand and specific to answer.
Do **not** ask vague questions like this.
The answers could be:

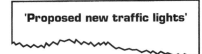

1 Did you see much of the Olympics on TV?

'Yes, a lot'
'Not much'
'Only the best bits'
'Once or twice a day'
You will find it hard to analyse this sort of data.

A **better** question is:

1 'How much of the Olympic coverage did you watch?' Tick one box			
Not at all	Up to 1 hour per day	1 to 2 hours per day	More than 2 hours per day
☐	☐	☐	☐

C Make sure that the questions are not **leading** questions. Remember that the survey is to find out opinions of other people, not to support your own.
Do **not** ask:

'Do you agree that BBC has the best sports coverage?'

A better question is:

'Which of these channels has the best sports coverage?'

BBC	ITV	Channel 4	Satellite TV
☐	☐	☐	☐

You might ask for one tick or possibly numbers 1, 2, 3, 4 to show an order of preference.

D If you are going to ask sensitive questions (about age or income, for example), design the question with care so as not to offend or embarrass.

Do **not** ask:

'How old are you?'

or 'Give your date of birth'

A better question is:

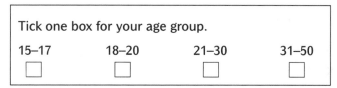

Tick one box for your age group.

15–17	18–20	21–30	31–50
☐	☐	☐	☐

E Do not ask more questions than necessary and put the easy questions first.

Exercise 3 ⓒ

In questions **1** to **7** explain why the question is not suitable for a questionnaire. Write an improved question in each case.

● Write some questions with 'yes/no' answers and some questions which involve multiple responses.
● Remember to word your questions simply.

1 Which sort of holiday do you like best?

2 What do you think of the new head teacher?

3 For how long do you watch television each day?

2–3 hours	3–4 hours	5–6 hours

4 How much would you pay to use the new car park?

☐ less than £1   ☐ more than £2·50

5 Do you agree that English and Maths are the most important subjects at school?

6 Do you or your parents often hire DVDs from a shop?

7 Do you agree that we get too much homework?

8 Some students designed a questionnaire to find out peoples' views about television. Comment on the questionnaire below and write an improved version.

Name _____   Sex   M/F

Age   _____

1 How much television do you watch?

☐   ☐   ☐

not much   quite a lot   a lot

2 What is your favourite programme on TV?

3 Do you agree that there should be more stations like MTV?

☐ agree   ☐ disagree

4 Do you like nature programmes?

☐   ☐   ☐   ☐

No   Not really   Sometimes   I love them

---

9 Write a suitable question to find out what **type** of TV programme people of your age watch most. For example: comedy, romance, sport etc.

10 Here is another style of question which can be useful.

Which of the following statements best describes your attitude to using a computer?

A I like using them for all sorts of things.
B I use them when I have to.
C I hate them.

Please circle: **A**   **B**   **C**

Write a similar style question about peoples' attitude to any topic of your own choice.

8.4.3 Bias

> ● A sample of data is biased when it does not give a true
> representation of the main population.

Bias can come from a variety of sources including non-random
sampling. For example, for data collected by a street survey, the
time of day and location may well mean that there is a sector or
sectors of the population who are not questioned.

A common cause of bias occurs when the questions asked in
the survey are not clear or are leading questions.

Exercise 4Ⓔ

In questions **1** to **4** decide whether the method of choosing
people to answer questions is satisfactory or not. Consider
whether or not the sample suggested might be **biased** in
some way. Where necessary, suggest a better way of finding
a sample.

1 A teacher, with responsibility for school meals,
   wants to hear students' opinions on the meals
   currently provided. She waits next to the dinner
   queue and questions the first 50 students as
   they pass.

2 An opinion pollster wants to canvass opinion about our
   European neighbours. He questions drivers as they are
   waiting to board their ferry at Dover.

3 A journalist wants to know the views of local people about
   a new one-way system in the town centre. She takes the
   electoral roll for the town and selects a random sample of
   200 people.

4 A pollster working for the BBC wants to know how many
   people are watching a new series. She questions 200 people as
   they are leaving a shop between 10:00 and 12:00 one
   Thursday.

8.5 GCSE coursework

8.5.1 Choosing your project

- Almost certainly the best idea for a statistics project or coursework will be an idea which **you** think of because **you** want to know the answer. You have to decide what data to collect and how you are going to collect them. You need to make a plan and then design an experiment or survey. You must decide what primary or secondary data will be suitable.

- At this point stop to consider two things.
  a Will the survey help to answer the question?
  b How many people do you need for the survey?
  Both of these questions can be answered by first doing a **pilot survey**, that is, a quick mini-survey of 5–10 people. Some of the questions may need changing straightaway.

- Having collected the data, you need to **process** them and then represent them, usually in pictorial form. You may need to calculate averages or percentages and then draw pie charts, scatter graphs, stem-and-leaf diagrams and so on.

  Do not be afraid to use colours!

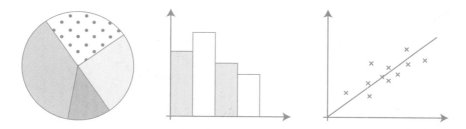

- Finally, interpret the results of your survey, making sure that you refer back to the initial question or problem. Write a **conclusion** in which you summarise the main findings of your work. Make sure that your conclusions are justified by the evidence in your report.
  You might suggest ways in which your work could be **extended** if more time was available.

- Remember: The best coursework will be on a topic which **you** find interesting. If you cannot think of a suitable idea you can use one from this list that some of the author's students have enjoyed investigating.

a 'Most people choose to shop in a supermarket where it is easy to park a car.'
b 'Children of school age watch more television than their parents.'
c 'Boys prefer action films and girls prefer some sort of story.'
d 'Most people who smoke have made at least one serious attempt to give it up.'
e 'Young people are more superstitious than old people.'
f 'Given a free choice, most girls would hardly ever choose to wear a dress in preference to something else.'
g 'More babies are born in the winter than in the summer.'
h 'Most cars these days use unleaded petrol.'
i 'The school day should start to 08:00 and end at 14:00.'
j 'Older people sleep less than younger people.'

Here is a project about people in a factory which provides practice in doing GCSE statistics coursework.

8.5.2 Statistics coursework practice

A sample of 50 employees from a factory were asked to anonymously complete a questionnaire which asked for their gender, their age and how much they were currently earning.

The results are summarised in the three tables on page 385.
Table 1 has been sorted by gender (female and males).
Table 2 has been sorted by age.
Table 3 has been sorted by wage.

Exercise 5Ⓔ

1 a Use **Table 1** to draw a pie chart showing the number of employees from the sample in each of these four categories
   i   females under 30
   ii   males under 30
   iii females 30 or over
   iv   males 30 and over

   b Summarise what the pie chart shows.

   c If the sample is representative of the whole factory population, how many people in each of the categories above would there be if the factory employed 1500 people?

2 a On two **separate** diagrams draw scatter graphs of wage against age for employees under 30 and employees over 30 (put age on the horizontal axis).
   Use **Table 2** to help you with this.

   b Comment on any correlation.

> There are 360° for 50 employees so for 1 employee the angle on the pie chart is 7·2°.

3 a With the help of **Table 3**, copy and complete this
frequency chart for wages. Note that a wage of exactly
£20 000 should be put in the 20 000–25 000 group.

Wage, £	Mid-value, x	Frequency, f	fx
15 000–20 000	17 500	4	70 000
20 000–25 000	22 500		
25 000–30 000	27 500		
30 000–35 000	32 500		
35 000–40 000	37 500	2	75 000
40 000–45 000	42 500		
125 000–130 000	127 500		
Total		50	A

b Calculate the mean wage from the frequency chart.
(mean = A ÷ 50)
c Calculate the mean wage from the raw data (the
50 individual wages in this table).
d Which of these two values for the mean is more reliable?
Why is that?

4 Use the frequency chart from question **3** to help you draw
a bar chart for these data.

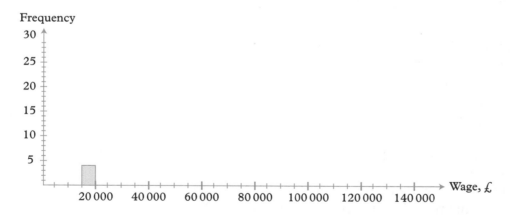

5 a Calculate the median wage with the help of **Table 3**.
b What is the range of the wages?
c Which do you think is a better estimate of the average
wage for the factory, the mean or the median?
d Explain your answer to part **c**.

Table 1

	Gender	Age	Wage, £
1	F	17	18500
2	F	18	19000
3	F	18	19500
4	F	20	21000
5	F	22	22000
6	F	23	21500
7	F	26	23000
8	F	28	23500
9	F	30	24000
10	F	31	25500
11	F	31	25000
12	F	33	27500
13	F	37	25000
14	F	42	29000
15	F	46	27000
16	F	48	29500
17	F	52	28500
18	F	53	28000
19	M	17	18000
20	M	18	20000
21	M	19	21000
22	M	19	20500
23	M	22	22000
24	M	24	23500
25	M	26	24000
26	M	27	25000
27	M	29	25000
28	M	30	27500
29	M	33	28000
30	M	35	28000
31	M	36	29500
32	M	37	42000
33	M	39	28000
34	M	40	27500
35	M	41	125000
36	M	43	28500
37	M	46	40000
38	M	46	27000
39	M	47	37000
40	M	48	27500
41	M	49	28000
42	M	50	27000
43	M	50	29000
44	M	52	28000
45	M	54	27500
46	M	58	36000
47	M	57	28000
48	M	59	29000
49	M	60	28500
50	M	62	29000

Table 2

	Gender	Age	Wage, £
1	M	17	18000
2	F	17	18500
3	F	18	19000
4	F	18	19500
5	M	18	20000
6	M	19	20500
7	M	19	21000
8	F	20	21000
9	F	22	22000
10	M	22	22000
11	F	23	21500
12	M	24	23500
13	F	26	23000
14	M	26	24000
15	M	27	25000
16	F	28	23500
17	M	29	25000
18	F	30	24000
19	M	30	27500
20	F	31	25000
21	F	31	25500
22	F	33	27500
23	M	33	28000
24	M	35	28000
25	M	36	29500
26	F	37	25000
27	M	37	42000
28	M	39	28000
29	M	40	27500
30	M	41	125000
31	F	42	29000
32	M	43	28500
33	F	46	27000
34	M	46	27000
35	M	46	40000
36	M	47	37000
37	M	48	27500
38	F	48	29500
39	M	49	28000
40	M	50	27000
41	M	50	29000
42	M	52	28000
43	F	52	28500
44	F	53	28000
45	M	54	27500
46	M	57	28000
47	M	58	36000
48	M	59	29000
49	M	60	28500
50	M	62	29000

Table 3

	Gender	Age	Wage, £
1	M	17	18000
2	F	17	18500
3	F	18	19000
4	F	18	19500
5	M	18	20000
6	M	19	20500
7	M	19	21000
8	F	20	21000
9	F	23	21500
10	F	22	22000
11	M	22	22000
12	F	26	23000
13	M	24	23500
14	F	28	23500
15	M	26	24000
16	F	30	24000
17	M	27	25000
18	M	29	25000
19	F	31	25000
20	F	37	25000
21	F	31	25500
22	F	46	27000
23	M	46	27000
24	M	50	27000
25	M	30	27500
26	F	33	27500
27	M	40	27500
28	M	48	27500
29	M	54	27500
30	M	33	28000
31	M	35	28000
32	M	39	28000
33	M	49	28000
34	M	52	28000
35	F	53	28000
36	M	57	28000
37	M	43	28500
38	F	52	28500
39	M	60	28500
40	F	42	29000
41	M	50	29000
42	M	59	29000
43	M	62	29000
44	M	36	29500
45	F	48	29500
46	M	58	36000
47	M	47	37000
48	M	46	40000
49	M	37	42000
50	M	41	125000

Test yourself

1 The table shows the heights of a group of Year 7 students together with the height of each of their fathers. All measurements are in centimetres.

Student	A	B	C	D	E	F	G	H	I
Height of student	138	141	145	148	149	154	155	161	162
Height of father	151	155	153	170	161	176	185	186	192

a Copy this scatter graph to show the information. The first three points are plotted for you.

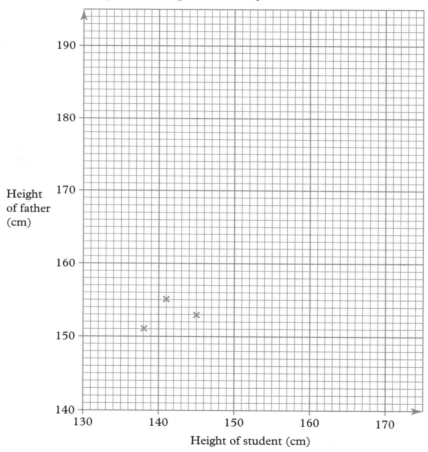

b Describe the correlation between the two sets of heights.
c Draw a line of best fit on your scatter diagram.
d A new student joins the group. His height is 151 cm. Use your line of best fit to estimate the height of his father.

(OCR, 2004)

2 The table lists the masses of twelve books and the number of pages in each one.

Number of pages	80	155	100	125	145	90	140	160	135	100	115	165
Mass (g)	160	330	200	260	320	180	290	330	260	180	230	350

a Draw a scatter graph to show the information in the table.

b Describe the correlation between the number of pages in these books and their masses.

c Draw a line of best fit on your scatter graph.

d Use your line of best fit to estimate
   i the number of pages in a book of mass 280 g
   ii the mass, in grams, of a book with 110 pages.

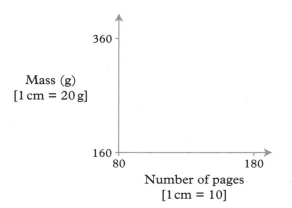

(Edexcel, 2003)

3 There is to be a survey about the need for a new leisure centre in a town.

a State why the following question is **not** suitable for use in a questionnaire.
'Do you agree that tennis courts are more important than squash courts?'

b Rewrite the question in a suitable form.

(AQA, 2003)

4 The manager of a school canteen has made some changes. She wants to find out what students think of these changes. She uses this question on a questionnaire.

'How much money do you normally spend in the canteen?'

A lot   Not much

Design a better question for the canteen manager to use. You should include some response boxes.

(Edexcel, 2004)

5 Howard is doing a survey about shops opening on a Sunday. Two of his questions are:

Question 1 How old are you?

Question 2 Everybody deserves to have a day off work to spend relaxing with their families, and going out. So shops shouldn't open on a Sunday, don't you agree?

a Explain why each question is not a good one for this questionnaire.

b Write a suitable question for Howard to ask to replace his question 2.

(AQA, 2004)

6 Kurt has to undertake a survey to find out the most popular drink in Wales. He carries out his survey by asking students at a school disco to answer this question.

Which is your favourite drink?
Tick the appropriate box.

Tea ☐   Coffee ☐   Cola ☐

a State **one** reason why the design of this question is unsuitable for his survey.

b State **two** reasons why his survey is likely to be biased.

(WJEC)

7 Sixty British students each visited one foreign country last week.

The two-way table shows some information about these students.

	France	Germany	Spain	Total
Female			9	34
Male	15			
Total		25	18	60

a Copy and complete the two-way table.

(AQA, 2004)

8 Each student at Redmond School studies one foreign language.
Students can choose from French, German and Spanish.

The headteacher wants to show how many boys and how many girls study each language.
a Draw a two-way table the headteacher could use to show this information.

23 boys study French.
b Write the number 23 in the correct place in your two-way table.

(Edexcel, 2003)

9 Jane conducts a survey of the favourite colours of the students in her class.
She records the results.

Male	Red	Female	Yellow
Male	Yellow	Female	Red
Male	Red	Female	Green
Female	Green	Female	Green
Female	Red	Male	Red
Male	Green	Male	Yellow
Male	Green		

Record the results in a two-way table.

(AQA, 2004)

9 Number 3

In this unit you will:
- revise how to convert between fractions, decimals and percentages
- learn how to calculate average speed
- learn about standard form
- revise metric and imperial units
- revise rounding and estimation.

9.1 Fractions, decimals and percentages

Percentages are simply a convenient way of expressing fractions or decimals. You should be able to convert readily from one form to another.

- To change a fraction to a decimal divide out the fraction.
- To change a fraction or a decimal to a percentage, multiply by 100%.

EXAMPLE

Change

a $\frac{7}{8}$ to a decimal

b 0.35 to a fraction

c $\frac{3}{8}$ to a percentage

d 0.85 to a percentage

e $\frac{5}{6}$ to a decimal.

....

a Divide 8 into 7 $\quad 8\overline{)7.000}^{0.875}$

b $0.35 = \frac{35}{100}$
$\qquad = \frac{7}{20}$

c $\frac{3}{8} = \frac{3}{8} \times 100\%$
$\qquad = 37\frac{1}{2}\%$

d $0.85 = 0.85 \times 100\%$
$\qquad = 85\%$

e $6\overline{)5.000}^{0.8333\ldots}$

> This is a **recurring** decimal. You write $\frac{5}{6} = 0.8\dot{3}$. Similarly, $0.525252\ldots = 0.\dot{5}\dot{2}$

Note that **all** recurring decimals are exact fractions but that not all exact fractions are recurring decimals.

Exercise 1 ⓒ

1 Change the fractions to decimals.

    a $\dfrac{1}{4}$ **b** $\dfrac{2}{5}$ **c** $\dfrac{3}{8}$ **d** $\dfrac{5}{12}$ **e** $\dfrac{1}{6}$ **f** $\dfrac{2}{7}$

2 Change the decimals to fractions and simplify.

    a 0·2   **b** 0·45   **c** 0·36   **d** 0·125   **e** 1·05   **f** 0·007

3 Change to percentages.

    a $\dfrac{1}{4}$ **b** $\dfrac{1}{10}$ **c** 0·72   **d** 0·075   **e** 0·02   **f** $\dfrac{1}{3}$

4 In July 2006, 360 000 people visited Bali for their holiday.

    a One-eighth of the people were American. Find the number of American visitors.

    b 11% of the people were French. How many people was that?

    c There were 12 000 people from Japan. What fraction of the total were from Japan?

5 Copy and complete the table.

	Fraction	Decimal	Percentage
a	$\dfrac{1}{4}$		
b		0·2	
c			80%
d	$\dfrac{1}{100}$		
e			30%
f	$\dfrac{1}{3}$		

6 Here are some fractions.

    $\dfrac{4}{10}$ $\dfrac{11}{33}$ $\dfrac{1}{5}$ $\dfrac{7}{12}$

    Write one that is

    a equal to 0·2   **b** equal to 40%

    c equal to $\dfrac{1}{3}$ **d** greater than $\dfrac{1}{2}$.

7 Max wants 3 bottles of Coke, which normally costs 90p per bottle. Which of the three offers shown is the cheapest for 3 bottles?

A   **B**   **C**

A *3* for the price of *2* !!!

B *30% off* *marked price*

C BUY ONE get the 2nd HALF PRICE !

8 Two shops had sale offers on an article which previously cost £69. One shop had '$\frac{1}{3}$ off' and the other had '70% of old price'. Which shop had the lower price?

9 Shareholders in a company can opt for either '$\frac{1}{6}$ of £5000' or '15% of £5000'. Which is the greater amount?

10 A photocopier increases the sides of a square in the ratio $4:5$. By what percentage are the sides increased?

*11 In an alloy the ratio of copper to iron to lead is $5:7:3$. What percentage of the alloy is lead?

9.2 Compound measures

9.2.1 Speed

When a train moves at a constant speed of 20 metres per second, it moves a distance of 20 metres in 1 second. In 2 seconds the train moves 40 metres, and so on.

- The distance moved is equal to the speed multiplied by the time taken.

$$\text{Distance} = \text{Speed} \times \text{Time} \qquad \text{Speed} = \frac{\text{Distance}}{\text{Time}} \qquad \text{Time} = \frac{\text{Distance}}{\text{Speed}}$$

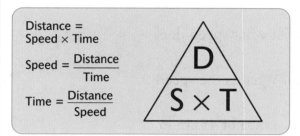

Distance =
Speed × Time

Speed = $\dfrac{\text{Distance}}{\text{Time}}$

Time = $\dfrac{\text{Distance}}{\text{Speed}}$

1 A bird takes 20 s to fly a distance of 100 m. Calculate the average speed of the bird.

$\left(S = \dfrac{D}{T}\right)$ Average speed = $\dfrac{100}{20}$

= 5 m/s

2 A car travels a distance of 200 m at a speed of 25 m/s. How long does it take?

$\left(T = \dfrac{D}{S}\right)$

Time taken = $\dfrac{200}{25}$ = 8 seconds

3 A boat sails at a speed of 12 km/h for 2 days. How far does it travel?

2 days = 48 hours

(D = S × T)

Distance travelled = 12 × 48
= 576 km

Exercise 2 C

1 Find the distance travelled.
   a 55 mph for 2 hours
   b 17 m/s for 20 seconds
   c 63 km/h for 5 hours
   d 5 cm/day for 12 days.

2 Find the speed.
   a 98 miles in 7 hours
   b 364 km in 8 hours
   c 250 m in 10 seconds
   d 63 cm in 6 minutes.

3 A car travels at a constant speed of 40 mph for three hours. How far does it go?

4 An athlete runs at a steady speed of 5 m/s for 100 s. How far does he run?

5 How far will a train travel in 15 s if it is going at a steady speed of 20 m/s?

6 A ball travels for 30 s at a speed of 12 m/s. Find the distance it covers.

7 An aircraft flies at a speed of 800 km/h for $2\frac{1}{2}$ h. How far does it fly?

8 How far will a ship sail in half an hour if it is going at a steady speed of 24 km/h?

9 A car travelling at a steady speed takes 4 hours to travel 244 km. What is the speed of the car?

10 Joseph runs 750 m in a time of 100 s. At what speed does he run?

11 A train takes 6 hours to travel 498 km. What is the speed of the train?

12 After a meal an earthworm moves a distance of 45 cm in 90 s. At what speed does the worm move?

13 A plane flies 720 miles at a speed of 240 mph. How long does it take?

14 An octopus swims 18 km at a speed of 3 km/h. How long does it take?

15 A rocket is flying at a speed of 1000 km/h. How far does it go in 15 minutes?

16 Find the time taken.
   a 360 km at 20 km/h   **b** 56 miles at 8 mph
   c 200 m at 40 m/s   **d** 60 km at 120 km/h.

17 A car takes 15 minutes to travel 22 miles. Find the speed in mph.

18 An athlete runs at 9 km/h for 30 minutes. How far does he run?

9.2.2 Other compound measures

Exercise 3Ⓔ

In questions **1**, **2** and **3** use this formula

$$\text{Density} = \frac{\text{Mass}}{\text{Volume}} \quad \text{or} \quad \text{Mass} = \text{Density} \times \text{Volume}$$

1 Find the density of each object.

  a

  b

  c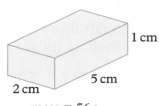

mass   = 150 g
volume = 100 cm³

mass   = 1800 g
volume = 200 cm³

1 cm
5 cm
2 cm
mass = 56 g

2 A steel ball has volume $1000 \, \text{cm}^3$. The density of steel is $8 \, \text{g/cm}^3$. Find the mass of the ball.

3 A gold ring has a volume of $3 \, \text{cm}^3$ and the density of gold is $12 \, \text{g/cm}^3$. Find the mass of the ring.

4 Silver plating costs £9 per cm^2. How much will it cost to plate this lid?

2 cm

5 cm

5 You can hire a satellite at £250 per second. How much will it cost to hire the satellite for one minute?

6 Telephone cable costs £3·50 per metre. Find the cost of 200 m of this cable.

7 A multi-millionaire earns £120 million in one year. On average how much did he earn per day? Give your answer to the nearest thousand pounds.

8 Rooney Farm has a rectangular field measuring 200 m by 500 m. The land can be sold at £3000 per hectare (1 hectare = $10\,000 \, \text{m}^2$). Find the price of the field.

*9 The town of Vernon has a population of 30 500 in an area of 33 square miles.
    a Work out the population density in people/square mile.
    b The area of the town is planned to increase to 47 square miles. Calculate the increased population of the town if the population density remains the same. Give your answer to a sensible degree of accuracy.

9.3 Standard form

9.3.1 Writing numbers in standard form

When dealing with either very large or very small numbers, it takes a long time to write them out in full in the normal way. It is better to use standard form, which uses powers of 10. Most calculators represent large and small numbers in this way.

A calculator may show

$2·3 \times 10^8$ as $\boxed{2.3 \quad ^{08}}$ or as $\boxed{2.3 \times 10 \quad ^{08}}$

> ● The number $a \times 10^n$ is in standard form when $1 \leqslant a < 10$ and n is a positive or negative integer.

> EXAMPLE
>
> Write these numbers in standard form.
> **a** 2000   **b** 150   **c** 0·0004
> ....
>
> **a** $2000 = 2 \times 1000 = 2 \times 10^3$
> **b** $150 = 1{\cdot}5 \times 100 = 1{\cdot}5 \times 10^2$
> **c** $0{\cdot}0004 = 4 \times \dfrac{1}{10\,000} = 4 \times 10^{-4}$

Exercise 4Ⓔ

Write these numbers in standard form.

1 4000	**2** 500	**3** 70 000
4 60	**5** 2400	**6** 380
7 46 000	**8** 46	**9** 900 000
10 2560	**11** 0·007	**12** 0·0004
13 0·0035	**14** 0·421	**15** 0·000 055
16 0·01	**17** 564 000	**18** 19 million

19 The population of China is estimated at 1 100 000 000.
Write this in standard form.

20 A hydrogen atom weighs
0·000 000 000 000 000 000 000 001 67 grams.
Write this mass in standard form.

21 The area of the surface of the earth
is about 510 000 000 km². Express
this in standard form.

22 A certain virus is 0·000 000 000 25 cm in diameter. Write
this in standard form.

23 Chemists use Avogadro's number, 602 300 000 000 000 000 000 000.
Express this number in standard form.

24 The speed of light is 300 000 km/s. Express this speed in
cm/s in standard form.

25 A very rich oil sheikh leaves his fortune of £3.6×10^8 to be divided between his 100 relatives.
How much does each relative receive? Give the answer in standard form.

9.3.2 Calculating with numbers in standard form

EXAMPLE

Work out $1500 \times 8\,000\,000$

$\begin{aligned}1500 \times 8\,000\,000 &= (1.5 \times 10^3) \times (8 \times 10^6)\\ &= 12 \times 10^9 \ (1.5 \times 8 = 12, \ 10^3 \times 10^6 = 10^9)\\ &= 1.2 \times 10^{10}\end{aligned}$

Note that you multiply the numbers and the powers of 10 separately.

Many calculators have an EXP button which is used for standard form.

a To enter 1.6×10^7 into the calculator:

press 1·6 EXP 7

Do **not** press the × button after 1·6 in part **a**.

b To enter 3.8×10^{-3}:

press 3·8 EXP − 3

c To calculate $(4.9 \times 10^{11}) \div (3.5 \times 10^{-4})$:

4·9 EXP 11 ÷ 3·5 EXP − 4 =

The answer is 1.4×10^{15}.

Exercise 5Ⓔ

In questions **1** to **22**, give the answer in standard form.

1 5000×3000	**2** $60\,000 \times 5000$	**3** $0.000\,07 \times 400$
4 $0.0007 \times 0.000\,01$	**5** $8000 \div 0.004$	**6** $(0.002)^2$
7 150×0.0006	**8** $0.000\,033 \div 500$	**9** $0.007 \div 20\,000$
10 $(0.0001)^4$	**11** $(2000)^3$	**12** $0.005\,92 \div 8000$

13 $(1.4 \times 10^7) \times (3.5 \times 10^4)$ **14** $(8.8 \times 10^{10}) \div (2 \times 10^{-2})$

15 $(1.2 \times 10^{11}) \div (8 \times 10^7)$ **16** $(4 \times 10^5) \times (5 \times 10^{11})$

17 $(2.1 \times 10^{-3}) \times (8 \times 10^{15})$ **18** $(8.5 \times 10^{14}) \div 2000$

19 $(3.3 \times 10^{12}) \times (3 \times 10^{-5})$ **20** $(2.5 \times 10^{-8})^2$

21 $(1.2 \times 10^5)^2 \div (5 \times 10^{-3})$ **22** $(6.2 \times 10^{-4}) \times (1.1 \times 10^{-3})$

23 A certain dinosaur laid its eggs 30 million years ago. How many days ago was that? Round off your answer to 2 significant figures.

24 A pile of ten thousand sheets of paper is 1·3 m high. How thick is each sheet of paper in metres?

25 These numbers are not in standard form. Write them in standard form.
$a = 512 \times 10^2$
$b = 0\cdot478 \times 10^6$
$c = 0\cdot0049 \times 10^7$

26 If the number $2\cdot74 \times 10^{15}$ is written out in full, how many zeros follow the 4?

27 If $m = 2 \times 10^3$ and $n = 4 \times 10^4$, find the value of

a $m + n$ **b** mn **c** $\dfrac{n}{m}$

28 If $x = 2 \times 10^5$ and $y = 5 \times 10^{-3}$, find the values of

a xy **b** $\dfrac{x}{y}$

*29 Oil flows through a pipe at a rate of 40 m³/s. How long will it take to fill a tank of volume $1\cdot2 \times 10^5$ m³?

$40\,\text{m}^3/\text{s}$

*30 Given that $L = 2\sqrt{\dfrac{a}{k}}$, find the value of L in standard form when $a = 4\cdot5 \times 10^{12}$ and $k = 5 \times 10^7$.

*31 A light year is the distance travelled by a beam of light in a year. Light travels at a speed of approximately 3×10^5 km/s.
a Work out the length of a light year in km.
b Light takes about 8 minutes to reach the earth from the sun. How far is the earth from the sun in km?

Use 'Distance = speed × time'.

9.4 Metric and imperial units

Years ago, measurements were made using parts of the body. The inch was measured using the thumb and the foot by using the foot. Even today many people, when asked their height or weight, will say '5 feet 3' or '9 stone' rather than '1 metre 60' or '58 kg'.

These are called imperial units.

9.4.1 Metric units

Length	Mass	Volume
10 mm = 1 cm	1000 g = 1 kg	1000 ml = 1 litre
100 cm = 1 m	1000 kg = 1 t	
1000 m = 1 km	(t for tonne)	

Exercise 6 C

Copy and complete.

1 4 m = __ cm
2 2·4 km = __ m
3 0·63 m = __ cm
4 25 cm = __ m
5 70 mm = __ cm
6 2 cm = __ mm
7 1·2 km = __ m
8 7 m = __ cm
9 0·5 km = __ m
10 815 cm = __ m
11 650 m = __ km
12 25 mm = __ cm
13 5 kg = __ g
14 4·2 kg = __ g
15 6·4 kg = __ g

16 3 kg = __ g
17 0·8 kg = __ g
18 400 g = __ kg
19 2 t = __ kg
20 250 g = __ kg
21 0·5 t = __ kg
22 0·62 t = __ kg
23 7 kg = __ g
24 1500 g = __ kg
25 8 litres = __ ml
26 2 litres = __ ml
27 1000 ml = __ litres
28 4·5 litres = __ ml
29 3·2 m = __ cm
30 55 mm = __ cm
31 1·4 kg = __ g
32 11 cm = __ mm

33 Write the most appropriate metric unit for measuring
    a the distance between Glasgow and Leeds   **b** the capacity of a wine bottle
    c the mass of raisins needed for a cake   **d** the diameter of a small drill
    e the mass of a car   **f** the area of a football pitch.

9.4.2 Imperial units

Length	Mass	Volume
Length	**Mass**	**Volume**
12 inches = 1 foot	16 ounces = 1 pound	8 pints = 1 gallon
3 feet = 1 yard	14 pounds = 1 stone	
1760 yards = 1 mile	2240 pounds = 1 ton	

Exercise 7 Ⓒ

1 How many inches are there in two feet?

2 How many ounces are there in a pound?

3 How many feet are there in ten yards?

4 How many pounds are there in two tons?

5 How many pints are there in six gallons?

6 How many yards are there in ten miles?

7 How many inches are there in one yard?

8 How many pounds are there in five stones?

9 How many pints are there in half a gallon?

10 How many yards are there in half a mile?

In questions **11** to **20** copy each statement and fill in the missing numbers.

11 9 feet = __ yards

12 16 pints = __ gallons

13 2 miles = __ yards

14 5 pounds = __ ounces

15 10 stones = __ pounds

16 4 yards = __ feet

17 4 feet = __ inches

18 10 tons = __ pounds

19 1 mile = __ feet

20 6 feet = __ yards

9.4.3 Changing units

Although the metric system is replacing the imperial system you still need to be able to convert from one set of units to the other.

Try to remember these approximate conversions.

1 inch ≈ 2·5 cm	1 gallon ≈ 5 litres
1 kg ≈ 2 pounds	1 km ≈ $\frac{5}{8}$ mile
1 ounce ≈ 30 g	(or 8 km ≈ 5 miles)

EXAMPLE

a Change 16 km into miles.   **b** Change 2 feet into cm.

$$1 \text{ km} \approx \frac{5}{8} \text{ mile}$$

$$\text{so } 16 \text{ km} \approx \frac{5}{8} \times 16$$

$$16 \text{ km} \approx 10 \text{ miles}$$

2 feet ≈ 24 inches

1 inch ≈ 2·5 cm

so 2 feet ≈ 2·5 × 24 cm
$$\approx 60 \text{ cm}$$

Exercise 8Ⓔ

Copy each statement and fill in the missing numbers.

1 10 inches = __ cm

2 10 gallons = __ litres

3 3 kg = __ pounds

4 4 kg = __ pounds

5 8 km = __ miles

6 2 pounds = __ kg

7 32 km = __ miles

8 4 inches = __ cm

9 8 pounds = __ kg

10 1 gallon = __ litres

11 40 km = __ miles

12 100 litres = __ gallons

13 3 kg = __ pounds

14 5 miles = __ km

15 400 litres = __ gallons

16 60 g = __ ounces

17 10 oz = __ g

18 5 litres = __ gallons

19 20 kg = __ pounds

20 25 cm = __ inches

21 1 foot = __ cm

22 A car handbook calls for the oil to be changed every 8000 km. How many miles is that?

23 On an Italian road the speed limit is 80 km/h. Convert this into a speed in mph.

24 Tomatoes are sold in Supermarket A at 85p per kilo and in Supermarket B at 30p per pound. Which supermarket has the lower price?

25 Here is the recipe for a pie.
Write the recipe with the correct metric quantities.

In questions **26** to **30** copy each sentence and choose the number which is the best estimate.

26 A one pound coin has a mass of about [1 g, 10 g, 1 kg].

27 The width of the classroom is about [100 inches, 7 m, 50 m].

28 A can of cola contains about [500 ml, $\frac{1}{2}$ gallon].

29 The distance from London to Birmingham is about [20 miles, 20 km, 100 miles].

30 The thickness of a one pound coin is about [3 mm, 6 mm, $\frac{1}{4}$ inch].

31 Here are scales for changing: **A** kilograms and pounds, **B** litres and gallons. In this question give your answers to the **nearest whole number**.
  a About how many kilograms are there in 6 pounds?
  b About how many litres are there in 3·3 gallons?
  c About how many pounds are there in 1·4 kilograms?

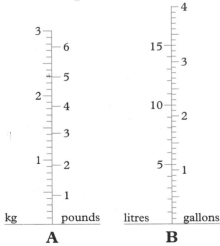

*32 A fitter is doing a job which requires a 3 mm drill. He has no metric drills but he does have the following imperial sizes (in inches): $\dfrac{1}{16}, \dfrac{1}{8}, \dfrac{3}{16}$. Which of these drills is the nearest in size to 3 mm?

9.5 Measurement is approximate

● Suppose you measure a line as 14 mm, correct to the nearest mm. The actual length could be from 13·5 mm to 14·5 m

> In these cases the measurement expressed to a given unit is in **possible error** of **half a unit**.

● If you measure the length of some fabric for a dress you might say the length is 145 cm **to the nearest** cm. The actual length could be anything from 144·5 cm to 145·49999... cm. This is effectively 145·5 and you could use this figure.

● Similarly if you say you weigh 57 kg to the nearest kg you could actually weigh anything from 56·5 kg to 57·5 kg. If your brother was weighed on more sensitive scales and the result was 57·2 kg, his actual weight could be from 57·15 kg to 57·25 kg.

> The 'unit' is 1 so half a 'unit' is 0·5.

> The 'unit' is 0·1 so half a 'unit' is 0·05.

● Here are some further examples.

	Lower limit	Upper limit
1 The diameter of a CD is 12 cm to the nearest cm.	11·5 cm	12·5 cm
2 The mass of a coin is 6·2 g to the nearest 0·1 g.	6·15 g	6·25 g
3 The length of a fence is 330 m to the nearest 10 m.	325 m	335 m

Exercise 9Ⓔ

1 In a DIY store the height of a door is given as 195 cm to the nearest cm. Write the greatest possible height of the door.

2 A vet weighs a sick goat at 37 kg to the nearest kg. What is the least possible weight of the goat?

3 A surveyor using a laser beam device can measure distances to the nearest 0·1 m. What is the least possible length of a warehouse which he measures at 95·6 m?

4 A cook's weighing scales weigh to the nearest 0·1 kg. What is the greatest possible weight of a chicken which she weighs at 3·2 kg?

5 In the county sports Jill was timed at 28·6 s for the 200 m. What is the greatest time she could have taken?

6 Copy and complete the table using this information.

	Lower limit	Upper limit

  a Temperature in a fridge = 2 °C, to nearest degree
  b Mass of an acorn = 2·3 g, to 1 dp
  c Length of telephone cable = 64 m, to nearest m
  d Time taken to run 100 m = 13·6 s, to nearest 0·1 s.

7 The length of a telephone is measured as 193 mm, to the nearest mm. The length lies between

> Choose A, B or C.

A	B	C
192 and 194 mm	192·5 and 193·5 mm	188 and 198 mm

8 The weight of a labrador is 35 kg, to the nearest kg. The weight lies between

A	B	C
30 and 40 kg	34 and 36 kg	34·5 and 35·5 kg

9 Liz and Julie each measure a different worm and they both say that their worm is 11 cm long to the nearest cm.
  a Does this mean that both worms are the same length?
  b If not, what is the maximum possible difference in the length of the two worms?

10 A card measuring 11·2 cm (to the nearest 0·1 cm) is
to be posted in an envelope which is 12 cm (to the
nearest cm).

11·2 cm   12 cm

Can you guarantee that the card will fit inside the envelope?
Explain your answer.

In questions **11** to **24** you are given a measurement. Write the
upper and lower bounds of the number. For example if you
are given a length as 13 cm you can write 'length is between
12·5 cm and 13·5 cm'.

11 mass = 17 kg   12 $d = 256$ km
13 length = 2·4 m   14 $m = 0·34$ grams
15 $v = 2·04$ m/s   16 $x = 8$ cm
17 $T = 81·4\,°C$ 18 $M = 0·3$ kg
19 $d = 4·80$ cm   20 $y = 0·07$ m
21 mass = 0·7 tonne   22 $t = 615$ seconds
23 $d = 7·13$ m   24 $n = 52\,000$
                                    (nearest thousand)

9.6 Solving numerical problems 3

Making a profit

EXAMPLE

A shopkeeper buys potatoes at a wholesale price of £180 per tonne and
sells them at a retail price of 22p per kg.
How much profit does he make on one kilogram of potatoes?

He pays £180 for 1000 kg
of potatoes.

So he pays £[180 ÷ 1000] for 1 kg
of potatoes.
So he pays 18p for 1 kg.

He sells at 22p per kg.
His profit = 4p per kg.

Exercise 10 C

Find the profit in each case.

	Commodity	Retail price	Wholesale price	Profit
1	cans of drink	15p each	£11 per 100	profit per can?
2	rulers	24p each	£130 per 1000	profit per ruler?
3	birthday cards	22p each	£13 per 100	profit per card?
4	soup	27p per can	£8·50 for 50 cans	profit per can?
5	newspapers	22p each	£36 for 200	profit per paper?
6	box of matches	37p each	£15·20 for 80	profit per box?
7	potatoes	22p per kg	£160 per tonne	profit per kg?
8	carrots	38p per kg	£250 per tonne	profit per kg?
9	T-shirts	£4·95 each	£38·40 per dozen	profit per T-shirt?
10	eggs	96p per dozen	£50 per 1000	profit per dozen?
11	oranges	5 for 30p	£14 for 400	profit per orange?
12	car tyres	£19·50 each	£2450 for 200	profit per tyre?
13	wine	55p for 100 ml	£40 for 10 litres	profit per 100 ml?
14	sand	16p per kg	£110 per tonne	profit per kg?
15	wire	23p per m	£700 for 10 km	profit per m?
16	cheese	£2·64 per kg	£87·50 for 50 kg	profit per kg?
17	copper tube	46p per m	£160 for 500 m	profit per m?
18	apples	9p each	£10·08 per gross	profit per apple?
19	carpet	£6·80 per m²	£1600 for 500 m²	profit per m²?
20	tin of soup	33p per tin	£72 for 400 tins	profit per tin?

Exercise 11 C (Cross squares)

Each empty square contains either a number or a mathematical symbol (+, −, ×, ÷). Copy each square and fill in the details.

1

5			→	60
×		÷		
	24	→	44	
↓		↓		
	×	½	→	50

2

	×	6	→	42
÷		÷		
14	−		→	
↓		↓		
		2	→	1

3

	×	2	→	38
−		÷		
			→	48
↓		↓		
7	−		→	6½

4

17	×		→	170
−		÷		
	÷		→	
↓		↓		
8	−	0·1	→	

5

0·3	×	20	→	
		−		
11	÷		→	
↓		↓		
11·3	−		→	2·3

6

	×	50	→	25
−		÷		
		½	→	0·6
↓		↓		
0·4	×		→	

7

7	×		→	0·7
÷		×		
	÷		→	
↓		↓		
1·75	+	0·02	→	

8

	+	8	→	9·4
−				
	×	0·1	→	
↓		↓		
1·3		0·8	→	2·1

9

	×		→	30
−				
	÷	10	→	0·25
↓		↓		
97·5	+	3	→	

10

3	÷	2	→	
÷		÷		
8	÷		→	
↓		↓		
	+	⅛	→	

11

	−	$\frac{1}{16}$	→	$\frac{3}{16}$
×				
	÷	4	→	
↓		↓		
⅛		¼	→	⅜

12

0·5	−	0·01	→	
		×		
	×		→	35
↓		↓		
4	÷	0·1	→	

Exercise 12 C

1 Write each calculation and find the missing digits.

a
```
  5 7 □ 2
+ □ 6 9 □
─────────
  8 □ 2 8
```

b
```
  8 □ 5
− 2 6 □
───────
  □ 7 3
```

c   □□□ ÷ 7 = 35

2 A hotel manager was able to buy loaves of bread at £4·44 per dozen, whereas the shop price was 43p per loaf. How much did he save on each loaf?

3 A high performance car uses one litre of petrol every 2·5 miles. How much petrol does it use on a journey of 37·5 miles?

4 John Lowe made darts history in 1984 with the first ever perfect game played in a tournament, 501 scored in just nine darts. He won a special prize of £100 000 from the sponsors of the tournament. His first eight darts were six treble 20s, treble 17 and treble 18.
  a What did he score with the ninth dart?
  b How much did he win per dart thrown, to the nearest pound?

5 An engineering firm offers all of its workers a choice of two pay rises. Workers can choose either an 8% increase on their salaries or they can accept a rise of £1600.
  a A fitter earns £10 400 a year. Which pay rise should he choose?
  b The personnel manager earns £23 000 a year. Which pay rise should he choose?

6 A map is 280 mm wide and 440 mm long. When reduced on a photocopier, the copy is 110 mm long. What is the width of the copy?

7 I have 213 mugs and one tray takes 9 mugs. How many trays do I need?

8 How many prime numbers are there between 120 and 130?

9 Write these answers, without a calculator.
   a 0.03×10 **b** $0.03 \div 10$ **c** $115 \div 1000$ **d** 0.07×1 million

10 Work out $\frac{2}{5} + 0.14 + \frac{3}{4}$, and write the answer as a decimal.

Exercise 13 Ⓒ

1 A maths teacher bought 40 calculators at £8·20 each and a number of other calculators costing £2·95 each. In all she spent £387. How many of the cheaper calculators did she buy?

2 The total mass of a jar one-quarter full of jam is 250 g. The total mass of the same jar three-quarters full of jam is 350 g.

   What is the mass of the empty jar?

$\frac{1}{4}$  **250 g**    $\frac{3}{4}$ **350 g**

3 I have lots of 1p, 2p, 3p and 4p stamps. How many different combinations of stamps can I make which total 5p?

4 8% of 2500 + 37% of P = 348. Find the value of P.

5 Eggs are packed twelve to a box. A farmer has enough eggs to fill 316 boxes with unbroken eggs and he has 62 cracked eggs left over. How many eggs had he to start with?

6 Booklets have a mass of 19 g each, and they are posted in an envelope of mass 38 g. Postage charges are shown in the table.

Mass (in grams) not more than	60	100	150	200	250	300	350	600
Postage (in pence)	24	30	37	44	51	59	67	110

   a A package consists of 15 booklets in an envelope. What is the total mass of the package?
   b The mass of a second package is 475 g. How many booklets does it contain?
   c What is the postage charge on a package of mass 320 g?
   d The postage on a third package was £1·10. What is the largest number of booklets it could contain?

7 A cylinder has a volume of 200 cm³ and a height of 10 cm. Calculate the area of its base.

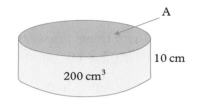

A

10 cm

200 cm³

8 You are told that $41 \times 271 = 11\,111$.
Use this to work out **in your head** the answer to 82×271.

9 Here is a subtraction using the digits 2, 3, 4, 5, 6.
Which subtraction using all the digits 2, 3, 4, 5, 6 has the smallest positive answer?

$$362 - 45$$

10 One litre of petrol costs 93·2p and one litre of oil costs 86p.
  a Find the cost of 100 litres of petrol.
  b Find the cost of 10 litres of oil.

Exercise 14 C

1 Copy and complete the additions.

a
```
    4 3 □
    2 □ 1
  + □ 3 5
  ─────────
  □ 3 9 8
```

b
```
    4 3 □
    5 □ 1
  + □ 1 5
  ─────────
  □ 0 7 2
```

2 Use the clues to find the mystery number
  ● the sum of the digits is 6
  ● the number reads the same forwards as backwards
  ● the number is less than 2000
  ● the number has four digits

3 The diagrams show magic squares in which the sum of the numbers in any row, column or diagonal is the same. Find the value of x in each square.

a

	x	6
3		7
		2

b

4		5	16
x		10	
	7	11	2
1			13

4 A 20p coin is 2 mm thick. Graham has a pile of 20p coins which is 18 cm tall. What is the value of Graham's pile?

5 The students in a school were given a general knowledge quiz. A mark of 20 or more was a 'pass'.
Some of the results are given in the table.
Copy and complete the table with the missing entries.

	Passed	Failed	Total
Boys		245	595
Girls	416		
Total	766		1191

6 **a** Find two consecutive numbers with a product of 342.

   b Find four consecutive numbers with a total of 102.

   c Find **any** two numbers with a product of 407.

   d Find a pair of numbers with a sum of 19 and a product of 48.

> Reminder:
> The **product** of 3 and 5 is $3 \times 5 = 15$.

7 In a supermarket five boxes of chocolates can be bought for £21·25. How many boxes can be bought for £34?

8 I think of a number. If I double the number and then take away ten the answer is 230. What number was I thinking of?

9 What is the fourth angle in a quadrilateral with angles 70°, 106° and 59°?

10 The test results of 100 students are shown in the table.

Result	5	6	7	8	9	10
Number of students	8	14	21	20	21	16

The pass mark in the test was 8. What percentage of the students passed?

Test yourself

You may use a calculator in this exercise.

1 Copy and complete this table.

Fraction	Decimal	Percentage
$\frac{1}{2}$	0·5	—
—	0·75	75%
$\frac{1}{4}$	—	—
—	—	9%

(OCR, 2003)

2 Write the metric unit that you would use to measure
  a the amount of butter in a pack
  b the length of a needle
  c the weight of a lorry.

(MEI, Spec)

3 What metric unit is used to measure each of these
  a the area of a carpet

  b the weight of an egg

  c the distance travelled on a train journey
  d the amount of petrol in the fuel
      tank of a car?

(OCR, 2003)

4 a Write $9·2 \times 10^7$ as an ordinary number.
  b A sheet of paper is $6·1 \times 10^{-2}$ mm thick.
     What is the total thickness of five sheets of this paper?
     Give your answer in standard form.

(OCR, 2005)

5 Work out $(3 \times 10^2) \times (4 \times 10^5)$
Give your answer in standard form.

(AQA, 2004)

6 Nassim buys petrol from his local garage.
On Monday, he filled up his tank.

On Tuesday, his tank was $\frac{3}{4}$ full.

  a What fraction of the full tank of petrol had he used?

  b Write $\frac{3}{4}$ as a decimal.

  c Write $\frac{3}{4}$ as a percentage.

The garage has a diagram for converting gallons to litres.
  d Use the diagram to convert
    i 2 gallons to litres
    ii 3·5 gallons to litres.

(Edexcel, 2003)

7 Change 5 kg to pounds.

(Edexcel, 2004)

8 Sharon travels from Leeds to London in her car.
The distance she travels is 200 miles.
The journey takes her 4 hours.
Find Sharon's average speed.

(AQA, 2004)

9 Mia drove a distance of 343 km.
She took 3 hours 30 minutes.
Work out her average speed.
Give your answer in km/h.

(Edexcel, 2003)

10 The video tape of *Jurassic Park* is 72 m long and can be
rewound in 4 minutes.
  a What is the speed of rewinding in cm per second?
The film *Gone with the Wind* is rewound at the same speed.
It takes 5 minutes to rewind.
  b How long in metres is the video tape of *Gone with the
Wind?*

(CCEA)

11 a Evaluate $\dfrac{1 \cdot 67 \times 10^{-7}}{4 \cdot 38 \times 10^{3}}$ giving your answer in standard form.

b A book published in 1993 contained the following statement.

> *At a constant 80 kilometres per hour it would take 35 years to travel over the entire road network of the world.*

Taking a year as 365 days, calculate the length of the road network of the world, giving your answer in standard form.

(WJEC)

12 The scales at an airport weigh luggage to the nearest kilogram.
What are the greatest and least possible weights of a case showing 25 kg on the scale?

(AQA, 2004)

13 The width of a wardrobe was given as 115 centimetres, measured to the nearest centimetre.
a Write the lower and upper limits of the width.

The wardrobe was bought to fit into a space between two walls. The width of the space had been measured as 1·2 metres, to the nearest tenth of a metre.

b Explain why the wardrobe might not fit into the space.

(CCEA)

14 Jo buys a new car.
a She fills it with 36 litres of petrol.
Roughly how many gallons is 36 litres?
b The distance from home to work is 5 miles.
She works 3 days a week.
Roughly how many kilometres will she drive to and from work each week?

(OCR, 2004)

10 Probability

In this unit you will:
- revise ideas of probability and chance
- use the probability scale
- use probability to list outcomes of two or more events
- learn how to find the probabilities of mutually exclusive events
- calculate probabilities.

10.1 Probability scale and experiments

10.1.1 How likely is it?

- The probability of something happening is the **likelihood** or **chance** that it might happen. It does **not** tell you what is definitely going to happen.

In probability you ask questions like
'What are the chances of Chelsea winning the F.A. cup?'
'How likely is it that I will pass my driving test first time?'

You do not know the answer to either of these questions yet.

Some events are **certain**. Some events are **impossible**.
Some events are in between certain and impossible.

- The probability of an event occurring is measured on a scale like this.

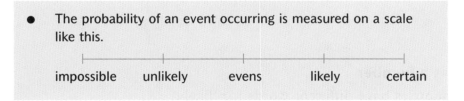

impossible   unlikely   evens   likely   certain

Think about where on this scale you would put these events.

1 The sun will rise tomorrow.

2 My teacher will win the National Lottery next week.

3 You spin a coin and get a head.

Exercise 1 C

Draw a probability scale like this.

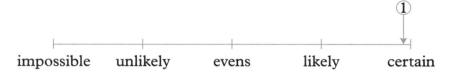

Draw an arrow to show the chance of these events happening.
(The arrow for question **1** has been done for you.)

 1 You will do some writing in this lesson.

 2 You will watch television this evening.

 3 You toss a coin and get a tail.

 4 You draw a card from a pack and get a king.

 5 You draw a card from a pack and get a heart.

 6 It will rain tomorrow.

 7 You will have a new pair of shoes in the next six months.

 8 You roll a dice and get a 1.

 9 You have the same birthday as the Prime Minister.

10 It will snow next week.

11 The next car to pass the school will be red.

12 The letter 'e' appears somewhere on the next page of this book.

10.1.2   Probability as a number

People in different countries have different words for 'likely' and 'unlikely'. All over the world people use probability as a way of avoiding confusion, using numbers on a scale instead of words.

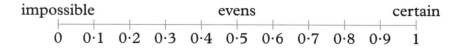

10.1.3 Experimental probability

The chance of certain events occurring can easily be predicted. For example, the chance of tossing a head with an ordinary coin. Many events cannot be so easily predicted, and you need to do an experiment to work out an experimental probability of the event occurring.

Experiment   To find the experimental probability of a drawing pin landing point up when dropped onto a hard surface.

Step 1   Do 50 **trials**.

Step 2   If the pin lands point up on a table this is a **success**.

Step 3   Make a tally chart like this.

Number of trials	Number of successes
LHT LHT II	LHT I

> Remember to put a mark in the 'trials' box every time you drop the pin.

● Experimental probability = $\dfrac{\text{Number of trials in which a success occurs}}{\text{Total number of trials made}}$

> Note that the experimental probability of an event is called its **relative frequency**.

Exercise 2 C

Carry out experiments to work out the experimental probability of some of these events.
Use a tally chart to record your results. Don't forget to record how many times you do the experiment (the number of trials).

1   Roll a dice. What is the chance of rolling a six?
Do 60 trials. Copy and complete this table and fill in the second row.

Number of trials (rolls of the dice)	12	24	36	48	60
Number of 6s you expect					10
Number of 6s you actually got					

a   What was the experimental probability of rolling a 6?
b   Would you expect to get the same result if you did the experiment again?

2 Toss two coins. What is the chance of tossing two tails?
Do 100 trials.

3 Pick a counter from a bag containing counters of different
colours. What is the chance of picking a red counter?
Do 100 trials.

4 Roll a pair of dice. What is the chance of rolling a double?
Do 100 trials.

5 This spinner has an equal chance of giving any digit from
0 to 9.
Four friends did an experiment when they spun the
pointer a different number of times and recorded the
number of zeros they got.
Here are their results.

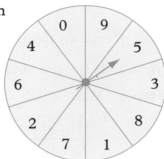

	Number of spins	Number of zeros	Relative frequency
Steve	10	2	0·2
Nick	150	14	0·093
Mike	200	41	0·205
Jason	1000	104	0·104

What is the chance
of spinning a zero
on this spinner?

One of the four recorded his results incorrectly. Say who
you think this was and explain why.

10.2 Working out probabilities

10.2.1 Expected probability

For simple events, like throwing a dice or tossing a coin, you
can work out the expected probability of an event occurring.

For a fair dice the **expected probability** of throwing a '3' is $\frac{1}{6}$.

For a normal coin the expected probability of tossing a 'head'
is $\frac{1}{2}$.

- Expected probability = $\dfrac{\text{The number of ways the event can happen}}{\text{The number of possible outcomes}}$

Random choice
If a card is chosen at random from a pack it means that every
card has an equal chance of being chosen.

EXAMPLE

Seven identical balls numbered 1, 2, 3, 4, 5, 6 and 7 are put
into a bag. One ball is selected at random.
What is the probability of selecting
a a '3'   **b** an even number?

There are 7 possible equally likely outcomes.

a The probability of selecting a '3' = $\frac{1}{7}$

   This may be written p (selecting a '3') = $\frac{1}{7}$

b p (selecting an even number) = $\frac{3}{7}$

EXAMPLE

A single card is drawn from a pack of 52 playing cards.
Find the probability of these results.
a The card is a queen.
b The card is a club.
c The card is the jack of hearts.

There are 52 equally likely outcomes of the trial
(drawing a card).

a p (queen) = $\frac{4}{52}$ = $\frac{1}{13}$ 13 Spades

b p (club) = $\frac{13}{52}$ = $\frac{1}{4}$ 13 Hearts
                                                   13 Diamonds
                                                   13 Clubs

c p (jack of hearts) = $\frac{1}{52}$ Total = 52

Exercise 3 C

1 A bag contains a red ball, a white ball and a black ball.
One ball is chosen at random.
Copy and complete these sentences.

a The probability that the red ball is chosen is $\frac{\boxed{}}{\boxed{3}}$.

b The probability that the white ball is chosen is $\frac{\boxed{}}{\boxed{}}$.

2 One ball is chosen at random from a bag which contains a red ball, a black ball, a grey ball and a white ball. Write the probability that the chosen ball will be
  a red
  b grey
  c black.

3 A bag contains 2 red balls and 1 white ball. One ball is chosen at random. Find the probability that it is
  a red
  b white.

4 Izzy rolls a fair dice. Write the chance of her getting
  a a 6
  b a 2
  c a 7.

5 Here is a spinner with 8 equal sectors. Write the probability of spinning
  a a 7
  b a 3
  c an even number.

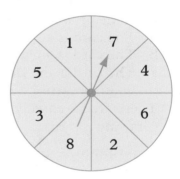

6 Write the probability of the arrow pointing at a shaded section when each of these spinners stops spinning.

a   **b**  **c**   **d**

e   **f**   **g**   **h**

7 Here are two spinners.
Say whether the following statements are
true or false.
Explain why in each case.
  a 'Dan is more likely to spin a 4 than
    Nick.'
  b 'Dan and Nick are equally likely to spin
    an odd number.'
  c 'If Nick spins his spinner six times he is
    bound to get at least one 6.'

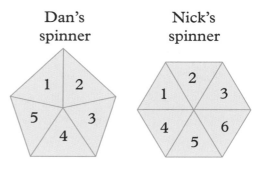

Dan's spinner   Nick's spinner

8 If one card is picked at random from a pack of 52 playing
cards, what is the probability that it is
  a a king
  b the ace of clubs
  c a heart?

9 Nine counters numbered 1, 2, 3, 4, 5, 6, 7, 8, 9 are placed
in a bag. One is taken out at random. What is the
probability that it is
  a a 5   **b** divisible by 3
  c less than 5 **d** divisible by 4?

10 A bag contains 5 grey balls, 2 red balls and 4 white balls.
One ball is taken out at random. What is the probability
that it is
  a grey
  b red
  c white?

11 A cash bag contains two 20p coins, four 10p coins,
five 5p coins, three 2p coins and three 1p coins.
Find the probability that one coin selected at random is
  a a 10p coin
  b a 2p coin
  c a silver coin.

12 A bag contains 8 orange balls, 5 green balls and 4 silver
balls. Find the probability that a ball picked out at
random is
  a silver
  b orange
  c green.

13 One card is selected at random from this hand.

You will not be tested on your knowledge of playing cards in your examination.

Find the probability of selecting
a a heart
b an ace
c the 10 of clubs
d a spade
e a heart or a diamond.

14 A pack of playing cards is well shuffled and a card is drawn. Find the probability that the card is
a a jack
b a queen or a jack
c the ten of hearts
d a club higher than the 9 (count the ace as high).

EXAMPLE

Cards with numbers 1, 2, 3, 4, 5, 6, 7, 8 are shuffled and then placed face down in a line. The cards are then turned over one at a time from the left. In this example the first card is a 4.

Find the probability that the next card turned over will be
a the 5
b a number less than 4
c the 4.

....

a The next card can be either 1, 2, 3, 5, 6, 7 or 8.

p (turning over a 5) $= \dfrac{1}{7}$

b You can turn over a 1, a 2 or a 3.

p (turning over a number less than 4) $= \dfrac{3}{7}$

c p (turning over a 4) $= 0$

Exercise 4 C

Questions **1**, **2** and **3** are about cards with numbers 1, 2, 3, 4, 5, 6, 7 and 8. The cards are shuffled and then placed face down in a line. The cards are then turned over one at a time.

1 In this case the first card is a 3.

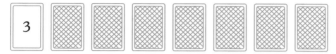

Find the probability that the next card will be
a the 6
b an even number
c higher than 1.

2 Suppose the second card is a 7.

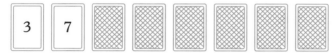

Find the probability that the next card will be
a the 5
b an odd number
c higher than 6.

3 Suppose the first three cards are

Find the probability that the next card will be
a the 8
b less than 4
c the 5 or the 6.

4 The numbers of matches in ten boxes are

48, 46, 45, 49, 44, 46, 47, 48, 45, 46

One box is selected at random. Find the probability of the box containing
a 49 matches
b 46 matches
c more than 47 matches.

5 One ball is selected at random from those below.

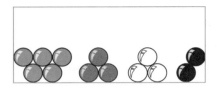

Find the probability of selecting
a a white ball
b a grey or a black ball
c a ball which is not red.

6 a A bag contains 5 red balls, 6 green balls and 2 black balls.
   Find the probability of selecting
   i a red ball   **ii** a green ball.

b One black ball is removed from the bag. Find the new
   probability of selecting
   i a red ball   **ii** a black ball.

7 A pack of playing cards is split so that all the picture cards
(kings, queens, jacks) are in Pile A and all the other cards
are in Pile B.
Find the probability
of selecting
a the queen of clubs
   from pile A
b the seven of spades
   from pile B
c any heart from pile B.

Pile A

Pile B

8 A bag contains 12 white balls, 12 green balls and 12 purple
balls. After 3 white balls, 4 green balls and 9 purple balls
have been removed, what is the probability that the next ball
to be selected will be white?

9 Jade has 3 queens and 1 king. She shuffles
the cards and takes one without looking.
Jade asks two of her friends about the
probability of getting a king.

Kim says: Megan says:
'It is $\frac{1}{3}$ because there   'It is $\frac{1}{4}$ because
are 3 queens and 1 king.'   there are 4 cards and only 1 king.'

Which of her friends is right?

10 A bag contains 9 balls, all of which are black or white. Jane selects a ball and then replaces it. She repeats this several times. Here are her results (B = black, W = white):

B W B W B B B W B B W B B W B
B B W W B B B B W B W B B W B

How many balls of each colour do you think there were in the bag?

11 A large firm employs 3750 people. One person is chosen at random.
What is the probability that that person's birthday is on a Monday in the year 2008?

12 There are eight balls in a bag. Asif takes a ball from the bag, notes its colour and then returns it to the bag.
He does this 25 times. The results are in the table.
 a What is the smallest number of red balls there could be in the bag?
 b Asif says 'There cannot be any blue balls in the bag because there are no blues in my table.' Explain why Asif is wrong.

Red	4
White	11
Black	10

13 The numbering on a set of 28 dominoes is like this.

6	6	6	6	6	6	6	5	5	5
6	5	4	3	2	1	0	5	4	3

5	5	5	4	4	4	4	4	3	3
2	1	0	4	3	2	1	0	3	2

3	3	2	2	2	1	1	0
1	0	2	1	0	1	0	0

 a What is the probability of drawing a domino from a full set with
  i at least one six on it?
  ii at least one four on it?
  iii at least one two on it?
 b What is the probability of drawing a 'double' from a full set?
 c If I draw a double five, which I do not return to the set, what is the probability of drawing another domino with a five on it?

14 A bag contains a number of green, red and blue discs. When a disc is selected at random, the probability of it being blue is 0·3. There are 50 discs in the bags. How many of the discs are blue?

10.2.2 Expectation

● Expected number of successes =
  (probability of a success) × (number of trials)

EXAMPLE

A fair dice is rolled 240 times. How many times would you expect to roll a number greater than 4?

....

You can roll a 5 or a 6 out of the six equally likely outcomes.

p (number greater than 4) $= \dfrac{2}{6} = \dfrac{1}{3}$

Expected number of scores greater than 4 $= \dfrac{1}{3} \times 240$

$= 80$

Exercise 5 C

1 A fair dice is rolled 300 times. How many times would you expect to roll
  a an even number
  b a six?

2 This spinner has four equal sectors. How many 3s would you expect in 100 spins?

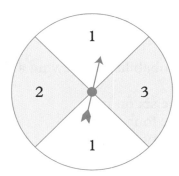

3 About one in eight of the population is left-handed. How many left-handed people would you expect to find in a firm employing 400 people?

4 A bag contains a large number of marbles of which one in five is red. If I randomly select one marble on 200 occasions how many times would I expect to select a red marble?

5 This spinner is used for a simple game. A player pays 10p and then spins the pointer, winning the amount shown.
   a What is the probability of winning nothing?
   b If the game is played by 200 people how many times would you expect the 50p to be won?

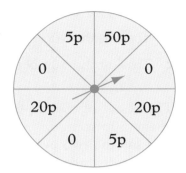

6 The numbered cards are shuffled and put into a pile.

One card is selected at random and not replaced. A second card is then selected.
   a If the first card was the '11', find the probability of selecting an even number with the second draw.
   b If the first card was the '8', find the probability of selecting a number higher than 9 with the second draw.

7 When the ball is passed to Vinnie, the probability that he kicks it is 0·2 and the probability that he heads it is 0·1. Otherwise he will miss the ball completely, fall over and claim a foul.
In one game his ever-optimistic team mates pass the ball to Vinnie 150 times.
   a How often would you expect him to head the ball?
   b How often would you expect him to miss the ball?

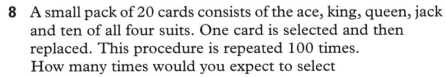

8 A small pack of 20 cards consists of the ace, king, queen, jack and ten of all four suits. One card is selected and then replaced. This procedure is repeated 100 times.
How many times would you expect to select
   a an ace
   b the queen of spades
   c a red card
   d any king or queen?

10.3 Listing possible outcomes, two or more events

When a 10p coin and a 50p coin are tossed
together two events are occurring.

1 Tossing the 10p coin

2 Tossing the 50p coin

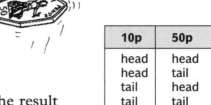

10p	50p
head	head
head	tail
tail	head
tail	tail

The result of tossing the 10p coin does not affect the result
of the 50p coin.
You say the two events are **independent**.
You can list all the possible outcomes as shown in the table.

A red dice and a black dice are thrown together. Show all the possible
outcomes.

....

You could list them in pairs with the red dice first.

$(1, 1)$, $(1, 2)$, $(1, 3)$, ... $(1, 6)$
$(2, 1)$, $(2, 2)$, ...
and so on.

In this example it is easier to see all the possible
outcomes when they are shown on a grid.
There are 36 possible outcomes.

The × shows 6 on the red dice and 2 on the black.
The o shows 3 on the red dice and 5 on the black.

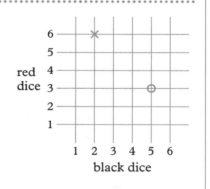

Exercise 6 C

1 Three coins (10p, 20p, 50p) are tossed together.
  a List all the possible ways in which they
     could land.
  b What is the probability of getting three heads?

2 List all the possible outcomes when four coins
   are tossed together. How many outcomes
   are there altogether?

3 A black dice and a white dice are thrown together.
  a Draw a grid to show all the possible outcomes (see the example above) and write how many possible outcomes there are.
  b How many ways can you get a total of nine on the two dice?
  c What is the probability that you get a total of nine?

4 A red spinner and a white spinner are spun together.

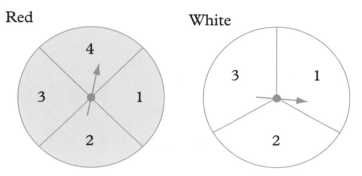

Red   White

  a List all the possible outcomes.
  b In how many ways can you get a total of 4?
  c What is the probability that you get a total of 4?

5 Four friends, Wayne, Xavier, Yves and Zara, each write their name on a card and the four cards are placed in a hat. Two cards are chosen to decide who does the maths homework that night.
List all the possible combinations.

6 The spinner is spun and the dice is thrown at the same time.

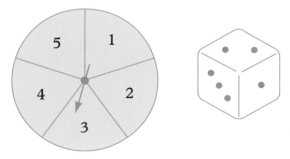

  a Draw a grid to show all the possible outcomes.
  b A 'win' occurs when the number on the spinner is greater than or equal to the number on the dice. Find the probability of a win.

7 The menu in a restaurant has two choices of starter, three choices of main course and two choices of dessert.

  a List all the different combinations I could choose from this menu.

Gourmet Greg always chooses his three courses at random.

  b What is the probability that he chooses Melon, Chicken pie and Gateau?

Starters	
Melon	(A)
or Pineapple	(B)
Main course	
Steak	(C)
or Cod fillet	(D)
or Chicken pie	(E)
Dessert	
Ice cream	(F)
or Gateau	(G)

8 By a strange coincidence the Branson family, the Green family and the Webb family all have the same first names for the five members of their families: James, Don, Samantha, Laura and Kate.

One year Father Christmas decides to give each person a embroidered handkerchief with two initials.

  a How many different handkerchiefs does he need for these three families?

  b How many different combinations are there if **any** first name and **any** surname is possible [for example 'Zak Quilfeldt']?

9 Keith, Len, Mike and Neil enter a cycling race.

  a List all the possible orders in which they could finish. State the number of different finishing orders.

  b In how many of these ways does Mike finish in front of Len?

10 Shirin and Dipika are playing a game in which three coins are tossed. Shirin wins if there are no heads or one head. Dipika wins if there are either two or three heads. Is the game fair to both players?

*11 Students Xavier, Yolanda and Zachary play a game in which four coins are tossed.

X wins if there is 0 or 1 head.
Y wins if there are 2 heads.
Z wins if there are 3 or 4 heads.

Is the game fair to all three players?

*12 Four cards numbered 2, 4, 5 and 7 are mixed up and placed face down.

In a game you pay 10p to select two cards. You win 25p if the total of the two cards is nine.

How much would you expect to win or lose if you played the game 12 times?

Exercise 7Ⓔ (A practical exercise)

1 The RAN # button on a calculator generates random numbers between 0·000 and 0·999. You can use it to simulate tossing three coins.

You could say any **odd** digit is a **tail** and any **even** digit is a **head**.

So the number 0·568 represents THH
and 0·605 represents HHT.

Use the RAN # button to simulate the tossing of three coins.

'Toss' the three coins 32 times and work out the relative frequencies of

a three heads

b two heads and a tail.

Compare your results with the values that you would expect to get theoretically.

2 In Gibson Academy there are six forms (A, B, C, D, E, F) in Year 10 and the Mathematics Department has been asked to work out a schedule so that, over 5 weeks, each team can play each of the others in a basketball competition.

The games are all played at lunch time on Wednesdays and three games have to be played at the same time.

For example in Week 1 you could have

| Form A v Form B |
| Form C v Form D |
| Form E v Form F |

Work out a schedule for the remaining four weeks so that each team plays each of the others. Check carefully that each team plays every other team just once.

10.4 Exclusive events

● Events are **mutually exclusive** if they cannot occur at the same time.

Example 1 Selecting an ace and selecting a ten from a pack of cards from one selection

Example 2 Tossing a head and tossing a tail from one throw

Example 3 Getting a total of 5 on two dice and getting a total of 7 on two dice from one throw

● The probability of something not happening is 1 minus the probability of it happening.

● Also, the sum of the probabilities of mutually exclusive events is 1.

EXAMPLE

A bag contains 3 red balls, 4 black balls and 1 white ball. Find the probability of selecting
a a white ball
b a ball which is not white
c a red ball
d a ball which is not red.

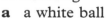

There are 8 balls in the bag.

a p (white ball) $= \dfrac{1}{8}$

b p (not white ball) $= 1 - \dfrac{1}{8} = \dfrac{7}{8}$

c p (red ball) $= \dfrac{3}{8}$

d p (not red ball) $= 1 - \dfrac{3}{8} = \dfrac{5}{8}$

Exercise 8 C

1 A bag contains a large number of balls including some red balls. The probability of selecting a red ball is $\frac{1}{5}$. What is the probability of selecting a ball which is not red?

2 A card is selected from a pack of 52. Find the probability of selecting
  a a king
  b a card which is not a king
  c any picture card (king, queen or jack)
  d a card which is not a picture card.

3 On a roulette wheel the probability of getting 21 is $\frac{1}{36}$.
What is the probability of not getting 21?

4 A motorist does a survey at some traffic lights on his way to work every day. He finds that the probability that the lights are red when he arrives is 0·24. What is the probability that the lights are not red?

5 Government birth statistics show that the probability of a woman giving birth to a boy is 0·506.
What is the probability of having a girl?

6 The spinner has 8 equal sectors.
Find the probability of
  a spinning a 5
  b not spinning a 5
  c spinning a 2
  d not spinning a 2
  e spinning a 7
  f not spinning a 7.

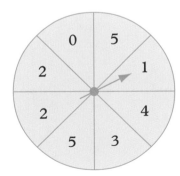

7 The king of clubs is removed from a normal pack of cards. One card is selected from the remaining cards. Find the probability of
  a selecting a king
  b not selecting a king
  c selecting a club.

8 A bag contains a large number of balls coloured red, white, black or green. The probabilities of selecting each colour are shown in the table.

Colour	red	white	black	green
Probability	0·3	0·1		0·3

Find the probability of selecting a ball
a which is black
b which is not white.

9 In a survey the number of people in cars is recorded. When a car passes the school gates the probability of it having 1, 2, 3, ... occupants is shown in the table.

Number of people	1	2	3	4	more than 4
Probability	0.42	0·23		0·09	0·02

a Find the probability that the next car past the school gates contains
i three people
ii less than 4 people.

b One day 2500 cars passed the gates. How many of the cars would you expect to have 2 people inside?

Test yourself

1 On a scale like the one below, draw and label arrows to show the probability of these events.
  a The sun will rise tomorrow.
  b The next baby born in a hospital will be a girl.
  c You will live to be 100.
  d You will get a number greater than 1 when you roll an ordinary dice.

0   1

(OCR, 2004)

2 a The diagram shows five probabilities marked on a probability line.

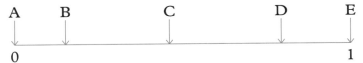

A   B   C   D   E

0   1

Copy these terms and match each one with the correct letter from the diagram.

Certain   Even   Impossible   Likely   Unlikely

  b Jade picks a ball at random from this box.

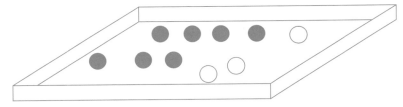

Which of these words describes the probability that she picks a red ball.

Certain   Even   Impossible   Likely   Unlikely
(MEI, Spec)

3 Mark throws a fair coin.
   He gets a head.

   Mark's sister then throws the same coin.
   a What is the probability that she will get a head?

   Mark throws the coin 30 times.
   b Explain why he may not get exactly 15 heads and 15 tails.
   (Edexcel, 2004)

4 a Ann has a spinner which has five equal sections.
Two sections are red, one is yellow, one is green and
one is white.

Ann spins the spinner once.
On what colour is the spinner most likely to land?

First spinner

b Ann has a second spinner which has six equal sections.
Three sections are red, one is yellow and two are blue.

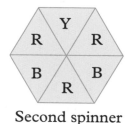

Ann spins this spinner once.
What is the probability that this spinner lands on red?

Second spinner

c Ann thinks that she has more chance of getting yellow on
the second spinner.

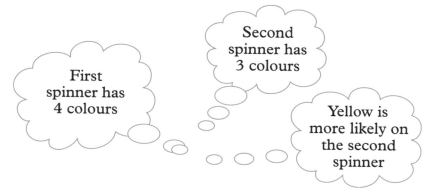

First
spinner has
4 colours

Second
spinner has
3 colours

Yellow is
more likely on
the second
spinner

Explain why Ann is wrong.

(AQA, 2004)

5 In a raffle 520 tickets are sold.
They are numbered from 1 to 520.

The winning number is chosen at random.

Work out the probability that it is
a 99
b greater than 240.

(MEI, Spec)

6 A spinner has a red sector (R) and a yellow sector (Y).
The arrow is spun 1000 times.

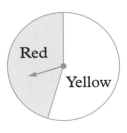

a The results for the first 20 spins are shown below.

R R Y Y Y R Y Y R Y Y Y Y Y R Y R Y Y Y

Work out the relative frequency of a red after 20 spins.
Give your answer as a decimal.

b The table shows the relative frequency of a red after
different numbers of spins.

Number of spins	Relative frequency of a red
50	0·42
100	0·36
200	0·34
500	0·3
1000	0·32

How many times was a red obtained after 200 spins?

(AQA, 2004)

7 Forty people take a driving test at Centre A on one day.
The table shows the results.

	Pass	Fail
Male	10	13
Female	6	11

a A person is chosen at random from the group.
What is the probability that the person is male?

b A person is chosen at random from the group.
What is the probability that the person passed the
test?

c It is known that throughout Britain the probability of a
person passing their test is 0·7.
John says it is easier to pass the test at Centre A.
Explain why John could be wrong.

(AQA, 2003)

8 **a** Marco is recycling his glass bottles.
He has one green (G), one brown (B) and one clear (C) bottle.

List the different orders he could recycle the three bottles.
The first one is done for you.

G	B	C

b i Jane has 11 green, 7 brown and 2 clear bottles to recycle.
She picks the first bottle at random.
What is the probability that it is brown?

ii The probability that the first bottle she picks is a juice bottle is 0·4.
What is the probability that the first bottle she picks is **not** a juice bottle?

(OCR, 2005)

9 Rory has two fair spinners.
The red spinner is numbered 1, 2, 3.
The white spinner is numbered 1, 2, 3, 4.

Rory spins them both.
He adds the numbers they land on.

a Copy and complete the table below to show all possible totals.

		Red spinner		
		1	2	3
White spinner	1			
	2			5
	3			
	4			

b Use the table to find the probability of obtaining a total of 5.

(OCR, 2005)

10 A red dice and a blue dice are rolled together. Both dice are unbiased.
The **difference** between the two scores is recorded in a table.

Score on red dice

		1	2	3	4	5	6
Score on blue dice	1	0	1	2	3	4	5
	2	1			2		
	3	2			1		
	4	3			0		
	5	4			1		
	6	5	4	3	2	1	0

   a Copy and complete the table.
   b Find the probability that the difference in the scores is 2.
   c Find the probability that the difference in the scores is
      greater than 3.

(OCR, 2004)

11 Rovers play Wanderers at football.
The probability that Rovers win the match is 0·55.
The probability that Wanderers win the match is 0·2.

Find the probability that the result is a draw.

(OCR, 2003)

12 A dice is biased.
The table below shows the probability of obtaining each
score when the dice is rolled.

Score	1	2	3	4	5	6
Probability	0·2	0·1	0·15	0·05	0·2	x

Find the value of x.

(OCR, 2005)

13 In a game at a fete a spinner is spun. The score obtained may be any one
of 2, 4, 6, 8, 10, 12, 14, the probabilities of which are given in the table.

Number	2	4	6	8	10	12	14
Probability	0·11	0·09	0·14	0·31	0·23	0·07	0·05

   a Calculate the probability that the score obtained will be
      greater than 8.
   b The spinner is spun 300 times. About how many times
      will the score be 6?

(WJEC)

11 Using and applying mathematics

11.1 Coursework tasks

There are a large number of possible starting points for investigations here so it may be possible to choose investigations which appeal to you. Alternatively the same investigation may be set to the whole class by your teacher.

Here are a few guidelines.
- If the set problem is too complicated try an easier case.
- Draw your own diagrams.
- Make tables of your results and be systematic.
- Look for patterns.
- Is there a rule or formula to describe the results?
- Can you **predict** further results?
- Can you **explain** any rules which you may find?
- Where possible extend the task further by asking questions like 'what happens if . . .'.

> Systematic means work through things in order.

11.1.1 Opposite corners

Here the numbers are arranged in 9 columns.

1	2	3	4	5	6	7	8	9
10	11	12	13	14	15	16	17	18
19	20	21	22	23	24	25	26	27
28	29	30	31	32	33	34	35	36
37	38	39	40	41	42	43	44	45
46	47	48	49	50	51	52	53	54
55	56	57	58	59	60	61	62	63
64	65	66	67	68	69	70	71	72
73	74	75	76	77	78	79	80	81
82	83	84	85	86	87	88	89	90

You can start with simpler cases:
In the 2×2 square . . .

6	7
15	16

$6 \times 16 = 96$
$7 \times 15 = 105$
. . . the difference between them is 9.

In the 3×3 square . . .

22	23	24
31	32	33
40	41	42

$22 \times 42 = 924$
$24 \times 40 = 960$
. . . the difference between them is 36.

Investigate to see if you can find any rules or patterns connecting the size of square chosen and the difference. If you find a rule, use it to **predict** the difference for larger squares.

Test your rule by looking at squares like 8 × 8 or 9 × 9.

Can you **generalise** the rule?
(What is the difference for a square of size $n \times n$?)

In a 3 × 3 square ...

Can you **prove** the rule?

What happens if the numbers are arranged in six columns or seven columns?

1	2	3	4	5	6
7	8	9	10	11	12
13	14	15	16	17	18
19	_	_	_	_	_
_	_	_	_	_	_

1	2	3	4	5	6	7
8	9	10	11	12	13	14
15	16	17	18	19	20	21
22	_	_	_	_	_	_
_	_	_	_	_	_	_

11.1.2   Hiring a car

You are going to hire a car for one week (7 days). Which of the firms below should you choose?

Gibson car hire	Snowdon rent-a-car	Hav-a-car
£270 per week unlimited mileage	£20 per day 10p per mile	£100 per week 500 miles without charge 30p per mile over 500 miles

Work out as detailed an answer as possible.

11.1.3   Half-time score

The final score in a football match was 3–2. How many different scores were possible at half-time?

Investigate for other final scores where the difference between the teams is always one goal (1–0, 5–4, etc.). Is there a pattern or rule which would tell you the number of possible half-time scores in a game which finished 8–7?

Suppose the game ends in a draw. Find a rule which would tell you the number of possible half-time scores if the final score was 10–10.

Investigate for other final scores (3–0, 5–1, 4–2, etc.). Find a rule which gives the number of different half-time scores for **any** final score (say *a–b*).

11.1.4 Squares inside squares

Here is a square drawn inside a square. The inside square is set at an angle to the outside square. Is there a connection between the **area** of the inside square and the numbers $\binom{2}{3}$ that describe the angle through which the square is rotated?

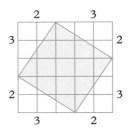

Method
Calculate the area of the inside square by subtracting the areas of triangles A, B, C and D from the area of the outside square.

Area of outside square $= 5 \times 5 = 25$ squares

Area of each triangle $= \frac{1}{2} \times 2 \times 3 = 3$ squares

So, area of inside square $= 25 - (4 \times 3)$
$= 13$ squares

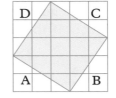

Your task is to investigate the areas of different squares drawn inside other squares in a similar way.

A Start with simple cases.

 2

 3

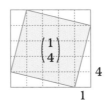 4

The outside squares are not always the same size.

Find the areas of the inside squares where the top number is 1. Can you find a connection or rule which you can use to find the area of the inside square **without** all the working with subtracting areas of triangles?

B Look at more difficult cases.

Again try to find a rule
which you can use to find
the area of the inside
square directly.

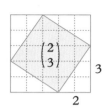

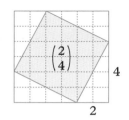

C Look at **any** size square.

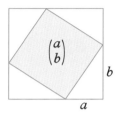

Try to find a rule and, if possible, use algebra to
show why it **always** works.

11.1.5 Maximum box

A You have a square sheet of card 24 cm by 24 cm.
You can make a box (without a lid) by cutting squares
from the corners and folding up the sides.
What size corners should you cut out so that the volume
of the box is as large as possible?
Try different sizes for the corners and record the results
in a table.

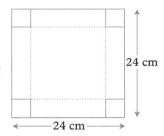

Length of the side of the corner square (cm)	Dimensions of the open box (cm)	Volume of the box (cm³)
1	22 × 22 × 1	484
2		
–		

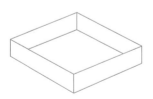

Now consider boxes made from different sized cards

15 cm × 15 cm and 20 cm × 20 cm.

What size corners should you cut out this time so that the
volume of the box is as large as possible?
Is there a connection between the size of the corners cut out
and the size of the square card?

B Investigate the situation when the card is not square.
Take rectangular cards where the length is twice the width

$(20 \times 10, \ 12 \times 6, \ 18 \times 9, \ \text{etc.})$.

Again, for the maximum volume, is there a connection
between the size of the corners cut out and the size of the
original card?

11.1.6 Diagonals

In a 4×7 rectangle, the diagonal passes through 10 squares.

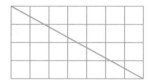

Draw rectangles of your own choice and count the number
of squares through which the diagonal passes.
A rectangle is 640×250. How many squares will the diagonal
pass through?

11.1.7 Painting cubes

The large cube on the right consists of 27 unit cubes.

All six faces of the large cube are painted red.
The large cube is dismantled into its unit cubes.

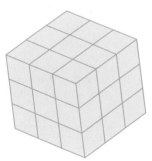

1 How many unit cubes have 3 red faces?
2 How many unit cubes have 2 red faces?
3 How many unit cubes have 1 red face?
4 How many unit cubes have 0 red faces?

Answer the four questions for the cube which is $n \times n \times n$.

11.2 Puzzles and games

11.2.1 Crossnumbers

Draw four copies of this crossnumber pattern and work out the
answers using the clues. You can check your working by doing
all the across and **all** the down clues.

There are four sets of clues for the same grid.

Part A

Across

1 $327 + 198$
3 $245 \div 7$
5 $3146 - 729$
6 $248 - 76$
7 2^6
8 $850 \div 5$
10 $10^2 + 1^2$
11 $3843 \div 7$
12 $1000 - 913$
13 $37 \times 5 \times 3$
16 $152\,300 \div 50$
19 3^6
20 $100 - \left(\dfrac{17 \times 10}{5}\right)$

Down

1 $3280 + 1938$
2 $65\,720 - 13\,510$
3 $3 \cdot 1 \times 1000$
4 $1284 \div 6$
7 $811 - 127$
9 65×11
10 $(12^2 - 8) \div 8$
11 $(7^2 + 1^2) \times 11$
12 $7 + 29 + 234 + 607$
14 $800 - 265$
15 $1 + 2 + 3 + 4 + 5 + 6 + 7 + 8 + 13$
17 $(69 \times 6) \div 9$
18 $3^2 + 4^2 + 5^2 + 2^4$

Part B

Draw decimal points on the lines between squares where necessary.

Across

1 $4 \cdot 2 + 1 \cdot 64$
3 $7 \times 0 \cdot 5$
5 $20 \cdot 562 \div 6$
6 $(2^3 \times 5) \times 10 - 1$
7 $0 \cdot 034 \times 1000$
8 $61 \times 0 \cdot 3$
10 $8 - 0 \cdot 36$
11 19×50
12 $95 \cdot 7 \div 11$
13 $8 \cdot 1 \times 0 \cdot 7$
16 $(11 \times 5) \div 8$
19 $(44 - 2 \cdot 8) \div 5$
20 Number of inches in a yard

Down

1 $62 \cdot 6 - 4 \cdot 24$
2 $48 \cdot 73 - 4 \cdot 814$
3 $25 + 7 \cdot 2 + 0 \cdot 63$
4 $2548 \div 7$
7 $0 \cdot 315 \times 100$
9 $169 \times 0 \cdot 05$
10 $770 \div 100$
11 $14 \cdot 2 + 0 \cdot 7 - 5 \cdot 12$
12 $11 \cdot 4 - 2 \cdot 64 - 0 \cdot 18$
14 $0 \cdot 0667 \times 10^3$
15 $0 \cdot 6 + 0 \cdot 7 + 0 \cdot 8 + 7 \cdot 3$
17 $0 \cdot 73$ m written in cm
18 $0 \cdot 028 \times 200$

Part C

Across

1 Eleven squared take away six
3 Next in the sequence
   21, 24, 28, 33
5 Number of minutes in a day
6 $2 \times 13 \times 5 \times 5$
7 Next in the sequence
   92, 83, 74
8 5% of 11 400
10 $98 + 11^2$
11 $(120 - 9) \times 6$
12 $1\frac{2}{5}$ as a decimal
13 $2387 \div 7$
16 9.05×1000
19 $8\,\text{m} - 95\,\text{cm}$ (in cm)
20 3^4

Down

1 Write 18·6 m in cm
2 Fifty-one thousand and fifty-one
3 Write 3·47 km in m
4 $1\frac{1}{4}$ as a decimal
7 $7\,\text{m} - 54\,\text{cm}$ (in cm)
9 0.0793×1000
10 2% of 1200
11 $\frac{1}{5}$ of 3050
12 $127 \div 100$
14 Number of minutes between
   12:00 and 20:10
15 4% of 1125
17 $7^2 + 3^2$
18 Last two digits of (67×3)

Part D

Across

1 $1\frac{3}{4}$ as a decimal
3 Two dozen
5 Forty less than ten thousand
6 Emergency
7 5% of 740
8 Nine pounds and five pence
10 1·6 m written in cm
11 $5649 \div 7$
12 One-third of 108
13 $6 - 0.28$
16 A quarter to midnight
   on the 24 h clock
19 $5^3 \times 2^2 + 1^5$
20 $3300 \div 150$

Down

1 Twelve pounds 95 pence
2 Four less than sixty thousand
3 245×11
4 James Bond
7 Number of minutes between
   09:10 and 15:30
9 $\frac{1}{20}$ as a decimal
10 Ounces in a pound
11 8·227 to two decimal places
12 $4\,\text{m} - 95\,\text{cm}$ (in cm)
14 Three to the power 6
15 20·64 to the nearest whole number
17 $\left(6\frac{1}{2}\right)^2$ to the nearest whole number
18 Number of minutes between
   14:22 and 15:14

11.2.2 Crossnumbers without clues

Here are four crossnumber puzzles with a difference. There are
no clues, only answers, and it is your task to find where the
answers go.

a Copy out the crossnumber pattern or ask your teacher for a
photocopy.

b Fit all the given numbers into the correct spaces.
Tick off the numbers from the lists as you write them
in the square.

Part A

2 digits	3 digits	4 digits	5 digits	6 digits
14	111	3824	24 715	198 264
25	127	4210	54 073	338 472
26	131	8072	71 436	387 566
30	249	8916	72 180	414 725
42	276	9603	82 125	
52	328			
53	571			*7 digits*
57	609			8 592 070
61	653			
99	906			
	918			

Part B

2 digits	3 digits	4 digits	5 digits	6 digits
19	106	1506	21 362	134 953
26	156	1624	21 862	727 542
41	180	2007	57 320	
63	215	2637	83 642	
71	234	4214		
76	263	4734		
	385	5216		
	427	5841		
	725	9131		
	872	9217		

The next two are more difficult but they are possible!
Don't give up.

Part C

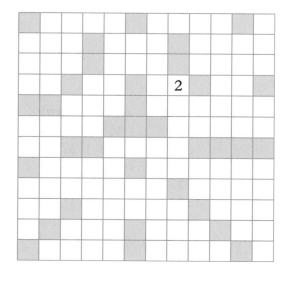

2 digits	3 digits	4 digits	5 digits	6 digits
26	306	3654	38 975	582 778
28	457	3735	49 561	585 778
32	504	3751	56 073	728 468
47	827	3755	56 315	
49	917	3819	56 435	
52	951	6426	57 435	*7 digits*
70		7214	58 535	8 677 056
74		7315	58 835	
		7618	66 430	
		7643	77 435	
		9847	77 543	

Part D

2 digits	3 digits	4 digits	5 digits	6 digits
11	121	2104	14 700	216 841
17	147	2356	24 567	588 369
18	170	2456	25 921	846 789
19	174	3714	26 759	861 277
23	204	4711	30 388	876 452
31	247	5548	50 968	
37	287	5678	51 789	
58	324	6231	78 967	*7 digits*
61	431	6789	98 438	6 645 678
62	450	7630		
62	612	9012		
70	678	9921		
74	772			
81	774			
85	789			
94	870			
99				

11.2.3 Designing square patterns

The aim is to design square patterns of different sizes. The patterns are all to be made from smaller tiles all of which are themselves square.

Designs for a 4 × 4 square:

This design consists of four tiles each 2 × 2.
The pattern is rather dull.

Suppose we say that the design must contain at least one 1 × 1 square.

This design is more interesting and consists of seven tiles.

1 Try a 5 × 5 square. Design a pattern which divides the 5 × 5 square into eight smaller squares.

2 Try a 6 × 6 square. Here you must include at least one 1 × 1 square. Design a pattern which divides the 6 × 6 square into nine smaller squares. Colour in the final design to make it look interesting.

3 The 7 × 7 square is more difficult. With no restrictions, design a pattern which divides the 7 × 7 square into nine smaller squares.

4 Design a pattern which divides an 8 × 8 square into ten smaller squares. You must not use a 4 × 4 square.

5 Design a pattern which divides a 9 × 9 square into ten smaller squares. You can use only one 3 × 3 square.

6 Design a pattern which divides a 10 × 10 square into eleven smaller squares. You must include a 3 × 3 square.

7 Design a pattern which divides an 11 × 11 square into eleven smaller squares. You must include a 6 × 6 square.

11.2.4 Calculator words

On a calculator the number 4615 looks like the word 'SIGH' when the calculator is held upside down.

Find the words given by these clues.

1 $221 \times 7 \times 5$ (Sounds like 'cell')

2 $5 \times 601 \times 5 \times 3$ (Wet and rainy)

3 $88^2 - 6$ (Ringer)

4 $0{\cdot}9 \times 5900 - 1$ (Leaves)

5 $62^2 - (4 \times 7 \times 5)$ (Nothing to it)

6 $0{\cdot}88^2 - \dfrac{1}{1000}$ (O Hell)

7 $(5 \times 7 \times 10^3) + (3 \times 113)$ (Gaggle)

8 $44^4 +$ Half of 67 682 (Readable)

9 $5 \times 3 \times 37 \times 1000 - 1420$ (Stick in mind)

10 $3200 - 1320 \div 11$ (Woodwind)

11 $48^4 + 8929$ (Deceitful dame)

12 $31^2 \times 32^2 - 276^2 + 30$ (Not a twig)

13 $(130 \times 135) + (23 \times 3 \times 11 \times 23)$ (Wobbly)

14 $164 \times 166^2 + 734$ (Almost big)

15 $8794^2 + 25 \times 342{\cdot}28 + 120 \times 25$ (Thin skin)

16 $0{\cdot}08 - (3^2 \div 10^4)$ (Ice house)

17 $235^2 - (4 \times 36{\cdot}5)$ (Shiny surface)

18 $(80^2 + 60^2) \times 3 + 81^2 + 12^2 + 3013$ (Ship gunge)

19 $3 \times 17 \times (329^2 + 2 \times 173)$ (Unlimbed)

20 $230 \times 230\tfrac{1}{2} + 30$ (Fit feet)

21 $33 \times 34 \times 35 + 15 \times 3$ (Beleaguer)

22 $0.32^2 + \dfrac{1}{1000}$ (Did he or didn't he?)

23 $(23 \times 24 \times 25 \times 26) + (3 \times 11 \times 10^3) - 20$ (Help)

24 $(16^2 + 16)^2 - (13^2 - 2)$ (Slander)

25 $(3 \times 661)^2 - (3^6 + 22)$ (Pester)

26 $(22^2 + 29.4) \times 10$; $(3.03^2 - 0.02^2) \times 100^2$ (Four words)
(Goliath)

27 $1.25 \times 0.2^6 + 0.2^2$ (Tissue time)

28 $(710 + (1823 \times 4)) \times 4$ (Liquor)

29 $(3^3)^2 + 2^2$ (Wriggler)

30 $14 + (5 \times (83^2 + 110))$ (Bigger than a duck)

31 $2 \times 3 \times 53 \times 10^4 + 9$ (Opposite to hello, almost!)

32 $(177 \times 179 \times 182) + (85 \times 86) - 82$ (Good salesman)

12 Revision

12.1 Revision exercises

Exercise 1

1 Copy this bill and complete it by filling in the four blank spaces.

8 rolls of wallpaper at £3·20 each = £ ⬚

3 tins of paint at £ ⬚ each = £ 20·10

⬚ brushes at £2·40 each = £ 9·60

Total = £ ⬚

2 Write each sequence and find the next two numbers.
  a 2, 9, 16, 23, ___, ___
  b 20, 18, 16, 14, ___, ___
  c −5, −2, 1, 4, ___, ___
  d 128, 64, 32, 16, ___, ___
  e 8, 11, 15, 20, ___, ___

3 Joshua buys 450 pencils at 3 pence each. What change does he receive from £20?

4 Every day at school Stephen buys a roll for 34p, crisps for 21p and a drink for 31p.
How much does he spend in pounds in the whole school year of 200 days?

5 An athlete runs 25 laps of a track in 30 minutes 10 seconds.
  a How many seconds does he take to run 25 laps?
  b How long does he take to run one lap, if he runs the 25 laps at a constant speed?

6 A pile of 250 tiles is 2 m thick. What is the thickness of one tile in cm?

7 Work out
  a 20% of £65
  b 37% of £400
  c 8·5% of £2000.

8 In a test, the marks of nine students were 7, 5, 2, 7, 4, 9, 7, 6, 6. Find
  a the mean mark
  b the median mark
  c the modal mark.

9 Work out
  a $-6 - 5$ **b** $-7 + 30$
  c $-13 + 3$ **d** -4×5
  e $-3 \times (-2)$ **f** $-4 + (-10)$

10 Given $a = 3$, $b = -2$ and $c = 5$, work out
  a $b + c$ **b** $a - b$
  c ab **d** $a + bc$

11 Solve the equations.
  a $x - 6 = 3$ **b** $x + 9 = 20$
  c $x - 5 = -2$ **d** $3x + 1 = 22$

12 Which of these nets can be used to make a cube?

  a

  b

  c

Cut out the nets on square grid paper.

13 a Draw the next diagram in this sequence.

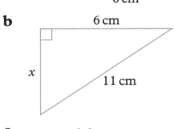

b Write the number of squares in each diagram.
c Describe in words the sequence you obtain in part **b**.
d How many squares will there be in the diagram which has 13 squares on the bottom row?

Exercise 2

1 Solve the equations.
a $3x - 1 = 20$ **b** $4x + 3 = 4$
c $5x - 7 = -3$

2 Copy the diagrams and then calculate x, correct to 3 sf.

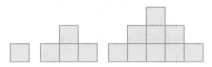

Use Pythagoras' theorem.

a

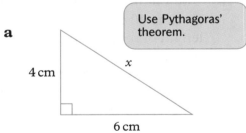

4 cm

x

6 cm

b

6 cm

x

11 cm

c

3·2 cm

x

8 cm

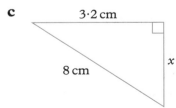

3 A bag contains 3 red balls and 5 white balls. Find the probability of selecting
a a red ball
b a white ball.

4 A box contains 2 yellow discs, 4 blue discs and 5 green discs. Find the probability of selecting
a a yellow disc
b a green disc
c a blue or a green disc.

5 Work out on a calculator, correct to 4 sf.
a $3·61 - (1·6 \times 0·951)$
b $\dfrac{(4·65 + 1·09)}{(3·6 - 1·714)}$

6 Look at this diagram.

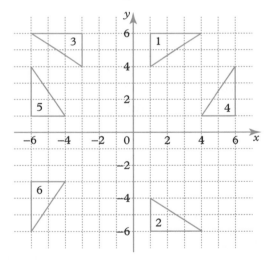

Describe fully these transformations.
a $\triangle 1 \rightarrow \triangle 2$
b $\triangle 1 \rightarrow \triangle 3$
c $\triangle 1 \rightarrow \triangle 4$
d $\triangle 5 \rightarrow \triangle 1$
e $\triangle 5 \rightarrow \triangle 6$
f $\triangle 4 \rightarrow \triangle 6$

7 Plot and label these triangles on an appropriate grid.
$\triangle 1$: $(-3, -6)$, $(-3, -2)$, $(-5, -2)$
$\triangle 2$: $(-5, -1)$, $(-5, -7)$, $(-8, -1)$
$\triangle 3$: $(-2, -1)$, $(2, -1)$, $(2, 1)$
$\triangle 4$: $(6, 3)$, $(2, 3)$, $(2, 5)$
$\triangle 5$: $(8, 4)$, $(8, 8)$, $(6, 8)$
$\triangle 6$: $(-3, 1)$, $(-3, 3)$, $(-4, 3)$

Describe fully these transformations.
a $\triangle 1 \rightarrow \triangle 2$
b $\triangle 1 \rightarrow \triangle 3$
c $\triangle 1 \rightarrow \triangle 4$
d $\triangle 1 \rightarrow \triangle 5$
e $\triangle 1 \rightarrow \triangle 6$
f $\triangle 3 \rightarrow \triangle 5$
g $\triangle 6 \rightarrow \triangle 2$

Exercise 3

1 The tables show the rail fares for adults and part of a rail timetable for trains between Cambridge and Bury St. Edmunds.

Fares for one adult

Cambridge

£2·00	Dullingham			
£2·40	80p	Newmarket		
£2·60	£2·00	£1·20	Kennett	
£4·00	£2·60	£2·40	£1·60	Bury St. Edmunds

Train times

Cambridge	11:20
Dullingham	11:37
Newmarket	11:43
Kennett	11:52
Bury St. Edmunds	12:06

a How much would it cost for four adults to travel from Dullingham to Bury St. Edmunds?
b How long does this journey take?

2 This is a sketch of a clock tower.

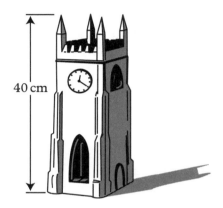

40 cm

Sarah makes a model of the tower using a scale of 1 to 20.
a The minute hand on the tower clock is 40 cm long. What is the length of the minute hand on the model?
b The height of the model is 40 cm. What is the height h, in metres, of the actual clock tower?

3 Look at this number pattern.
$(2 \times 1) - 1 = 2 - 1$
$(3 \times 3) - 2 = 8 - 1$
$(4 \times 5) - 3 = 18 - 1$
$(5 \times 7) - 4 = 32 - 1$
$(6 \times a) - 5 = b - 1$
i What number does the letter a stand for?
ii What number does the letter b stand for?
iii Write the next line in the pattern.

4 a Plot and label these triangles.
$\triangle 1$: $(-3, 4)$, $(-3, 8)$, $(-1, 8)$
$\triangle 5$: $(-8, -2)$, $(-8, -6)$, $(-6, -2)$

b Draw $\triangle 2$, $\triangle 3$, $\triangle 4$, $\triangle 6$ and $\triangle 7$ using these transformations.

i $\triangle 1 \rightarrow \triangle 2$: translation
$$\begin{pmatrix} 9 \\ -4 \end{pmatrix}$$

ii $\triangle 2 \rightarrow \triangle 3$: translation
$$\begin{pmatrix} -4 \\ -8 \end{pmatrix}$$

iii $\triangle 3 \rightarrow \triangle 4$: reflection in the line $y = x$

iv $\triangle 5 \rightarrow \triangle 6$: rotation 90° anticlockwise, centre $(-4, -1)$

v $\triangle 6 \rightarrow \triangle 7$: rotation 180°, centre $(0, -1)$

c Write the coordinates of $\triangle 2$, $\triangle 3$, $\triangle 4$, $\triangle 6$, and $\triangle 7$.

5 The faces of a round and square clock are exactly the same area. If the round clock has a radius of 10 cm, how wide is the square clock?

6 A metal ingot is in the form of a solid cylinder of length 7 cm and radius 3 cm.
a Calculate the volume, in cm^3, of the ingot. Give your answer to 1 dp.
The ingot is to be melted down and used to make cylindrical coins of thickness 0·3 cm and radius 1·2 cm.
b Calculate the volume, in cm^3, of each coin. Give your answer to 1 dp.
c Calculate the number of coins which can be made from the ingot, assuming that there is no wastage of metal.

Exercise 4

1 In December 2006, a factory employed 220 men, each man being paid £300 per week.
a Calculate the total weekly wage bill for the factory.
b In January 2007, the work force of 220 was reduced by 10 per cent.
Find the number of men employed at the factory after the reduction.
c Also in January 2007, the weekly wage of £300 was increased by 10 per cent. Find the new weekly wage.
d Calculate the total weekly wage bill in January 2007.

2 $1 + 3 = 2^2 \qquad 1 + 3 + 5 = 3^2$
a $1 + 3 + 5 + 7 = x^2$
Calculate x.
b $1 + 3 + 5 + \ldots + n = 100$
Calculate n.

3 A motorist travelled 800 miles during May, when the cost of petrol was 90 pence per litre. In June the cost of petrol increased by 10% and he reduced his mileage for the month by 5%.

   a What was the cost, in pence per litre, of petrol in June?

   b How many miles did he travel in June?

4 The distance–time graphs for several moving objects are shown. Decide which line represents each of these.

   ● Hovercraft from Dover
   ● Car ferry from Dover
   ● Cross-channel swimmer
   ● Marker buoy outside harbour
   ● Train from Dover
   ● Car ferry from Calais

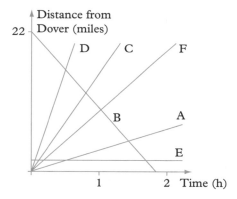

5 The mean mass of 10 boys in a class is 56 kg.

   a Calculate the total mass of these 10 boys.

   b Another boy, whose mass is 67 kg, joins the group. Calculate the mean mass of the 11 boys.

6 This electricity bill is not complete.

METER READING on	
07-11-05	26819 units
METER READING on	
04-02-06	$\boxed{A}$ units
ELECTRICITY USED	1455 units
1455 units at 5·44 pence per unit	£ $\boxed{B}$
Quarterly charge	£ 6·27
TOTAL (now due)	£ $\boxed{C}$

   a Find the correct amounts, A, B and C, to be placed in each box.

   b In 2005, in what month was the meter read?

Exercise 5

1 $a = \dfrac{1}{2}$, $b = \dfrac{1}{4}$. Which one of these expressions has the greatest value?

   a ab **b** $a + b$ **c** $\dfrac{a}{b}$

   d $\dfrac{b}{a}$ **e** $(ab)^2$

2 **a** Calculate the speed (in metres per second) of a slug which moves a distance of 300 cm in 1 minute.

   b Calculate the time taken for a bullet to travel 8 km at a speed of 5000 m/s.

   c Calculate the distance flown, in a time of four hours, by a pigeon which flies at a speed of 12 m/s.

3 Given $a = 3$, $b = 4$ and $c = -2$, evaluate
  a $2a^2 - b$ **b** $a(b - c)$
  c $2b^2 - c^2$

4 In the diagram, the equations of the lines are $y = 3x$, $y = 6$, $y = 10 - x$ and $y = \frac{1}{2}x - 3$.

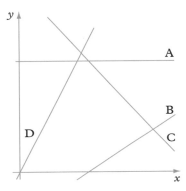

Find the equation corresponding to each line.

5 Given that $s - 3t = rt$, express
  a s in terms of r and t
  b r in terms of s and t.

6 Find the area of this shape.

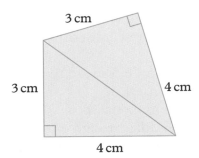

3 cm

3 cm   4 cm

4 cm

7 A cylinder of radius 8 cm has a volume of 2 litres. Calculate the height of the cylinder. (1 litre = 1000 cm³) Give your answer to 1 dp.

Exercise 6

1 The pump shows the price of petrol in a garage.

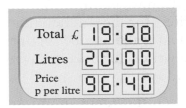

Total £	19·28
Litres	20·00
Price p per litre	96·40

One day I buy £20 worth of petrol. How many litres do I buy?

2 Given that $x = 4$, $y = 3$, $z = -2$, evaluate
  a $2x(y + z)$ **b** $(xy)^2 - z^2$
  c $x^2 + y^2 + z^2$ **d** $(x + y)(x - z)$

3 Twenty-seven small wooden cubes fit exactly inside a cubical box without a lid. How many of the cubes are touching the sides or the bottom of the box?

4 Each diagram in the sequence below consists of a number of dots.

Diagram number	1	2	3
	• • • • • •	• • • • • • • • • •	• • • • • • • • • • • • • •

  a Draw diagram number 4, diagram number 5 and diagram number 6.
  b Copy and complete this table.

Diagram number	Number of dots
1	6
2	10
3	
4	
5	
6	

c Without drawing the diagrams, state the number of dots in
  i diagram number 10
  ii diagram number 15.
d Write x for the diagram number and n for the number of dots and write a formula involving x and n.

5 Write each statement and make corrections where necessary.
a $t + t + t = t^3$
b $a^2 \times a^2 = 2a^2$
c $2n \times n = 2n^2$

6 The table shows the number of students in a class who scored marks 3 to 8 in a test.

Marks	3	4	5	6	7	8
Number of students	2	3	6	4	3	2

Find
a the mean mark
b the modal mark
c the median mark.

7 The diagram shows a regular octagon with centre O.
Find angles a and b.

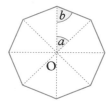

8 a Given that $x - z = 5y$, express z in terms of x and y.
b Given that $mk + 3m = 11$, express k in terms of m.
c For the formula $T = \dfrac{C}{z}$ express, z in terms of T and C.

Exercise 7

1 This graph shows a car journey from Gateshead to Middlesbrough and back again.

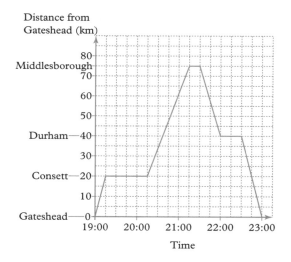

a Where is the car
  i at 19:15
  ii at 22:15
  iii at 22:45?
b How far is it
  i from Consett to Middlesbrough
  ii from Durham to Gateshead?
c At what speed does the car travel
  i from Gateshead to Consett
  ii from Consett to Middlesbrough
  iii from Middlesbrough to Durham
  iv from Durham to Gateshead?
d For how long is the car stationary during the journey?

2 Work out the difference between one ton and one tonne.

Give your answer to the nearest kg.

> 1 tonne = 1000 kg
> 1 ton = 2240 lb
> 1 lb = 454 g

3 A motorist travelled 200 miles in five hours. Her average speed for the first 100 miles was 50 mph. How long did she take for the second 100 miles?

4 Evaluate these and give the answers to 3 significant figures.

a $\sqrt{(9\cdot61 \times 0\cdot0041)}$

b $\dfrac{1}{9\cdot5} - \dfrac{1}{11\cdot2}$

c $\dfrac{15\cdot6 \times 0\cdot714}{0\cdot0143 \times 12}$

d $\sqrt[3]{\left(\dfrac{1}{5 \times 10^3}\right)}$

5 Throughout his life Mr Cram's heart has been beating at an average rate of 72 beats per minute. Mr Cram is exactly sixty years old. How many times has his heart beat during his life? Give the answer in standard form correct to two significant figures.

6 **Estimate** the answer correct to one significant figure. Do not use a calculator.

a $(612 \times 52) \div 49\cdot2$

b $(11\cdot7 + 997\cdot1) \times 9\cdot2$

c $\sqrt{\left(\dfrac{91\cdot3}{10\cdot1}\right)}$

> For **a** write $(600 \times 50) \div 50$

d $\pi\sqrt{(5\cdot2^2 + 18\cdot2^2)}$

7 In the quadrilateral PQRS, PQ = QS = QR, PS is parallel to QR and ∠QRS = 70°. Calculate

a ∠RQS

b ∠PQS.

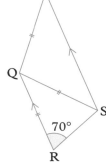

8 A bag contains x green discs and 5 blue discs. A disc is selected. Find, in terms of x, the probability of selecting a green disc.

9 In the diagram, the equations of the lines are $2y = x - 8$, $2y + x = 8$, $4y = 3x - 16$ and $4y + 3x = 16$, but not in that order.

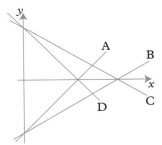

Find the equation corresponding to each line.

Exercise 8

1 Asda sell their 'own-label' raspberry jam in two sizes.

Which jar represents the better value for money? (1 kg = 2·2 lb)

2 A photo 21 cm by 12 cm is enlarged as shown.

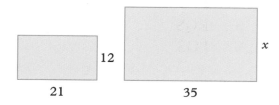

12

21

35

x

  a What is the scale factor of the enlargement?

  b Work out the length *x*.

3 Nadia said: 'I thought of a number, multiplied it by 6, then added 15. My answer was less than 200.'

  a Write Nadia's statement in symbols, using *x* as the starting number.

  b Nadia actually thought of a prime number. What was the largest prime number she could have thought of?

4 Find the area of the red part of the flag. All lengths are in cm.

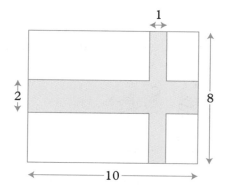

1

2

8

10

5 Point B is on a bearing 120° from point A. The distance from A to B is 110 km.

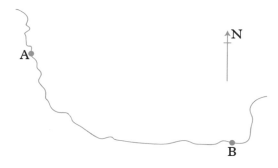

A

N

B

  a Draw a diagram showing the positions of A and B.
Use a scale of 1 cm to 10 km.

  b Ship S is on a bearing 072° from A.
Ship S is on a bearing 325° from B.
Show S on your diagram and state the distance from S to B.

6 Evaluate these using a calculator (answers to 4 sf).

  a $\dfrac{0.74}{0.81 \times 1.631}$ **b** $\sqrt{\left(\dfrac{9.61}{8.34 - 7.41}\right)}$

  c $\left(\dfrac{0.741}{0.8364}\right)^{3}$ **d** $\dfrac{8.4 - 7.642}{3.333 - 1.735}$

7 The mean of four numbers is 21.

  a Calculate the sum of the four numbers.

Six other numbers have a mean of 18.

  b Calculate the mean of the ten numbers.

8 Tins of peaches are packed 24 to a box. How many boxes are needed for 1285 tins?

12.2 Multiple choice tests

Test 1

1 How many mm are there in 1 m 1 cm?

A 1001
B 1110
C 1010
D 1100

2 If $x = 3$, then $x^2 - 2x = ?$

A 3
B 4
C 0
D 5

3 The gradient of the line $y = 2x - 1$ is

A 2
B −1
C $\frac{1}{2}$
D −2

4 The mean weight of a group of 11 men is 70 kg. What is the mean weight of the remaining group when a man of weight 90 kg leaves?

A 80 kg
B 72 kg
C 68 kg
D 62 kg

5 A, B, C and D are points on the sides of a rectangle. Find the area in cm² of quadrilateral ABCD.

A $27\frac{1}{2}$
B 28
C $28\frac{1}{2}$
D cannot be found

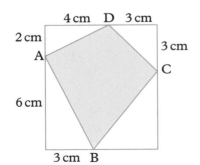

6 The area of the triangle is

A 6
B 10
C 12
D depends on x

7 The formula $\frac{x}{a} + b = c$ is rearranged to make x the subject. What is x?

A $a(c - b)$
B $ac - b$
C $\frac{c - b}{a}$
D $ac + ab$

8 In standard form the value of $2000 \times 80\,000$ is

A 16×10^6
B $1{\cdot}6 \times 10^9$
C $1{\cdot}6 \times 10^7$
D $1{\cdot}6 \times 10^8$

9 The sum of the lengths of the edges of a cube is 36 cm. The volume, in cm³, of the cube is

A 36
B 27
C 64
D 48

10 In the triangle the size of angle x is

A 35°
B 70°
C 110°
D 40°

11 A man paid tax on £9000 at 30%. He paid the tax in 12 equal payments. Each payment was

A £2·25
B £22·50
C £225
D £250

12 The approximate value of
$$\frac{3 \cdot 96 \times (0 \cdot 5)^2}{97 \cdot 1}$$ is

  A 0·01
  B 0·02
  C 0·04
  D 0·1

13 Given that $\dfrac{3}{n} = 5$, then $n =$

  A 2
  B −2
  C $1\frac{2}{3}$
  D 0·6

14 Cube A has side 2 cm. Cube B has side 4 cm.
$$\frac{\text{Volume of B}}{\text{Volume of A}} =$$

  A 2
  B 4
  C 8
  D 16

15 How many tiles of side 50 cm will be needed to cover this floor?

  A 16
  B 32
  C 64
  D 84

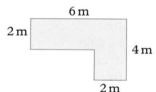

16 The equation $ax + 6 = 0$ has a solution $x = 3$. What is a?

  A 1
  B −2
  C $\sqrt{2}$
  D 2

17 Which of these is/are correct?
1 $\sqrt{0 \cdot 16} = 0 \cdot 4$
2 $0 \cdot 2 \div 0 \cdot 1 = 0 \cdot 2$
3 $\dfrac{4}{7} > \dfrac{3}{5}$

  A 1 only
  B 2 only
  C 3 only
  D 1 and 2

18 How many prime numbers are there between 30 and 40?

  A 0
  B 1
  C 2
  D 3

19 Kevin is paid £180 per week after a pay rise of 20%. What was he paid before?

  A £144
  B £150
  C £160
  D £164

20 A car travels for 20 minutes at 45 mph. How far does the car travel?

  A 900 miles
  B $2\frac{1}{4}$ miles
  C 9 miles
  D 15 miles

21 The point $(3, -1)$ is reflected in the line $y = 2$. The new coordinates are

  A $(3, 5)$
  B $(1, -1)$
  C $(3, 4)$
  D $(0, -1)$

22 Given the equation $5^x = 120$, the best approximate solution is $x =$

  A 2
  B 3
  C 4
  D 22

23 Here are four statements about the diagonals of a rectangle. The statement which is not **always** true is
  A They are equal in length.
  B They divide the rectangle into four triangles of equal area.
  C They cross at right angles.
  D They bisect each other.

24 The shaded area in cm^2 is

  A $16 - 2\pi$
  B $16 - 4\pi$
  C $\dfrac{4}{\pi}$
  D $64 - 8\pi$

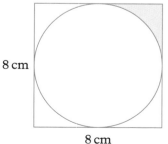

25 An estimate of the value of
$$\frac{204 \cdot 7 \times 97 \cdot 5}{1064 \cdot 2},$$
to one significant figure, is

A 2
B 20
C 200
D 2000

Test 2

1 What is the value of the expression
$(x - 2)(x + 4)$
when $x = -1$?

A 9
B −9
C 5
D −5

2 The perimeter of a square is 36 cm. What is its area?

A 36 cm²
B 324 cm²
C 81 cm²
D 9 cm²

3 The gradient of the line $2x + y = 3$ is

A 3
B −2
C $\dfrac{1}{2}$
D $-\dfrac{1}{2}$

4 The shape consists of four semicircles placed round a square of side 2 m. The area of the shape, in m², is

A $2\pi + 4$
B $2\pi + 2$
C $4\pi + 4$
D $\pi + 4$

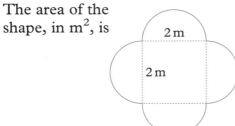

5 A firm employs 1200 people, of whom 240 are men. The percentage of employees who are men is

A 40%
B 10%
C 15%
D 20%

6 A car is travelling at a constant speed of 30 mph. How far will the car travel in 10 minutes?

A $\dfrac{1}{3}$ mile
B 3 miles
C 5 miles
D 6 miles

7 What are the coordinates of the point $(1, -1)$ after reflection in the line $y = x$?

A $(-1, 1)$
B $(1, 1)$
C $(-1, -1)$
D $(1, -1)$

8 $\dfrac{1}{3} + \dfrac{2}{5} =$
[no calculators!]

A $\dfrac{2}{8}$
B $\dfrac{3}{8}$
C $\dfrac{3}{15}$
D $\dfrac{11}{15}$

9 In the triangle the size of the largest angle is

A 30°
B 90°
C 120°
D 80°

10 800 decreased by 5% is

A 795
B 640
C 760
D 400

11 What is the area, in m², of a square with each side 0·02 m long?

A 0·0004
B 0·004
C 0·04
D 0·4

12 Given $a = \dfrac{3}{5}$,
$b = \dfrac{1}{3}$, $c = \dfrac{1}{2}$ then

A $a < b < c$
B $a < c < b$
C $a > b > c$
D $a > c > b$

13 The **larger** angle between south-west and east is

A 225°
B 240°
C 135°
D 315°

14 In a triangle PQR, $\angle$PQR = 50° and point X lies on PQ such that QX = XR. Calculate $\angle$QXR.

A 100°
B 50°
C 80°
D 65°

15 What is the value of $1 - 0.05$ as a fraction?

A $\dfrac{1}{20}$

B $\dfrac{9}{10}$

C $\dfrac{19}{20}$

D $\dfrac{5}{100}$

16 I start with x, then square it, multiply by 3 and finally subtract 4. The final result is

A $(3x)^2 - 4$
B $(3x - 4)^2$
C $3x^2 - 4$
D $(3x - 4)^2$

17 Given that $m = 2$ and $n = -3$, what is mn^2?

A -18
B 18
C -36
D 36

18 How many prime numbers are there between 50 and 60?

A 1
B 2
C 3
D 4

19 £240 is shared in the ratio $2 : 3 : 7$. The largest share is

A £130
B £140
C £150
D £160

20 Adjacent angles in a parallelogram are x and $3x$. The smallest angles in the parallelogram are each

A 30°
B 45°
C 60°
D 120°

21 A square has sides of 10 cm. When the sides of the square are increased by 10% the area is increased by

A 10%
B 20%
C 21%
D 15%

22 The volume, in cm^3, of the cylinder is

A 9π
B 12π
C 600π
D 900π

6 cm

1 m

23 $3a + 2a - 7 - a = 21$ The value of a is

A $3\dfrac{1}{2}$

B 7

C $4\dfrac{2}{3}$

D 8

24 Four people each toss a coin. What is the probability that the fourth person will toss a tail?

A $\dfrac{1}{2}$

B $\dfrac{1}{4}$

C $\dfrac{1}{8}$

D $\dfrac{1}{16}$

25 What is next in the sequence $1, 3, 7, 15$?

A 23
B 21
C 31
D 24

Answers

1   Number 1

page 1 **Exercise 1 Ⓒ**

1 a 3   **b** 4   **c** 539   **d** 2000 + 400 + 10 + 6
2 a 80   **b** 3000
3 a Five hundred and twenty three
  b Six thousand, four hundred and ten
  c Twenty five thousand
4 a 217   **b** 4250   **c** 5 000 000   **d** 6020   **5** Roydon, Penton, Quarkby
6 a 29, 85, 290, 314
  b 564, 645, 1666, 2000, 5010
  c 7510, 8888, ten thousand, 60 000
7 a 258, 285, 528, 582   **b** 852   **8 a** 1348   **b** 8431   **c** 483   **9** 2 × 20p, 2 × 2p
10 25 000   **11** 4110   **12** 510 212   **13** 70   **14 a** 7320   **b** 2037
15 a 75 423   **b** 23 574   **16 a** 257   **b** 3221   **c** 704   **17 a** 1392   **b** 26 611   **c** 257 900
18 a 0   **b** 52 000   **19 a** 2058, 2136, 2142, 2290   **b** 5029, 5299, 5329, 5330
  c 25 000, 25 117, 25 171, 25 200, 25 500
20 a 96 + 84 + 73 *or* 94 + 86 + 73 *or* 96 + 83 + 74
    or 94 + 83 + 76 *or* 93 + 86 + 74 *or* 93 + 84 + 76
  b 974 + 863 *or* 964 + 873 *or* 973 + 864 *or* 963 + 874
21 100   **22** 10

page 4 **Exercise 2 Ⓒ**

1 a 1, 3, 9   **b** 1, 2, 5, 10   **c** 1, 2, 3, 4, 6, 12
2 a 1, 2, 7, 14 **b** 1, 2, 4, 5, 10, 20 **c** 1, 2, 3, 5, 6, 10, 15, 30   **3** 1 and 5
4 a

Number	Factors
12	1, 2, 3, 4, 6, 12
20	1, 2, 4, 5, 10, 20

  b 1, 2, 4

5 b i 3, 6, 9, 12   **ii** 4, 8, 12, 16   **iii** 10, 20, 30, 40
6 16   **7** 76   **8** 22
9 34   **10** 2   **11 a** even   **b** odd

page 6 **Exercise 3 Ⓔ**

1 a 2, 4, 6, 8, 10, 12   **b** 5, 10, 15, 20, 25, 30   **c** 10
2 a 4, 8, 12, 16   **b** 12, 24, 36, 48   **c** 12
3 a 18   **b** 24   **c** 70   **d** 12   **e** 10   **f** 63
4 12   **5** 6
6 a 6   **b** 11   **c** 9   **d** 6   **e** 12   **f** 10
7 a 6   **b** 40   **c** Many possible answers, e.g. 22 and 33   **d** 2 and 5 *or* 1 and 10
8 15   **9** 21

page 7 **Exercise 4 C**

1

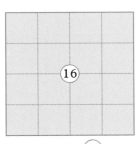

$4 \times 4 = \boxed{16}$ $5 \times 5 = \boxed{25}$

2 a 25   **b** 100   **c** 36   **d** 13   **e** 50   **f** 96
3 a 4   **b** 9   **c** 1   **4 a** 3   **b** 5   **c** 10   **d** 7
5 a 8   **b** 7   **c** 36   **6 a** True   **b** True   **c** False   **d** True
7 a 104   **b** 28   **c** 18   **d** 900   **e** 6   **f** 10 001
8 a 4 9 ⑷⑻ 64   **b** 16 25 36 ⑸⑸   **c** 1 ⑻ 16 100

   d 4 ⑽ 49 81   **e** 25 64 ⑴⑵⓪ 144   **f** 16 36 ⑴⓪⑻ 121
9 a 9   **b** 25   **c** 36   **d** 16   **e** 4   **f** 100   **g** 49   **h** 81
10 a 25   **b** 27   **c** 125   **11** 13

page 9 **Exercise 5 E**

1 2, 7, 17   **2** 2, 3, 5, 7, 11, 13, 17, 19, 23, 29   **3** 2   **4** 31, 37
5 a 11   **b** 17   **c** 29   **d** 23   **e** 37   **f** 47   **g** 59   **h** 41
6 a 7 ⑼ 13 17   **b** 2 13 ⑵⑴ 23   **c** 11 13 19 ⑵⑺

   d ⑴⑸ 19 29 31   **e** 31 37 ⑶⑼ 41   **f** 23 43 47 ⑷⑼

7 a 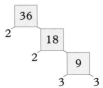   **b** 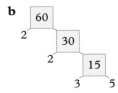   **c**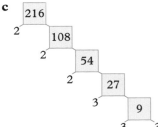

   d 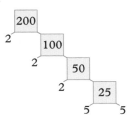   **e**

8 $2 \times 2 \times 2 \times 2 \times 3 \times 5 \times 5$
9 a $2 \times 2 \times 7$ **b** $2 \times 2 \times 2 \times 2 \times 2$ **c** 2×17 **d** $3 \times 3 \times 3 \times 3$ **e** $2 \times 3 \times 7 \times 7$

10

	Even number	Square number	Prime number
Factor of 14	2	1	7
Multiple of 3	6	9	3
Between 3 & 9	8	4	5

11

	Factor of 10	More than 3	Factor of 18
Square number	1	4	9
Even number	2	8	6
Prime number	5	7	3

12 Multiple solutions   **13** 29, 31   41, 43   59, 61   71, 73

14 a = 9 = 3^2 **b** $1 + 3 + 5 + 7 + 9 = 25 = 5^2$
    = $16 = 4^2$ $1 + 3 + 5 + 7 + 9 + 11 = 36 = 6^2$
          $1 + 3 + 5 + 7 + 9 + 11 + 13 = 49 = 7^2$
          $1 + 3 + 5 + 7 + 9 + 11 + 13 + 15 = 64 = 8^2$
          $1 + 3 + 5 + 7 + 9 + 11 + 13 + 15 + 17 = 81 = 9^2$

15 $13 + 15 + 17 + 19 = 64 = 4^3$ **16** No, e.g. 2 + 3 = 5, an odd number.
    $21 + 23 + 25 + 27 + 29 = 125 = 5^3$
    $31 + 33 + 35 + 37 + 39 + 41 = 216 = 6^3$

17 For example, $7 = 3 + 2^2$, $9 = 5 + 2^2$, $11 = 7 + 2^2$, $13 = 5 + 2^3$,
          $15 = 7 + 2^3$ $17 = 13 + 2^2$, $19 = 11 + 2^3$, $21 = 17 + 2^2$

page 12 **Exercise 6** ⒸC

 1 67   **2** 96   **3** 13   **4** 121   **5** 144   **6** 83   **7** 225   **8** 23   **9** 231   **10** 333

page 13 **Exercise 7** ⒸC

1 58	**2** 67	**3** 251	**4** 520	**5** 961	**6** 337	**7** 496	**8** 511
9 320	**10** 992	**11** 647	**12** 1071	**13** 328	**14** 940	**15** 197	**16** 2384
17 3312	**18** 5335	**19** 7008	**20** 8193	**21** 1031	**22** 3121	**23** 3541	**24** 827
25 6890	**26** 1021	**27** 13011	**28** 21844	**29** 115387	**30** 19885		

page 14 **Exercise 8** ⒸC

1 34	**2** 28	**3** 23	**4** 82	**5** 111	**6** 204	**7** 57	**8** 15
9 56	**10** 23	**11** 137	**12** 461	**13** 381	**14** 542	**15** 301	**16** 113
17 533	**18** 123	**19** 522	**20** 81	**21** 265	**22** 5646	**23** 4819	**24** 6388
25 7832	**26** 384	**27** 399	**28** 5804	**29** 1361	**30** 548	**31** 50	

page 14 **Exercise 9** ⒸC

1

3	8	1
2	4	6
7	0	5

2

4	11	6
9	7	5
8	3	10

3

8	1	6
3	5	7
4	9	2

4

7	2	9
8	6	4
3	10	5

5

12	7	14
13	11	9
8	15	10

6

17	10	15
12	14	16
13	18	11

7

1	12	7	14
8	13	2	11
10	3	16	5
15	6	9	4

8

15	6	9	4
10	3	16	5
8	13	2	11
1	12	7	14

9

3	10	12	17
14	15	5	8
9	4	18	11
16	13	7	6

10

18	9	12	7
13	6	19	8
11	16	5	14
4	15	10	17

11

11	24	7	20	3
4	12	25	8	16
17	5	13	21	9
10	18	1	14	22
23	6	19	2	15

12

16	23	10	17	4
3	15	22	9	21
20	2	14	26	8
7	19	6	13	25
24	11	18	5	12

page 16 **Exercise 10 C**

1 63	**2** 96	**3** 252	**4** 140	**5** 639	**6** 230
7 1230	**8** 168	**9** 1477	**10** 2114	**11** 1065	**12** 1923
13 168	**14** 1884	**15** 1179	**16** 1712	**17** 4920	**18** 3684
19 12 846	**20** 15 125	**21** 2592	**22** 4501	**23** 2655	**24** 6410
25 8460	**26** 2200	**27** 4417	**28** 7965	**29** 3976	**30** 12 918

page 16 **Exercise 11 C**

1 23	**2** 143	**3** 211	**4** 115	**5** 178	**6** 232
7 527	**8** 528	**9** 83	**10** 497	**11** 273	**12** 6024
13 604	**14** 271	**15** 415	**16** 383	**17** 824	**18** 936
19 321	**20** 2142	**21** 9486	**22** 2314	**23** 241	**24** 7005
25 837	**26** 6145	**27** 2638	**28** 415	**29** 15 r 1	**30** 21 r 3
31 28 r 1	**32** 62 r 1	**33** 41 r 1	**34** 24 r 4	**35** 56 r 1	**36** 130 r 3
37 535 r 3	**38** 1283 r 3	**39** 1506 r 3	**40** 689 r 1		

page 17 **Speed tests**

Test 1

1 22	**2** 45	**3** 8	**4** 58	**5** 77	**6** 48	**7** 36	**8** 9
9 110	**10** 42	**11** 48	**12** 7	**13** 116	**14** 21	**15** 900	

Test 2

1 22	**2** 27	**3** 54	**4** 45	**5** 143	**6** 9	**7** 5	**8** 1300
9 198	**10** 50	**11** 57	**12** 21	**13** 49	**14** 37	**15** 12	

Test 3

1 40	**2** 40	**3** 10	**4** 81	**5** 98	**6** 90	**7** 6	**8** 35
9 52	**10** 190	**11** 5	**12** 8	**13** 110	**14** 195	**15** 32	

Test 4

1 35	**2** 18	**3** 83	**4** 8	**5** 32	**6** 89	**7** 29	**8** 12
9 100	**10** 154	**11** 55	**12** 11	**13** 5000	**14** 225	**15** 63	

page 17 **Exercise 12 C**

1 135 cm **2** 20p, 10p, 5p, 2p **3** £27 **4** 57
5 a £8·37 **b** £1·63 **6** 893 **7** £6·44 **8** 900 g
9 42 **10 a** 145 **b** 135 **c** 145 + 135 = 280 **11** *e.g.* $a = 30$, $b = 25$ There are other possible
$a = 225$, $b = 1$ answers.

page 18 **Exercise 13 C**

1

11	+	4	→	15
×		÷		
6	÷	2	→	3
↓		↓		
66	×	2	→	132

2

9	+	17	→	26
×		−		
5	×	8	→	40
↓		↓		
45	÷	9	→	5

3

14	+	17	→	31
×		+		
4	×	23	→	92
↓		↓		
56	−	40	→	16

4

15	÷	3	→	5
+		×		
22	×	5	→	110
↓		↓		
37	−	15	→	22

5

9	×	10	→	90
+		÷		
11	÷	2	→	$5\frac{1}{2}$
↓		↓		
20	×	5	→	100

6

26	×	2	→	52
−		×		
18	×	4	→	72
↓		↓		
8	÷	8	→	1

7

5	×	12	→	60
×		÷		
20	+	24	→	44
↓		↓		
100	×	$\frac{1}{2}$	→	50

8

7	×	6	→	42
÷		÷		
14	−	3	→	11
↓		↓		
$\frac{1}{2}$	×	2	→	1

9

19	×	2	→	38
−		÷		
12	×	4	→	48
↓		↓		
7	−	$\frac{1}{2}$	→	$6\frac{1}{2}$

page 20 **Exercise 14** C

1 a
```
    2  8  5
 + [5] 1  4
 ─────────
    7 [9][9]
```

b
```
    6  3 [7]
 + [2] 5  2
 ─────────
    8 [8] 9
```

c
```
   [6] 3  5
 +  3  4 [4]
 ─────────
    9 [7] 9
```

2 a
```
    3  5  6
 +5 [2] 6
 ─────────
   [8] 8 [2]
```

b
```
    2 [2] 4
 +  5  3  7
 ─────────
   [7] 6  1
```

c
```
    3  8  8
 + [4] 2 [5]
 ─────────
    8 [1] 3
```

3 a
```
      4 [8]
   ×     3
 ─────────
    1  4  4
```

b
```
      3 [3]
   ×     7
 ─────────
    2  3  1
```

c
```
   [3][2] 1
   ×      5
 ─────────
  1 6  0  5
```

4 a [1][5][0] ÷ 3 = 50   **b** [1][5] × 4 = 60

c 9 × [9] = 81   **d** [1][1][5][2] ÷ 6 = 192

5 a
```
    4 [4] 5
 +2  8 [5]
 ─────────
   [7] 3  0
```

b
```
    4 [2] 7
 + [1] 7 [7]
 ─────────
    6  0  4
```

c
```
   [5] 3 [5]
 +  2 [6] 4
 ─────────
    7  9  9
```

6 a [3][5] × 7 = 245   **b** [5][8] × 10 = 580

c 32 ÷ [4] = 8   **d** [9][5][0] ÷ 5 = 190

7 a [7][2] + 29 = 101   **b** [1][0][8] − 17 = 91

c
```
   [8] 8  9
 − 3 [4] 6
 ─────────
    5  4 [3]
```

d
```
    3  3  5
 −2  1 [8]
 ─────────
   [1][1] 7
```

8 Many possible answers

9 a [4] × [4] − [4] = 12   **b** [8] ÷ [8] + [8] = 9   **c** [8] × [8] + [8] = 72

10 a 5 → ×6 → +9 → 39   **b** 6 → +2 → ×5 → 40

c 2 → +7 → ×4 → 36   **d** 7 → ×3 → −11 → 10

11 a 7 × 4 (−) 3 = 25   **b** 8 × 5 (÷) 2 = 20   **c** 7 (×) 3 − 9 = 12

d 12 (÷) 2 + 4 = 10   **e** 75 ÷ 5 (+) 5 = 20

12 a 5 × 4 × 3 (+) 3 = 63   **b** 5 + 4 (−) 3 (−) 2 = 4   **c** 5 × 2 × 3 (+) 1 = 31

page 22 **Exercise 15 ⓒ**

1 a $\frac{3}{10}, 0.3$ **b** $\frac{1}{10}, 0.1$ **c** $\frac{1}{5}, 0.2$ **d** $\frac{7}{10}, 0.7$ **e** $\frac{4}{5}, 0.8$ **f** $\frac{3}{5}, 0.6$ **g** $\frac{2}{5}, 0.4$ **h** $\frac{9}{10}, 0.9$

2 a $\frac{1}{2}$ **b** 20 **c** 7 **d** $\frac{9}{10}$ **e** $\frac{1}{100}$ **f** $\frac{7}{10}$ **g** 30 **h** $\frac{8}{100} = \frac{2}{25}$

3 a 0·6, 0·7 **b** 0·8, 1·0 **c** 2·5, 3·0 **d** 0·9, 1·1
4 a 0·2 **b** 0·5 **c** 0·9 **d** 1·1 **e** 1·3 **f** 1·7
5 a 0·2 **b** 0·13 **c** 0·02 **d** 0·15 **e** 0·155 **f** 0·227

page 23 **Exercise 16 ⓒ**

1 1·5, 3·7, 4, 12   **2** 1·3, 3·1, 13, 31   **3** 0·2, 5·2, 6, 11
4 0·11, 0·4, 1, 1·7   **5** 2, 2·2, 20, 22   **6** 0·12, 0·21, 0·31
7 0·04, 0·35, 0·4   **8** 0·67, 0·672, 0·7   **9** 0·045, 0·05, 0·07
10 0·089, 0·09, 0·1   **11** 0·57, 0·705, 0·75   **12** 0·041, 0·14, 0·41
13 0·8, 0·809, 0·81   **14** 0·006, 0·059, 0·6   **15** 0·143, 0·15, 0·2
16 0·04, 0·14, 0·2, 0·53   **17** 0·12, 0·21, 1·12, 1·2   **18** 0·08, 0·75, 2·03, 2·3
19 0·26, 0·3, 0·602, 0·62   **20** 0·5, 1·003, 1·03, 1·3   **21** 0·709, 0·79, 0·792, 0·97
22 0·312, 0·321, 1·04, 1·23   **23** 0·007, 0·008, 0·09, 0·091   **24** 2, 2·046, 2·05, 2·5
25 CARWASH   **26 a** 32·51 **b** 0·853 **c** 1·16
27 a £3·50 **b** £0·15 **c** £0·03 **d** £0·10 **e** £12·60 **f** £0·08
28 a True **b** False **c** True **d** True

page 24 **Exercise 17 ⓒ**

1 c **2** a **3** c **4** b **5** b **6** a **7** False
8 True **9** True **10** True **11** True **12** False **13** True **14** True
15 True **16** False **17** True **18** False **19** True **20** False **21** False
22 True **23** True **24** True **25** True **26** False **27** False **28** True
29 False **30** False **31** True **32** True **33** True **34** True

35 **36**

37 **38**

page 25 **Exercise 18 ⓒ**

1 4·3 **2** 0·7 **3** 9·4 **4** 1·2 **5** 16 **6** 10·7 **7** 17·4 **8** 128 **9** 375
10 0·24 **11** 1·92 **12** 5·2 **13** 0·06 **14** 1·76 **15** 3·16 **16** 105 **17** 50 **18** 125

page 26 **Exercise 19 ⓒ**

1 6·8 **2** 8·8 **3** 11·7 **4** 14·9 **5** 6·6 **6** 12·81 **7** 8·77 **8** 10·19
9 14·54 **10** 9·07 **11** 10·14 **12** 20·94 **13** 26·71 **14** 216·95 **15** 9·6 **16** 23·1
17 12·25 **18** 17·4 **19** 0·062 **20** 85·47 **21** 28 **22** 27·02 **23** 2·44 **24** 275·9

25 96·8	**26** 12·078	**27** 5·83	**28** 22·61	**29** 560·357	**30** 55·2	**31** 58·4	**32** 20·055
33 35·24	**34** 1008·09	**35** 22·03	**36** 257·4	**37** 35·5	**38** 29·66	**39** 43·55	**40** 583·6

page 26 Exercise 20 Ⓒ

1 7·2	**2** 5·4	**3** 7·1	**4** 5·5	**5** 6·8	**6** 2·7	**7** 4·5
8 2·7	**9** 9·11	**10** 7·08	**11** 1·11	**12** 4·36	**13** 2·41	**14** 10·8
15 1·36	**16** 6·23	**17** 2·46	**18** 12·24	**19** 8·4	**20** 15·96	**21** 2·8
22 2·2	**23** 10·3	**24** 21·8	**25** 0·137	**26** 0·054	**27** 6·65	**28** 4·72
29 0·57	**30** 3·6	**31** 7·23	**32** 6·53	**33** 0·43	**34** 48·01	**35** 1·3
36 7·96	**37** 9·34	**38** 2·43	**39** 16·65	**40** 7·53	**41** 105·2 cm	**42** £8·96

page 27 Exercise 21 Ⓒ

1 6·34	**2** 8·38	**3** 81·5	**4** 7·4	**5** 7245	**6** 32	**7** 6·3
8 142	**9** 4·1	**10** 30	**11** 710	**12** 39·5	**13** 0·624	**14** 0·897
15 0·175	**16** 0·236	**17** 0·127	**18** 0·705	**19** 1·3	**20** 0·08	**21** 0·007
22 21·8	**23** 0·035	**24** 0·0086	**25** 95	**26** 111·1	**27** 32	**28** 70
29 5·76	**30** 9·99	**31** 660	**32** 1	**33** 0·42	**34** 6200	**35** 0·009
36 0·0555	**37 a** 0	**b i** 5 and 2	**ii** 5, 2 and 0	**iii** 0, 0, 5 and 2		

page 28 Exercise 22 Ⓒ

1 3·9	**2** 4·8	**3** 20·4	**4** 30·5	**5** 28·8	**6** 7·8	**7** 4·9	**8** 9·9
9 3·36	**10** 8·15	**11** 21·7	**12** 0·84	**13** 81·9	**14** 17·12	**15** 8·65	**16** 8·16
17 200·4	**18** 494·9	**19** 23·8	**20** 8·36	**21** £16·25	**22** £31	**23** £66·15	**24** 12 cm^2

page 29 Exercise 23 Ⓔ

1 0·06	**2** 0·15	**3** 0·12	**4** 0·006	**5** 1·8	**6** 3·5	**7** 1·8	**8** 0·8
9 0·36	**10** 0·014	**11** 1·26	**12** 2·35	**13** 8·52	**14** 3·12	**15** 0·126	**16** 127·2
17 0·170	**18** 0·327	**19** 0·126	**20** 0·34	**21** 0·055	**22** 0·52	**23** 1·30	**24** 0·001

page 30 Exercise 24 Ⓔ

1 2·1	**2** 2·3	**3** 2·5	**4** 1·5	**5** 45·7	**6** 3·45	**7** 3·8	**8** 2·7
9 2·15	**10** 1·2	**11** 1·84	**12** 1·3	**13** 2·4	**14** 4·2	**15** 2·6	**16** 6·2
17 6·8	**18** 6·4	**19** 4·6	**20** 3·6	**21** 0·18	**22** 0·14	**23** 0·35	**24** 0·15

page 30 Exercise 25 Ⓔ

1 2·1	**2** 3·1	**3** 4·36	**4** 4	**5** 4	**6** 2·5	**7** 16
8 200	**9** 70	**10** 9·2	**11** 30·5	**12** 6·2	**13** 12·5	**14** 122
15 212	**16** 56	**17** 60	**18** 1500	**19** 0·3	**20** 0·7	**21** 0·5
22 3·04	**23** 5·62	**24** 0·78	**25** 0·14	**26** 3·75	**27** 0·075	**28** 0·15
29 1·22	**30** 163·8	**31** 1·75	**32** 18·8	**33** 8	**34** 88	**35** 580

page 31 Exercise 26 Ⓔ

1 4·32	**2** 5·75	**3** 9·16	**4** 1·008	**5** 0·748	**6** 20·24	**7** 10·2
8 2·95	**9** 4·926	**10** 34	**11** 0·621	**12** 8·24	**13** 0·1224	**14** 12·15

15 2·658 **16** 66·462 **17** 34 100 **18** 0·0041 **19** 2·104 **20** 0·285 **21** 0·25884
22 3·27 **23** 2·247 **24** 0·54 **25** 0·027 **26** 6·6077 **27** 6·56 **28** 7·84

page 31 Exercise 27Ⓔ Crossnumbers

A

¹9	1		²2	1	³4	0
4		⁴1	7		5	
	⁵6	4		⁶8		⁷9
⁸7	0		⁹3	1	¹⁰8	
4			8		¹¹3	6
¹²4	3	¹³7		¹⁴8	1	
0		¹⁵2	3		¹⁶6	3

B

¹9	6		²1	5	³7	3
3		⁴5	6		1	
	⁵8	3		⁶3		⁷8
⁸6	6		⁹5	9	¹⁰4	
3			6		¹¹7	0
¹²1	7	¹³4		¹⁴8	9	
8		¹⁵7	0		¹⁶9	3

C

¹2	6		²6	3	³1	4
0		⁴3	7		8	
	⁵2	5		⁶2		⁷6
⁸8	0		⁹3	1	¹⁰5	
4			2		¹¹7	7
¹²6	0	¹³2		¹⁴1	0	
3		¹⁵5	7		¹⁶6	4

page 32 Exercise 28Ⓔ

1 a 608 **b** 6080 **c** 6·08   **2 a** 203·04 **b** 203·04 **c** 20304
3 a 1892·9 **b** 189·29 **c** 18·929   **4 a** 36·173 **b** 3·6173 **c** 0·36173
5 a 485 080 **b** 48·508 **c** 36·2   **6 a** 17 544 **b** 175·44 **c** 215

page 33 Exercise 29 Ⓒ

1 $+5°$ **2** $-1°$ **3** $-4°$ **4** $-7°$ **5** $-4°$ **6** $-4°$
7 $+4°$ **8** $+12°$ **9** $-5°$ **10** $0°$ **11** $-11°$ **12** $-15°$
13 a $-7°C, -2°C, 5°C$ **b** $-1°C, 0°C, 2°C$ **c** $-11°C, -8°C, 3°C$
    d $-7°C, -4°C, -2°C, -1°C$ **e** $-5°C, -4°C, -2°C, 4°C$
14 a $-5, -3, 0, 2, 6$ **b** $-8, -6, -1, 3, 11$
    c $-20, -15, -10, -2, 5$ **d** $-15, -10, -5, 18, 23$
15 a $2, 0, -2$ **b** $0, -3, -6$ **c** $-2, -3, -4$ **d** $-4, -6, -8$
    e $-6, -12, -18$ **f** $0, 1, 2$ **g** $-2, 0, 2$ **h** $-2, -6, -10$

page 34 Exercise 30 Ⓒ

1 -4 **2** -12 **3** -11 **4** -3 **5** -5 **6** 4 **7** -5
8 -8 **9** 19 **10** -17 **11** -4 **12** -5 **13** -11 **14** 6
15 -4 **16** 6 **17** 0 **18** -18 **19** -3 **20** -11 **21** -12
22 4 **23** 4 **24** 0 **25** -8 **26** -3 **27** 3 **28** -12
29 18 **30** -5 **31** -66 **32** 98

page 35 Exercise 31 Ⓒ

1 -8 **2** -7 **3** 1 **4** 1 **5** 9 **6** 11 **7** -8
8 42 **9** 4 **10** 15 **11** -7 **12** -9 **13** -1 **14** -7
15 0 **16** 11 **17** -14 **18** 0 **19** 17 **20** 3 **21** -1
22 -3 **23** 12 **24** -9 **25** 3 **26** 0 **27** 8 **28** 2

29 a

+	−5	1	6	−2
3	−2	4	9	1
−2	−7	−1	4	−4
6	1	7	12	4
−10	−15	−9	−4	−12

b

+	−3	2	−4	7
5	2	7	1	12
−2	−5	0	−6	5
10	7	12	6	17
0	−3	2	−4	7

page 36 **Exercise 32 C**

1 −6	**2** −4	**3** −15	**4** 9	**5** −8	**6** −15	**7** −24	**8** 6
9 12	**10** −18	**11** −21	**12** 25	**13** −60	**14** 21	**15** 48	**16** −16
17 −42	**18** 20	**19** −42	**20** −66	**21** −4	**22** −3	**23** 3	**24** −5
25 4	**26** −4	**27** −4	**28** −1	**29** −2	**30** 4	**31** −16	**32** −2
33 −4	**34** 5	**35** −10	**36** 11	**37** 16	**38** −10	**39** 1	**40** −5
41 64	**42** −27	**43** −600	**44** 40	**45** 2	**46** 36	**47** −2	**48** −8

49 a

×	4	−3	0	−2
−5	−20	15	0	10
2	8	−6	0	−4
10	40	−30	0	−20
−1	−4	3	0	2

b

×	−2	5	−1	−6
3	−6	15	−3	−18
−3	6	−15	3	18
7	−14	35	−7	−42
2	−4	10	−2	−12

page 37 **Test 1**

1 −16	**2** 64	**3** −15	**4** −2	**5** 15	**6** 18	**7** 3	**8** −6	**9** 11	**10** −48
11 −7	**12** 9	**13** 6	**14** −18	**15** −10	**16** 8	**17** −6	**18** −30	**19** 4	**20** −1

page 37 **Test 2**

1 −16	**2** 6	**3** −13	**4** 42	**5** −4	**6** −4	**7** −12	**8** −20	**9** 6	**10** 0
11 36	**12** −10	**13** −7	**14** 10	**15** 6	**16** −18	**17** −9	**18** 15	**19** 1	**20** 0

page 37 **Test 3**

1 100	**2** −20	**3** −8	**4** −7	**5** −4	**6** 10	**7** 9	**8** −10	**9** 7	**10** 35
11 −20	**12** −24	**13** −10	**14** −7	**15** −19	**16** −1	**17** −5	**18** −13	**19** 0	**20** 8

page 38 **Exercise 33 C**

1 11	**2** 1	**3** −5	**4** 12	**5** 21	**6** 2	**7** 17	**8** 24	**9** 9	**10** 30
11 30	**12** 25	**13** 8	**14** 5	**15** 6	**16** 8	**17** 6	**18** 3	**19** 7	**20** −2
21 −4	**22** 14	**23** 13	**24** 0	**25** 52	**26** 11	**27** 10	**28** 20	**29** 5	**30** 5

page 39 **Exercise 34** Ⓒ

1 15 **2** 10 **3** 5 **4** 9 **5** 11 **6** 1 **7** 7 **8** 8 **9** 8
10 4 **11** 0 **12** 1 **13** 18 **14** 18 **15** 12 **16** 27 **17** 8 **18** 6
19 1 **20** 22 **21** 9 **22** 0 **23** 5 **24** 0 **25** 20 **26** 10 **27** 16
28 52 **29** 40 **30** 111 **31** 51 **32** 30 **33** 11 **34** 9 **35** 28 **36** 106
37 54 **38** 4 **39** 4 **40** 153 **41** 59 **42** 165 **43** 85 **44** 12 **45** 33
46 64 **47** 67 **48** 1172 **49** 52 **50** 5 **51** 4 **52** 16 **53** 8 **54** 2

page 40 **Exercise 35** Ⓒ

1 3^4 **2** 5^2 **3** 6^3 **4** 10^5 **5** 1^7 **6** 8^4 **7** 7^6
8 $2^3 \times 5^2$ **9** $3^2 \times 7^4$ **10** $3^2 \times 10^3$ **11** $5^4 \times 11^2$ **12** $2^2 \times 3^3$ **13** $3^2 \times 5^3$ **14** $2^2 \times 3^3 \times 11^2$
15 a 16 **b** 36 **c** 100 **d** 27 **e** 1000
16 a 81 **b** 441 **c** 1·44 **d** 0·04 **e** 9·61 **f** 10 000 **g** 625 **h** 75·69 **i** 0·81 **j** 6625·96
17 a 169 cm^2 **b** 6·25 cm^2 **c** 129·96 cm^2 **18 a** a^3 **b** n^4 **c** s^5 **d** p^2q^3 **e** b^7
19 a 216 **b** 256 **c** 243 **d** 100 000 **e** 64 **f** 0·001 **g** 8·3521 **h** 567 **i** 1250
20 10^{10} **21** 2^7 **22 a** $2^1, 2^2, 2^3, 2^4$ **b** $2^{25}p = £335\,544·32$ **23** Sean is right.

page 42 **Exercise 36** Ⓔ

1 a 4 **b** 6 **c** 1 **d** 10 **2 a** 9 cm **b** 7 cm **c** 12 cm
3 a 3·2 **b** 5·4 **c** 10·3 **d** 4·4 **e** 49·1 **f** 7·7 **g** 0·4 **h** 0·9
4 12·2 cm **5** 447 m **6** 7·8 cm **7 a** 4 **b** 5 **c** 10 **8** 5·8 cm **9** 10

page 43 **Exercise 37** Ⓔ

1 $\frac{1}{3}$ **2** $\frac{1}{4}$ **3** $\frac{1}{10}$ **4** 1 **5** $\frac{1}{9}$ **6** $\frac{1}{16}$ **7** $\frac{1}{100}$ **8** 1
9 $\frac{1}{49}$ **10** 1 **11** $\frac{1}{81}$ **12** 1 **13** True **14** False **15** True **16** True
17 False **18** False **19** False **20** True **21** True **22** True **23** False **24** False
25 False **26** True **27** True **28** True **29** True **30** True **31** True **32** False

page 44 **Exercise 38** Ⓔ

1 5^6 **2** 6^5 **3** 10^9 **4** 7^8 **5** 3^{10} **6** 8^6 **7** 2^{13} **8** 3^4 **9** 5^3
10 7^4 **11** 5^2 **12** 3^{-4} **13** 6^5 **14** 5^{-10} **15** 7^6 **16** 7^2 **17** 6^5 **18** 8
19 5^8 **20** 10^2 **21** 9^{-2} **22** 3^{-2} **23** 2^4 **24** 3^{-2} **25** 7^{-6} **26** 3^{-4} **27** 5^{-5}
28 8^{-5} **29** 5^{-5} **30** 6^4 **31** 1 **32** 1 **33** 3^7 **34** 2^7 **35** 7^2 **36** 5^{-1}

page 45 **Exercise 39** Ⓔ

1 3^6 **2** 5^{12} **3** 7^{10} **4** 8^{20} **5** x^6 **6** a^{15} **7** n^{14} **8** y^9
9 2^{-2} **10** 3^{-4} **11** 7^2 **12** x^{-3} **13** $6a^5$ **14** $20n^4$ **15** $14x^5$ **16** $24y^7$
17 $5n^7$ **18** $12y^2$ **19** $9p^5$ **20** $10p^6$ **21** $8x^6$ **22** $27a^6$ **23** $16y^6$ **24** $25x^8$
25 ± 3 **26** 1 **27** 3 **28** 0 **29** 3 **30** 1 **31** 2 **32** 3
33 -1 **34** -1 **35** 0 **36** 2 **37** 4 **38** 0 **39** -1 **40** 0

page 46 **Exercise 40** Ⓒ

1 a £6·10   **b** £6·40   **c** £4·70   **d** £116   **e** £129·30   **f** £0·04
2 a 2·5h   **b** 4·25h   **c** 3·75 h   **d** 0·1h   **e** 0·2h   **f** 0·25h
3 a 24·75 h   **b** 22·75h   **c** 2·9 h   **d** 2·75 h   **e** 2·5h   **f** 1·75h

page 47 **Exercise 41** Ⓒ

1 4·2	**2** 15·9	**3** 0·6	**4** 5·3	**5** 4·0	**6** 12·7	**7** 0·5	**8** 5·6
9 14·0	**10** 2·1	**11** 14·1	**12** 1·2	**13** 9·9	**14** 9·1	**15** 9·5	**16** 0·6
17 23·0	**18** 11·4	**19** 7·4	**20** 5·5	**21** 11·5	**22** 11·7	**23** 10·9	**24** 1·9
25 13·0	**26** 4·9	**27** 18·8	**28** 3·4	**29** 2·4	**30** 2·9		

page 48 **Exercise 42** Ⓒ

1 9·1	**2** 11·4	**3** 4·9	**4** 12·4	**5** 1·5	**6** 4·7	**7** 2·2	**8** 2·6
9 0·7	**10** 1·4	**11** 3·7	**12** 15·1	**13** 9·3	**14** 10·0	**15** 6·0	**16** 6·9
17 1·0	**18** 0·2	**19** 5·4	**20** 80·6	**21** 7·8	**22** 16·6	**23** 7·3	**24** 12·5
25 64·1	**26** 1·6	**27** 14·1	**28** 2·5	**29** 0·6	**30** 2·7	**31** 86·6	**32** 44·9
33 1·038	**34** 1·0	**35** 6·3	**36** 0·8	**37** 2·3	**38** 9·9	**39** 13·4	**40** 1·5

page 49 **Puzzle Exercise 43** Ⓒ

1 SOIL	**2** ISLES	**3** HE LIES	**4** SOS
5 HOHOHO	**6** ESSO OIL	**7** SOLIDS	**8** SOLO
9 BIGGLES	**10** HE IS BOSS	**11** LODGE	**12** SIGH
13 HEDGEHOG	**14** GOSH	**15** GOBBLE	**16** BEG
17 BIG SLOB	**18** SID	**19** HILL	**20** LESLIE
21 HOBBIES	**22** GIGGLE	**23** BIBLE	**24** BOILED EGGS
25 BOBBLE	**26** HEIDI	**27** BOBBIE	**28** HIGH
29 HELLS BELLS	**30** GOD BLESS	**31** SHE DIES	**32** SOLEIL

page 50 **Exercise 44** Ⓒ

1 0·7, 2·44, 3·5   **2** 23   **3** 130   **4** £28·50   **5** 14·55   **6** 15h 5mins   **7** £10·35
8 a   **b**   **9** 128 cm^2 **10** 48

a

6	13	8
11	9	7
10	5	12

b

11	8	5	10
2	13	16	3
14	1	4	15
7	12	9	6

page 52 **Exercise 45** Ⓒ

1 a
```
    5 [1] 3
  + 3  4 [6]
  ─────────
  [8] 5  9
```

b
```
    3  3  4
  + [3] 4 [6]
  ─────────
    6 [8] 0
```

c [1] [6] [0] ÷ 5 = 32

2 a 16 384 **b** 4096 **3** £4550 **4 a** 3·32 **b** 1·61 **c** 1·46 **d** 4·4 **e** 6·2 **f** 2·74

5 18 **6 a**

−1	−2	3
4	0	−4
−3	2	1

b

4	3	−1
−3	2	7
5	1	0

c

0	1	−4
−5	−1	3
2	−3	−2

7 0·05, 0·2, 0·201, 0·21, 0·5

8 13°C, −2°C, 22°C, − 21°C **9** 9 h 15 mins **10** 96 523

page 53 Exercise 46 Ⓒ

1 £2·60; £3·60; 7 jars; Total = £16·45 **2 a** 3^4 **b** 1^7 **c** 7^5 **d** 2^4 **e** 10^6 **f** 10^4

3 a 9 **b** −5 **c** −6 **d** −10 **e** 7 **f** −10 **4** 20

5 a 50p, 20p, 5p, 2p **b** 50p, 20p, 10p, 5p, 1p **c** £1, 50p, 5p, 1p, 1p *or* 50p, 50p, 50p, 5p, 2p

6 m, 9, z **7** 24

8

	Prime Number	Multiple of 3	Factor of 16
Number >5	7	9	8
Odd number	5	3	1
Even number	2	6	4

9 1·50 m **10 a** What time do we finish **b** Spurs are rubbish **c** We are under attack

page 55 Exercise 47 Ⓒ

1 £3·26 **2** £1·70 **3 a** 108 m² **b** 3

4 1 3 9 27 81

4 12 36 108

16 48 144

64 192

256

5 100 m

6 a 5 **b** 27 **c** 2, 5 **d** 25, 49 **e** 8, 27 **7** 10 h 30 mins

8 a 0·54 **b** 40 **c** 0·004 **d** 2·2 **e** £9 **f** £40

9 a 782 or 827 or 872 **b** 278 or 287 **c** 287 or 827

10 a 3 and 4 **b** 6 and 7 **c** 4 and 8 **d** 2 and 24

page 56 Exercise 48 Ⓒ

1 a 69 **b** 65 **c** Many possibilities e.g. SAT **2** 120 **3** 360 000 kg **4** 16

5

1	2	3	4
2	3	4	1
3	4	1	2
4	1	2	3

6 64 mph

7 a Yes **b** No **c** Yes **d** Yes **e** Yes **f** Yes **g** No **h** No

8 £5·12

Other solutions are possible

page 58 **Exercise 49** Ⓒ

1 a 45 mins   **b** 30 mins   **c** 30 mins   **d** 20 mins   **2** 2 h 10 mins   **3** 20·05
4 The London Blackout Murders   **5** 15 mins   **6** 3 h 45 mins   **7** 11 h 05 mins
8 21·10   **9** 5   **10** 1 h 45 mins   **11** 18·00   **12** 15 mins
13 13 h 25 mins excluding CEEFAX

page 59 **Test yourself**

1 a $2 \times 2 \times 7$ **b** 84   **2 a** 16, 4   **b** 27, 3   **3 a** 847   **b** 328   **c** 1056   **d** 0·08
4 0·075, 0·7, 0·705, 0·75   **5 a** $-32°C$ **b** 42°C   **c** 15°C
6 Positive, as a positive × a positive is positive and a negative × a negative is positive.
7 18   **8** £1028·06
9 a i Multiple answers e.g. 10   **ii** Multiple answers e.g. 16, 18
    iii Multiple answers e.g. 17, 10   **iv** 16
   b i 10 or 15   **ii** 5 divides into it exactly   **c** 16
10 a i 14   **ii** 56   **iii** 15   **iv** 125   **b** 8 cm   **11** 40 cm

12 a

−1	4	−3
−2	0	2
3	−4	1

     b 12

13 a 11·23   **b** 11·25   **c** 43 mins   **d** 24 mins   **14 a** 250 000   **b** 7 million
15 a 119·31   **b** 119 310   **c** 1·23
16 Large pot costs 0·107 pence per g and small pot costs 0·104 pence per g
   ∴ Small pot is a better value.
17 a i 146·41 = 146·4 (1 dp)   **ii** 6·8 (1 dp)   **iii** 25   **b** 6   **18** 17·9867
19 a 7 DVDs £5·50 change   **b** 6
20 a 17·9867   **b** $(1·6 + 3·8 \times 2·4) \times 4·2$

2   Algebra 1

page 63 **Exercise 1** Ⓒ

1 $2n$ **2** $y + 4$ **3** $x - 7$ **4** $x + 100$ **5** $5y$ **6** $100s$
7 $3t$ **8** $z + 11$ **9** $p - 9$ **10** $n + x$ **11** $4n$ **12** $2x + 3$
13 $2n - 12$ **14** $3m + 2$ **15** $20y$ **16** $3x + 3$ **17** $2y - 7$ **18** $3k + 10$

page 64 **Exercise 2** Ⓒ

1 $3(x + 4)$ **2** $5(x + 3)$ **3** $6(y + 11)$ **4** $\dfrac{x + 3}{4}$ **5** $\dfrac{x - 7}{3}$

6 $\dfrac{y - 8}{5}$ **7** $\dfrac{2(4a + 3)}{4}$ or $\dfrac{4a + 3}{2}$ **8** $\dfrac{3(m - 6)}{4}$ **9** $x^2 - 6$ **10** $\dfrac{x^2 + 3}{4}$

11 $(n + 2)^2$ **12** $(w - x)^2$ **13** $\dfrac{x^2 - 7}{3}$ **14** $(x - 9)^2 + 10$ **15** $\dfrac{(y + 7)^2}{x}$

16 $\dfrac{(a-x)^3}{y}$ **17** $l-3$ cm **18** $15-x$ cm **19** $l+200-m$ kg **20** $4(n+2)$

21 $6w$ kg **22** xl kg **23** $\dfrac{n}{6}$ pence **24** $£\dfrac{p}{5}$ **25** $\dfrac{12}{n}$ kg

page 66 **Exercise 3 C**

1 $5a$ **2** $11a$ **3** $9a$ **4** $6n$ **5** $5n$
6 $9n$ **7** $4x$ **8** $16x$ **9** $3x$ **10** $6a+4b$
11 $9a+6b$ **12** $5x+5y$ **13** $9x+7y$ **14** $5m+3n$ **15** $16m+2n$
16 $5x+11$ **17** $13x+8$ **18** $8x+10$ **19** $16x+4y$ **20** $5x+8$
21 $9x+5$ **22** $7x+4$ **23** $7x+4$ **24** $7x+7$ **25** $8x+12$
26 $12x-6$ **27** $2x+5$ **28** $2x-5$ **29** $2x-5$ **30** $13a+3b-1$
31 $10m+3n+8$ **32** $3p-2q-8$ **33** $2s-7t+14$ **34** $2a+1$ **35** $x+y+7z$
36 $5x-4y+4z$ **37** $5k-4m$ **38** $4a-9+5b$ **39** $a-4x-5e$ **40** $3n+3$
41 $2x+8$ **42** $2x+7$ **43** $2x+14$ **44** $3a+b+3$ **45** $6x+2y+12$
46 $10x+8$

page 68 **Exercise 4 C**

1 a ab **b** nm **c** xy **d** ht **e** adn **f** $3ab$ **g** $4nm$ **h** $3abc$
2 a $6ab$ **b** $15ab$ **c** $12cd$ **d** $10nm$ **e** $14pq$ **f** $30an$ **g** $20st$ **h** $36ab$ **i** $32uv$
3 a $3y^2$ **b** y^2 **c** $6x^2$ **d** $24t^2$ **e** $6a^2$ **f** $10y^2$ **g** $2x^2$ **h** $3y^2$ **i** $100x^2$

page 68 **Exercise 5 C**

1 $2x+6$ **2** $3x+15$ **3** $4x+24$ **4** $4x+2$ **5** $10x+15$
6 $12x-4$ **7** $12x-12$ **8** $15x-6$ **9** $15x-20$ **10** $14x-21$
11 $4x+6$ **12** $6x+3$ **13** $5x+20$ **14** $12x+12$ **15** $8x-2$
16 $2a+6b$ **17** $6a+15b$ **18** $10m+15n$ **19** $14a-21b$ **20** $11a+22b$
21 $24a+16b$ **22** x^2+5x **23** x^2-2x **24** x^2-3x **25** $2x^2+x$
26 $3x^2-2x$ **27** $3x^2+5x$ **28** $2x^2-2x$ **29** $2x^2+4x$ **30** $6x^2+9x$
31 $7x+10$ **32** $8x+2$ **33** $5a-3$ **34** $11a+17$ **35** $8a-10$
36 $8t+4$ **37** $5x+8$ **38** $7x+18$ **39** $8x-16$ **40** $8a-6$
41 $2x^2+4x+6$ **42** $2x^2+2x+5$ **43** $3a^2+6a-4$ **44** $5y^2+4y-3$ **45** $5x^2+2x$
46 $3a^2+4a$

page 69 **Exercise 6 C**

1 a $-m-n$ **b** $-a-b$ **c** $-2a-b$ **d** $-m+n$ **e** $-a+b$ **f** $-3a+b$ **g** $-a-b+c$
h $-2a-b+2c$ **i** $-3x+y+2$ **2** $a+4b$ **3** $4a+b$ **4** $4a+7b$
5 $a+4b$ **6** $a+3b$ **7** $5a+b$ **8** $m+2n$ **9** $4m+11n$
10 $11m+4n$ **11** $6m+8n$ **12** $x-2y$ **13** $4x-3y$ **14** $9a+35b$
15 $3x+11y$ **16 a** $4a+8$ **b** $10x+4$ **c** $8a-2$ **17 a** $6x+2y+4$ **b** $6x+2y$
18 a $l=2x-2$ **b** $l=2x+2$ **c** $l=1$ **d** $l=x-1$ **e** $l=x+2$ **f** $l=x-3$
g $l=x+3$ **h** $l=x+3$

page 71 **Exercise 7 C**

1 True **2** False $\left(\text{unless } n=1\dfrac{1}{2}\right)$ **3** True

4 True **5** False (unless $n=\sqrt[\pm]{3}$ or 0) **6** True

7 False (unless $m = n$)   **8** True   **9** False (unless $n = 0$ or $\sqrt[+]{3}$)

10 False (unless $m = 0$)   **11** False $\left(\text{unless } c = 1\frac{1}{2}\right)$ **12** True

13 True   **14** False (unless $n = \pm 2$)   **15** True

16 False (unless $n = 0$)   **17** True   **18** False (unless $n = 0$ or 1)

19 **a** $n + n, 4n - 2n$ **b** $n \times n^2, n \times n \times n$ **c** $3n \div 3, n^2 \div n$

   d $4 \div n$ **e** Many possible answers e.g. $\boxed{n \times n + n \times n}$

20 **a** $n \rightarrow \boxed{\times 6} \rightarrow \boxed{-1} \rightarrow 6n - 1$ **b** $n \rightarrow \boxed{\times 8} \rightarrow \boxed{+10} \rightarrow 8n + 10$

   c $n \rightarrow \boxed{\div 2} \rightarrow \boxed{+3} \rightarrow \frac{n}{2} + 3$ **d** $n \rightarrow \boxed{\times 2} \rightarrow \boxed{+5} \rightarrow \boxed{\times 3} \rightarrow 3(2n + 5)$

   e $n \rightarrow \boxed{\times 2} \rightarrow \boxed{-4} \rightarrow \boxed{\times 5} \rightarrow 5(2n - 4)$ **f** $n \rightarrow \boxed{+4} \rightarrow \boxed{\div 7} \rightarrow \frac{(n + 4)}{7}$

21 3   **22** 1   **23** n **24** n **25** $2a + b + c$ **26** $2n^2$ **27** $2mn$ **28** n^2

29 3   **30** a^3 **31** n **32** $3t - 3p + 3$ **33** 1   **34** $4n + 2$ **35** $2n + 8$

page 72 **Exercise 8 C**

1 4 kg   **2** 3 kg   **3** 3 kg   **4** 2 kg   **5** 3 kg   **6** 4 kg   **7** 4 kg   **8** 3 kg

9 2 kg   **10** 2 kg

page 73 **Exercise 9 C**

1 3   **2** 17   **3** 14   **4** 16   **5** 7   **6** 7   **7** 3   **8** 13

9 4   **10** 0   **11** 31   **12** 8   **13** 10   **14** 8   **15** 8   **16** 3

17 5   **18** 3   **19** 2   **20** 4   **21** 1   **22** 0   **23** 2   **24** 1

page 74 **Exercise 10 C**

1 $\frac{4}{5}$ **2** $2\frac{1}{3}$ **3** $7\frac{1}{2}$ **4** $1\frac{5}{6}$ **5** 0   **6** $\frac{5}{9}$ **7** 1   **8** $\frac{1}{5}$

9 $\frac{2}{7}$ **10** $\frac{3}{4}$ **11** $\frac{2}{3}$ **12** $1\frac{1}{4}$ **13** $1\frac{1}{5}$ **14** $1\frac{5}{9}$ **15** $\frac{1}{3}$ **16** $\frac{1}{2}$

17 $\frac{1}{10}$ **18** $-\frac{3}{8}$ **19** $\frac{9}{50}$ **20** $\frac{1}{2}$ **21** $\frac{3}{5}$ **22** $-\frac{4}{9}$ **23** 0   **24** $4\frac{5}{8}$

25 $-1\frac{3}{7}$ **26** $2\frac{1}{3}$ **27** $\frac{3}{4}$ **28** 1   **29** $3\frac{3}{5}$ **30** $\frac{1}{3}$ **31** $2\frac{1}{14}$ **32** -1

33 $-\frac{5}{6}$ **34** $8\frac{1}{4}$ **35** -55

page 74 **Exercise 11 E**

1 $2\frac{3}{4}$ **2** $1\frac{2}{3}$ **3** 2   **4** $\frac{1}{5}$ **5** $\frac{1}{2}$ **6** 2   **7** $5\frac{1}{3}$ **8** $1\frac{1}{5}$

9 0   **10** $\frac{2}{9}$ **11** $1\frac{1}{2}$ **12** $\frac{1}{6}$ **13** $1\frac{1}{3}$ **14** $\frac{6}{7}$ **15** $\frac{4}{7}$ **16** 7

17 $\frac{5}{8}$ **18** 5   **19** $\frac{2}{5}$ **20** $\frac{1}{3}$ **21** 4   **22** -1 **23** 1   **24** $\frac{6}{7}$

25 $1\frac{1}{4}$ **26** 1   **27** $\frac{7}{9}$ **28** $-1\frac{1}{2}$ **29** $\frac{2}{9}$ **30** $-1\frac{1}{2}$

page 75 **Exercise 12Ⓔ**

1 3　　　　**2** 5　　　　**3** $10\frac{1}{2}$　　**4** -8　　**5** $\frac{1}{3}$　　**6** $-4\frac{1}{2}$　　**7** $3\frac{1}{3}$　　**8** $3\frac{1}{2}$

9 $3\frac{2}{3}$　　**10** -2　　**11** $-5\frac{1}{2}$　　**12** $4\frac{1}{5}$　　**13** $\frac{3}{7}$　　**14** $\frac{7}{11}$　　**15** $4\frac{4}{5}$　　**16** 5

17 9　　**18** $-2\frac{1}{3}$　　**19** $\frac{2}{5}$　　**20** $\frac{3}{5}$　　**21** -1　　**22** 13　　**23** 9　　**24** $4\frac{1}{2}$

25 $3\frac{1}{3}$

page 76 **Exercise 13Ⓔ**

1 $\frac{3}{5}$　　**2** $\frac{4}{7}$　　**3** $\frac{11}{12}$　　**4** $\frac{6}{11}$　　**5** $\frac{2}{3}$　　**6** $\frac{5}{9}$　　**7** $\frac{7}{9}$　　**8** $1\frac{1}{3}$

9 $\frac{1}{2}$　　**10** $\frac{2}{3}$　　**11** 3　　**12** $1\frac{1}{2}$　　**13** 24　　**14** 15　　**15** -10　　**16** 21

17 21　　**18** $2\frac{2}{3}$　　**19** $4\frac{3}{8}$　　**20** $1\frac{1}{2}$　　**21** $3\frac{3}{4}$　　**22** 11　　**23** 13　　**24** 10

25 $\frac{1}{3}$　　**26** 1　　**27** $3\frac{2}{3}$　　**28** 28　　**29** $4\frac{1}{2}$　　**30** 1　　**31** 220　　**32** -500

33 $-\frac{98}{99}$　　**34** 6　　**35** 30　　**36** $1\frac{1}{2}$　　**37** 84　　**38** 6　　**39** $\frac{5}{7}$　　**40** $\frac{3}{5}$

page 76 **Exercise 14Ⓔ**

1 3　　**2** $\frac{3}{4}$　　**3** $4\frac{1}{2}$　　**4** $-\frac{3}{10}$　　**5** $-\frac{1}{2}$　　**6** $17\frac{2}{3}$　　**7** $\frac{1}{6}$　　**8** 5

9 12　　**10** $3\frac{1}{3}$　　**11** $4\frac{2}{3}$　　**12** -9

page 78 **Exercise 15Ⓔ**

1 $\frac{3}{4}$　　**2** $1\frac{1}{2}$　　**3** $1\frac{3}{8}$ cm　　**4** $1\frac{1}{4}$ cm　　**5** 7 cm　　**6 a** $3\frac{3}{5}$　**b** 0　**c** 7

7 a 41　**b** 31　**c** 65　　**8** 29　　　　**9** 55, 56, 57　　**10** 41, 42, 43, 44

11 a i $x - 3$ cm　**ii** $2x - 6$ cm　**b** 12·5　　**12** $x = 8$; perimeter $= 60$

13 a -6　**b** 2　**c** A and C　　　　**14** 11　　　　**15** £6　　　　**16** 3

page 80 **Exercise 16Ⓒ**

1 a £210　**b** £360　　**2** 60 cm　　**3 a** 15 cm　**b** 33 cm　　**4 a** 30 cm²　**b** 16 cm²　**c** 35 cm²

5 a 4000 grams　**b** 6100 grams　**c** 7750 grams　　**6 a** £20　**b** £60

page 82 **Exercise 17Ⓒ**

1 a 14 cm　**b** 16 cm　　**2 a** 10 cm²　**b** 21 cm²　　**3 a** 13　**b** 53　**c** 33

4 a 60　**b** 140　**c** 88　　**5 a** 56　**b** 17　**c** 120　**d** 74　　**6 a** 600　**b** 300　　**7** $T = 8n + 50$

page 83 **Exercise 18Ⓔ**

1 36　**2** 29　**3** 8　　**4** 18　**5** 84　**6** 52　**7** 165　**8** 181　**9** 1·62 (2 dp)　**10** 650

page 84 **Exercise 19Ⓒ**

1 11　**2** 29　**3** 33　**4** 3　**5** 45　**6** 6　**7** 9　　**8** 46　**9** 304　**10** 10

11 a 11　**b** 18　**c** 27　　**12 a** 3　**b** 3·5　**c** 3　　**13 a** 2　**b** 6　**c** 12　　**14 a** 9　**b** 0　**c** 1　**d** 5　**e** 1　**f** 10

page 85 **Exercise 20 C**

1 2 **2** 14 **3** 0 **4** 4 **5** 6 **6** −6 **7** 17 **8** 2 **9** 4 **10** 1 **11** 24 **12** 13
13 3 **14** −1 **15** 17 **16** −9 **17** 30 **18** 10 **19** 3 **20** −3 **21** 45 **22** 23 **23** 41 **24** 9
25 4 **26** 8 **27** 21 **28** −12 **29** 14 **30** 27 **31** −7 **32** 8 **33** 9 **34** 25

page 86 **Exercise 21 E**

1 4 **2** 4 **3** 9 **4** 16 **5** 8 **6** −8 **7** −27 **8** 64 **9** 8 **10** 16
11 8 **12** 16 **13** 18 **14** 36 **15** 48 **16** 16 **17** 20 **18** 54 **19** 144 **20** 24
21 13 **22** 10 **23** 1 **24** 18 **25** 13 **26** 19 **27** 10 **28** 32 **29** 16 **30** 144
31 36 **32** 36 **33** 4 **34** 1 **35** 2 **36** −14 **37** −5 **38** −5 **39** −10 **40** 10
41 0 **42** 4 **43** 50 **44** 4 **45** −10 **46** −4 **47** −6 **48** −16 **49** 28 **50** 44

page 86 **Exercise 22 E**

1 2 **2** $\frac{1}{2}$ **3** 0 **4** 18 **5** −3 **6** 8 **7** 26 **8** −24
9 2 **10** 0 **11** $\frac{1}{4}$ **12** $-\frac{2}{3}$ **13** 17 **14** $\frac{1}{3}$ **15** −3 **16** $\frac{3}{4}$

page 87 **Exercise 23 C**

1 a 16 **b** 14 **c** 3 **d** 17 **2** 13 **3** 20 **4** 14 **5** 5 **6** 25 **7** 7 **8** 26 **9** 45
10 11 **11** 32 **12** 81 **13** 20 000 **14** 5 **15** 12·5 **16** 121 **17** 27
18 18 **19** 6 **20** 16 **21** 9 **22** 25 **23** 3, 81 **24** 0, 32 **25** 0·1, 1000

page 87 **Exercise 24 C**

1 a 2, 5, 8, 11, 14 **b** 30, 25, 20, 15, 10 **c** 1, 2, 4, 8, 16 **d** 1, 10, 100, 1000, 10 000
e 35, 28, 21, 14, 7 **f** 64, 32, 16, 8, 4 **g** −10, −8, −6, −4, −2
2 11, 16, 21, 26, 31
3 a 96, 98, 100, 102, 104 **b** 100, 99, 88, 77, 66 **c** 10, 20, 40, 80, 160 **d** −6, −3, 0, 3, 6
4 a Add 3 **b** Subtract 5 **c** Multiply by 2 **d** Add 7 **e** Subtract 4
5 a 16 **b** 10 **c** 3000 **d** 2 **e** 16 **f** 0 **g** 14 **h** 8
6 a Add 4 **b** Subtract 10 **c** Multiply by 4 **d** Divide by 2
e Subtract 5 **f** Add 1 **g** Divide by 2 **h** Divide by 10

page 89 **Exercise 25 E**

1 a 15 **b** 30 **c** 19 **d** 16 **e** 40 **f** 13 **g** $\frac{1}{2}$ **h** 14 **2** 13, 21, 34, 55
3 a 11, 18, 29, 47, 76, 123 **b** 12, 19, 31, 50 **4 a** $6 \times 7 = 6 + 6^2$ **b** $10 + 10^2$
$7 \times 8 = 7 + 7^2$ $30 + 30^2$
5 $5 + 9 \times 1234 = 11\,111$ **6** 63, 3968
$6 + 9 \times 12\,345 = 111\,111$ **7** 3, 5, 5
$7 + 9 \times 123\,456 = 1\,111\,111$ **8 a** 16 **b** 15 **c** 26 **d** 25
9 a $1^3 + 2^3 + 3^3 + 4^3 = (1 + 2 + 3 + 4)^2 = 100$ **b** 3025
$1^3 + 2^3 + 3^3 + 4^3 + 5^3 = (1 + 2 + 3 + 4 + 5)^2 = 225$
$1^3 + 2^3 + 3^3 + 4^3 + 5^3 + 6^3 = (1 + 2 + 3 + 4 + 5 + 6)^2 = 441$
10 a Yes **b i** 5 **ii** 10 **iii** 1331

page 92 **Exercise 26Ⓔ**

1 a $2n$ **b** $10n$ **c** $3n$ **d** $11n$ **e** $100n$ **f** $6n$ **g** $22n$ **h** $30n$

2 a $11 \rightarrow 55$ **b** $20 \rightarrow 180$ **c** $12 \rightarrow 1200$
        $n \rightarrow 5n$ $n \rightarrow 9n$ $n \rightarrow 100n$

3 a $10 \rightarrow 20 \rightarrow 23$ **b** $20 \rightarrow 60 \rightarrow 61$
        $n \rightarrow 2n \rightarrow 2n + 3$ $n \rightarrow 3n \rightarrow 3n + 1$

4 a $3 \rightarrow 18 \rightarrow 20$ **b** $3 \rightarrow 15 \rightarrow 13$
        $4 \rightarrow 24 \rightarrow 26$ $4 \rightarrow 20 \rightarrow 18$

5 a

Term number (n)		$5n$		Term
1	$\rightarrow$	5	$\rightarrow$	6
2	$\rightarrow$	10	$\rightarrow$	11
3	$\rightarrow$	15	$\rightarrow$	16
4	$\rightarrow$	20	$\rightarrow$	21
5	$\rightarrow$	25	$\rightarrow$	26
$\vdots$		$\vdots$		$\vdots$
n	$\rightarrow$	$5n$	$\rightarrow$	$5n + 1$

b i 51
   iii $5n + 1$

6 a A

Term number (n)		$3n$		Term
1	$\rightarrow$	3	$\rightarrow$	1
2	$\rightarrow$	6	$\rightarrow$	4
3	$\rightarrow$	9	$\rightarrow$	7
$\vdots$				
n	$\rightarrow$	$3n$	$\rightarrow$	$3n - 2$

B

Term number (n)		$4n$		Term
1	$\rightarrow$	4	$\rightarrow$	6
2	$\rightarrow$	8	$\rightarrow$	10
3	$\rightarrow$	12	$\rightarrow$	14
$\vdots$				
n	$\rightarrow$	$4n$	$\rightarrow$	$4n + 2$

C

Term number (n)		$7n$		Term
1	$\rightarrow$	7	$\rightarrow$	5
2	$\rightarrow$	14	$\rightarrow$	12
3	$\rightarrow$	21	$\rightarrow$	19
$\vdots$				
n	$\rightarrow$	$7n$	$\rightarrow$	$7n - 2$

b (A) $3n - 2$ (B) $4n + 2$ (C) $7n - 2$

7 a A10 = 80   **b** B10 = 76   **c** $8n$ **d** $8n - 4$
8 a 79   **b** 81   **c** $4n - 1$ **d** $4n + 1$
9 a $6n + 2$ **b** $6n - 1$

page 95 **Exercise 27Ⓔ**

1 $w = r + 4$ **2** $w = 2r + 6$

3

r	w
8	4
9	6
10	8
11	10

$w = 2r - 12$

4

t	m
1	3
2	5
3	2
4	9
5	11
6	13
7	15
8	17
9	19
10	21

$m = 2t + 1$

5

t	m
1	5
2	8
3	11
4	14

$m = 3t + 2$

6 $t = s - 2$ or $s = t + 2$
7 a $p = 5n - 2$
   b $k = 7n + 3$
   c $w = 2n + 11$
8 $m = 8c + 4$

page 98 **Test yourself**

1 a $4x$ **b** $3x + 7y$ **c** $12a$

2 $y + y \to 2y$ $5y - y \to 4y$ $y + 2y \to 2y + y$ $y \div 2 \to \dfrac{y}{2}$

3 a 3 **b** -5 **c** $2\frac{1}{2}$ **4 a** $9p + 2q$ **b i** $w = 3$ **ii** $z = 3\frac{1}{2}$

5 a 4 **b** 6 **c** 2 **d** 5 **6 a i** $4c$ **ii** p^4 **iii** $8g$ **iv** $10\,r\,p$ **b** $10y - 15$

7 a $3\frac{4}{5}$ **b** $17\frac{1}{2}$ **8 a** $3a + 2b$ **b i** x^2y **ii** $2x^2 + 4xy$

9 a $y = 3k - 1$ **b** 5 **10 a** $3a + 2b$ **b** $6\frac{1}{2}$ **c i** -7 **ii** $4\frac{1}{2}$

11 a 14 **b** 53 **c** $k = 2 \times$ any even number will demonstrate this, e.g. $k = 20 \Rightarrow \frac{1}{2}k + 1 = 11$ which is odd.

12 $x = 1\frac{1}{2}$, A = 7

13 a Bryani, because $4x^2$ means square x, then multiply by 4. **b** 64

14 a £70 **b** £104 **c** 16 miles **15** $w = 6, x = 8, y = 5, z = 4$

16 a $4c$ **b** p^4 **c** $10rp$

17 a i 16, 8 **ii** The sequence continually repeats as 2, 1, 4, 2, 1, 4 etc. **b** 0, 3, 8

18 a $6x$ **b** $x - 4$ **c** $12(x - 4) = 12x - 48$ **d** $18x - 48$

3 Number 3

page 103 **Exercise 1 C**

1 A(5, 1) B(1, 4) C(4, 4) D(1, 2) E(7, 3) F(3, 0)
G(2, 1) H(0, 3) I(6, 5) J(6, 0) K(3, 5)

2 2 : (1, 6), (2, 7), (4, 7), (5, 6), (5, 4), (1, 0), (5, 0)
S : (10, 2), (11, 1), (12, 1), (13, 2), (13, 3), (12, 4), (11, 4), (10, 5), (10, 6), (11, 7), (12, 7), (13, 6)

3 A Parrot

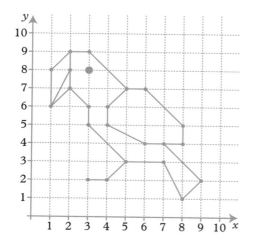

4 A Mouse

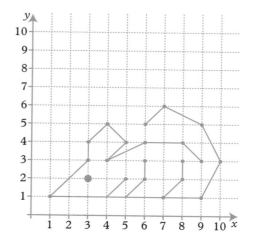

5 A hollowed out cuboid

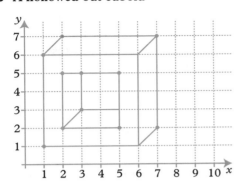

6 A car

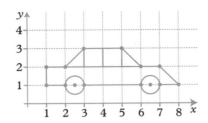

7 What do you call a man with a spade? Doug.

page 104 **Exercise 2 C**

1 A(3, −2) B(4, 3) C(−2, 3) D(−4, −2) E(5, −3) F(−4, 2) G(1, −4) H(−2, −4) I(−2, 1)

2 A face

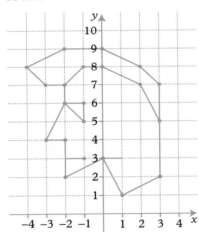

3

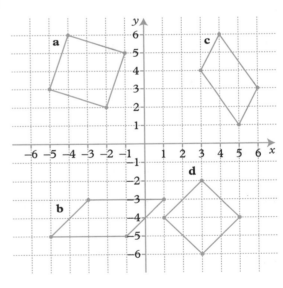

4 **a** (−4, 0) **b** (1, 0) **c** (3, 1) **d** (2, 1). Another possible answer is (0, 7).
5 (4, 6) **6** **a** (5, 2) **b** (3, 1½) **c** (6, 5) **d** (2, 5) **e** (4, 1) **f** (2, 2)

page 107 **Exercise 3 C**

1 True	**2** True	**3** False, 70°	**4** False, 80°	**5** True
6 True	**7** False, 135°	**8** False, 75°	**9** True	**10** True
11 True	**12** True	**13** True	**14** True	**15** False, 110°
16 True	**17** False, 60°	**18** False, 330°	**19** True	**20** False, 120°

page 108 **Exercise 4** Ⓒ

1 Acute	**2** Right angle	**3** Acute	**4** Acute	**5** Obtuse
6 Reflex	**7** Obtuse	**8** Acute	**9** Obtuse	**10** Reflex
11 Obtuse	**12** Reflex	**13** Right angle	**14** Reflex	**15** Obtuse
16 Acute	**17** Acute	**18** Reflex	**19** Acute	**20** Obtuse

page 109 **Exercise 5** Ⓒ

1 $70°$ **2** $100°$ **3** $70°$ **4** $100°$ **5** $55°$

6 $70°$ **7** $70°$ **8** $33\frac{1}{3}°$ **9** $30°$ **10** $35°$

11 $155°$ **12** $125°$ **13** $44°$ **14** $80°$ **15** $40°$

16 $48°$ **17** $40°$ **18** $35°$ **19** $a = 40°, b = 140°$ **20** $y = 72°, x = 108°$

page 111 **Exercise 6** Ⓒ

1 $50°$ **2** $70°$ **3** $40°$ **4** $26°$ **5** $130°$ **6** $73°$

7 $18°$ **8** $75°$ **9** $29°$ **10** $30°$ **11** $70°$ **12** $42°$

13 $120°$ **14** $100°$ **15** $45°$ **16** $72°$ **17** $40°$ **18** $s = 55°, t = 70°$

19 $u = 72°, v = 36°$ **20** $w = z = 55°$ **21** $140°$ **22** $75°$ **23** $60°$ **24** $d = 122°, y = 116°$

25 $135°$ **26** $75°$ **27** $30°, 60°, 90°$ **28** $28°$

page 113 **Exercise 7** Ⓔ

1 $72°$ **2** $98°$ **3** $80°$ **4** $74°$

5 $86°$ **6** $88°$ **7** $x = 95°, y = 50°$ **8** $a = 87°, b = 74°$

9 $a = 65°, c = 103°$ **10** $a = 68°, b = 42°$ **11** $y = 65°, z = 50°$ **12** $a = 55°, b = 75°, c = 50°$

page 113 **Exercise 8** Ⓔ

1 $42°$ **2** $68°$ **3** $100°$ **4** $73°$ **5** $120°$

6 $64°$ **7** $20°$ **8** $a = 70°, b = 60°$ **9** $x = 58°, y = 109°$ **10** $66°$

11 $65°$ **12** $e = 70°, f = 75°$ **13** $x = 72°, y = 36°$ **14** $a = 68°, b = 72°, c = 68°$

15 $4°$ **16** $28.5°$ **17** $x = 60°, y = 48°$ **18** $a = 65°, b = 40°$

19 $x = 49°, y = 61°$ **20** $a = 60°, b = 40°$ **21** $136°$ **22** $80°$

page 115 **Exercise 9** Ⓔ

1, 2, 3 – Check students' work.

page 117 **Exercise 10** Ⓒ

1 $40°$ **2** $65°$ **3** $110°$ **4** $72°$

5 A $= 60°$, B $= 66°$, c $= 54°$
   AB $= 4.8$ cm, BC $= 5.1$ cm AC $= 5.4$ cm

6 D $= 38°$, E $= 112°$, F $= 30°$
   DE $= 5.4$ cm, EF $= 6.6$ cm, DF $= 10$ cm

7 G $= 48°$, I $= 54°$, H $= 78°$
   GH $= 8.1$ cm, HI $= 7.5$ cm, IG $= 9.8$ cm

8 J $= 108°$, K $= 35°$, L $= 37°$ **9** Check students' work.
   JK $= 5.4$ cm, KL $= 8.5$ cm, LJ $= 5.1$ cm

page 119 **Exercise 11 C**

1 7·4 cm   **2** 7·9 cm   **3** 8·0 cm   **4** 10·3 cm   **5** 6·4 cm   **6** 6·8 cm
7 9·0 cm   **8** 9·6 cm   **9** 7 cm   **10** 7·4 cm   **11** 7·6 cm   **12** 9·1 cm

page 119 **Exercise 12 C**

2 Check students' work.   **3** 61°   **4** 85°   **5** 72°   **6** 121°

page 120 **Exercise 13 E**

1 Check students' work.

2 Two triangles are possible with the angle opposite the 8 cm side being 59° and 121° in each case.

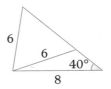

3 Two triangles are possible with the angle opposite the 7 cm side being 63° and 117° in each case.

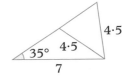

4 AAS, SSA or RHS.

page 121 **Exercise 14 E**

1 a, b, d

2

3

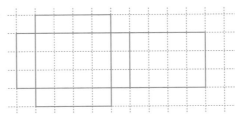

4 The net can be completed by any one of squares ①–④

5 a A   6 vertices   5 faces
     B   5 vertices   5 faces
   b A is a prism and B a pyramid.
6 Check students' work.

page 124 **Exercise 15 E**

1–8 Check student's work.

page 126 **Exercise 16** C

1 A and G, B and O, C and F, H and P, I and N, J and K

2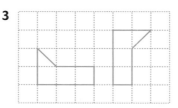

Shapes labelled with same
numbers are congruent

3

Many other answers are possible.

4 A and G, C and J, K and M, L and F   **5** Any enlargement of I

6 The ratio of the sides for the first figure is 3 : 2 whereas for the second
figure it is 4 : 1.

page 128 **Exercise 17** C

1 a 1	**b** 1	**2 a** 1	**b** 1	**3 a** 4	**b** 4	**4 a** 2	**b** 2
5 a 0	**b** 6	**6 a** 0	**b** 2	**7 a** 0	**b** 2	**8 a** 4	**b** 4
9 a 0	**b** 4	**10 a** 4	**b** 4	**11 a** 6	**b** 6	**12 a** infinite	**b** infinite

13 a   or   or   **b**

14 a 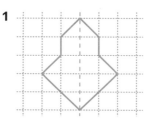 **b**

Other answers are possible.

15 a 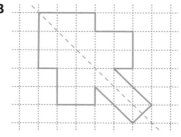 **b**

Other answers are possible.

page 129 **Exercise 18** C

1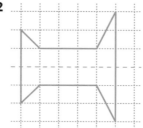

2

3

4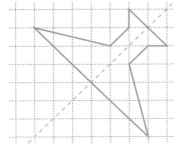

5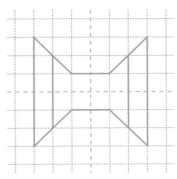

6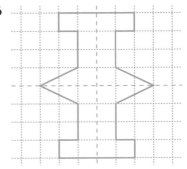

7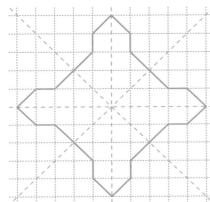

8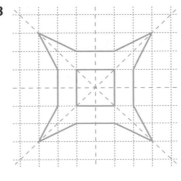

9 Check students' work.

page 130 **Exercise 19**Ⓔ

1 3   **2 a** 1   **b** 1   **c** 2   **3 a** Multiple answers – check students' work   **b** 9

4 2 planes of symmetry are shown.
There are another 2 formed by
joining the diagonals of the base to
the vertex of the pyramid.

                                                5 4

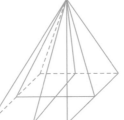

page 132 **Exercise 20**Ⓒ

1 A Isosceles trapezium   **B** Rhombus   **C** Rectangle   **D** Trapezium   **E** Parallelogram

2

	Diagonals always equal	Diagonals always perpendicular	Diagonals always bisect the angles	Diagonals always bisect each other
Square	✓	✓	✓	✓
Rectangle	✓			✓
Parallelogram				✓
Rhombus	✓	✓	✓	✓
Trapezium				

3 a 30°   **b** 90°   **c** 115°

4 a

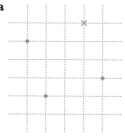

b

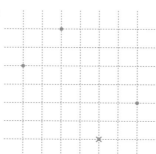

c

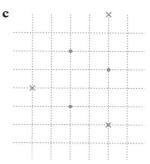

5 Quadrilaterals   **6** Square   **7** Rectangle   **8** Trapezium
9 Equilateral triangle   **10** Isosceles triangle **11** Rhombus   **12** True
13 False   **14 a** Rectangle   **b** Square   **c** Trapezium   **d** Parallelogram
15 a Trapezium   **b** Rectangle   **c** Square   **d** Parallelogram   **e** Isosceles triangle
16 Any trapezium with 2 right angles   **17** (5, 1) and (0, 0)

page 134 **Exercise 21Ⓔ**

1 a 34°   **b** 56°   **2 a** 35°   **b** 35°   **3 a** 72°   **b** 108°   **c** 80°   **4 a** 40°   **b** 30°   **c** 110°
5 a 116°   **b** 32°   **c** 58°   **6 a** 55°   **b** 55°   **7 a** 26°   **b** 26°   **c** 77°

page 135 **Exercise 22Ⓔ**

1–3 Check students' work.
4 Multiple answers – check students' work.
5 Multiple answers – check students' work.

6

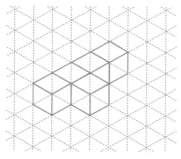

7

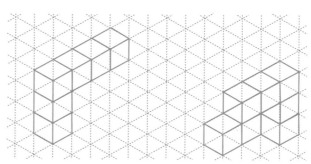

8 Check students' work.

9 B

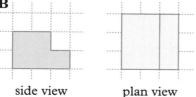

side view   plan view

C

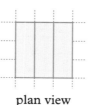

side view   plan view

page 137 **Exercise 23 Ⓒ**

1 $8\,\text{cm}^2$ **2** $14\,\text{cm}^2$ **3** $10\,\text{cm}^2$ **4** $6\,\text{cm}^2$ **5** $8\,\text{cm}^2$

6 $7\,\text{cm}^2$ **7** $9\frac{1}{2}\,\text{cm}^2$ **8** $12\,\text{cm}^2$ **9** $34\,\text{cm}^2$

10 a $19\,\text{cm}^2$ **b** $15\,\text{cm}^2$ **c** $16\,\text{cm}^2$ **d** $21\,\text{cm}^2$ **e** $23\,\text{cm}^2$ **f** $17\,\text{cm}^2$ **g** $16\,\text{cm}^2$ **h** $17\,\text{cm}^2$

page 139 **Exercise 24 Ⓒ**

1 $24\,\text{cm}^2$ **2** $24\,\text{cm}^2$ **3** $20\,\text{cm}^2$ **4** $8\,\text{cm}^2$ **5** $10\,\text{cm}^2$ **6** $25\,\text{cm}^2$

7 $35\,\text{cm}^2$ **8** $6000\,\text{cm}^2$ **9** $40\,000\,\text{m}^2$ **10** $12\,\text{cm}^2$ **11** $28\,\text{cm}^2$

page 140 **Exercise 25 Ⓒ**

1 $17\,\text{cm}^2$ **2** $34\,\text{cm}^2$ **3** $33\,\text{cm}^2$ **4** $54\,\text{cm}^2$

5 $18\,\text{cm}^2$ **6** $20\,\text{cm}^2$ **7** $25\,\text{cm}^2$ **8** $23\,\text{cm}^2$

9 $31\,\text{cm}^2$ **10** $35\,\text{cm}^2$ **11** $54\,\text{cm}^2$ **12** $51\,\text{cm}^2$

13 a B   **b** $A = 6\,\text{cm}^2$, $C = 4\,\text{cm}^2$ $D = 6\,\text{cm}^2$ **c** Multiple answers – check students' work.

14 $36\,\text{cm}^2$ **15 a** $4\,\text{cm}$ **b** $5\,\text{cm}$ **c** $5\,\text{cm}$ **16** $248\,\text{cm}^2$

page 142 **Exercise 26 Ⓒ**

1 $5\,\text{cm}^2$ **2** $14\,\text{cm}^2$ **3** $20\,\text{cm}^2$ **4** $9\,\text{cm}^2$ **5** $12\,\text{cm}^2$ **6** $25\,\text{cm}^2$ **7** $14\,\text{cm}^2$ **8** $33\,\text{cm}^2$

9 $39\,\text{cm}^2$ **10** $39\,\text{cm}^2$ **11** $15\,\text{cm}^2$ **12** $23\frac{1}{2}\,\text{cm}^2$ **13** $33\,\text{cm}^2$ **14 a** $6\,\text{cm}$ **b** $5\,\text{cm}$ **c** $7\,\text{cm}$

15 a b Multiple answers – check students' work.   **16** Mark

page 144 **Exercise 27 Ⓔ**

1 b $A = 10$, $B = 6$, $C = 3$ square units   **c** 36 square units   **d** 17 square units

2 b $A = 5$, $B = 14$, $C = 6$ square units   **c** 42 square units   **d** 17 square units

3 $13\frac{1}{2}$ square units   **4** $14\frac{1}{2}$ square units   **5** 24 square units   **6** 21 square units   **7** 21 square units

page 145 **Exercise 28 Ⓒ**

1 a $20\,\text{cm}$ **b** $30\,\text{cm}$ **c** $38\,\text{cm}$ **d** $28\,\text{cm}$

2 a $12\,\text{cm}$ **b** $12\,\text{cm}$ **c** $13\,\text{cm}$

3 a $16\,\text{cm}$ (A) $14\,\text{cm}$ (B) $18\,\text{cm}$ (C)   **b** Check students' work.

4 a $12\,\text{cm}$ **b** $12\,\text{cm}$ **c** $12\,\text{cm}$ **5** $50\,\text{cm}$ **6** $8\,\text{m}$ **7** $26\,\text{cm}$ **8** $34\,\text{cm}$

9 $30\,\text{cm}$ **10** $30\,\text{cm}$ **11** $40\,\text{cm}$ **12** $32\,\text{cm}$ **13** $28\,\text{cm}$ **14** $28\,\text{cm}$

page 148 **Test yourself**

1 a $18\,\text{cm}$ **b** $12\,\text{cm}^2$ **c** any copy of the shape with 3 squares shaded

2 a 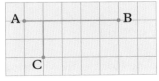   **b**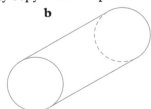

3 a $80°$ **b** $b = 60°$, $c = 110°$ **4** $142°$ **5 a** South   **b** Bampton Hill   **c** $1\cdot75\,\text{km}$ **d** $12\,\text{km}^2$

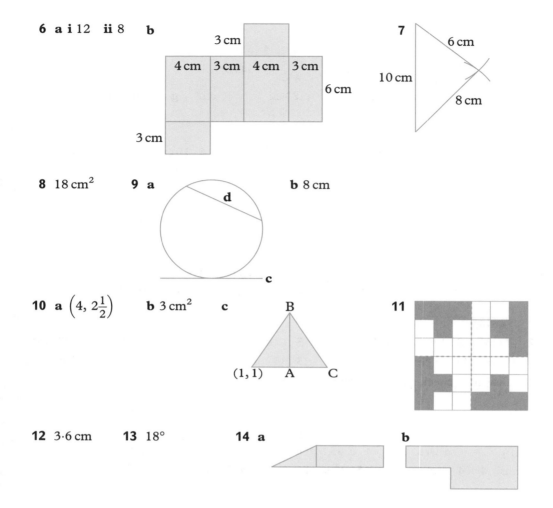

6 a i 12 **ii** 8 **b**

3 cm

4 cm | 3 cm | 4 cm | 3 cm

6 cm

3 cm

7

6 cm

10 cm

8 cm

8 18 cm² **9 a** **b** 8 cm

d

c

10 a $\left(4, 2\frac{1}{2}\right)$ **b** 3 cm² **c**

B

(1, 1) A C

11

12 3·6 cm **13** 18° **14 a** **b**

4 Handling data 1

page 153 **Exercise 1 C**

1 a i 10·00 **ii** 08·30 **iii** 10·45 **b i** Farnham **ii** Dorking **iii** Reigate
2 a i 15·00 **ii** 13·30 **iii** 15·15
 b i Westbury **ii** Southampton **iii** Bristol **iv** Porstmouth
 c 40 km **d** 1 hour **e** 40 km/hour
3 a i 13·30 **ii** 11·00 **b i** Stump Cross **ii** Mountfitchet **c** 20 km
 d 1 hour **e** 20 km/hour
4 a i 16·30 **ii** 19·00 **iii** 16·07 **b i** Sevenoaks **ii** Chiddingstone Hoath **iii** Maidstone
 c 80 km **d** 80 km/hour **e** 40 km **f** 20 km/hour
5 a i 15·30 **ii** 16·30 **iii** 14·30 **b i** Chipping Norton **ii** Furzy Leaze **iii** Long Compton
 c 30 km **d** 20 km/hour **e** 20 km **f** 10 km/hour

6 a i 08·45   **ii** 09·15   **iii** 11·00   **b i** Great Bricett **ii** Bury St. Edmunds **iii** Shimpling
  c 30 km   **d** 20 km/hour   **e** 20 km   **f** 10 km/hour   **g** 14·3 km/hour (1 dp)

page 155 Exercise 2 C

1 a 30 minutes **b i** 30 km/hour **ii** 40 km/hour **iii** 70 km/hour **c i** 12·00 **ii** 10·30 **iii** 09·30
2 a 45 mins   **b** 15·15   **c** 14·15   **d i** 40 km/hour **ii** 50 km/hour **iii** 70 km/hour **e** 16·15
3 a 40 km   **b** 60 km   **c** York and Scarborough   **d** 15 minutes **e i** 11·00   **ii** 13·45
  f i 40 km/hour   **ii** 60 km/hour   **iii** 100 km/hour
4 a 25 km   **b** 15 km   **c** 09·45   **d** 1 hour
  e i 26·7 km/hour (1 dp)   **ii** 5 km/hour **iii** 30 km/hour **iv** 40 km/hour
5 a i 14·00   **ii** 13·45   **b i** 15·45   **ii** Towards Aston
  c i 15 km/hour   **ii** 40 km/hour   **iii** 40 km/hour **iv** 20 km/hour
6 a 45 mins   **b** 09·15   **c** 60 km/hour   **d** 100 km/hour   **e** 57·1 km/hour (1 dp)

page 160 Exercise 3 C

1 a i £2·80   **ii** £2·30   **b i** €4·25   **ii** €2   **c** £35 000
2 a i £25   **ii** £44   **iii** £30   **iv** £40   **v** £10   **vi** £15
  b i $32   **ii** $80   **iii** $10   **iv** $26   **v** $58   **vi** $54   **c** £45
3 a i £3·60   **ii** £8·50   **iii** £5·40   **iv** £4·90   **v** £7·70   **vi** £0·50
  b i 620 rupees **ii** 580 rupees **iii** 360 rupees **iv** 160 rupees **v** 780 rupees **vi** 515 rupees
  c 700 rupees
4 a 11·2 cm   **b** May and October
  c November, could have been heavy rainfall, or snow followed by a thaw.
5 a 15·2 °C   **b** October   **c** April and November   **d** September and October   **e** 21 °C
6 a 30 litres   **b i** 8 miles per litre   **ii** 6 miles per litre
7 a

x	0	50	100	150	200	250	300
C	35	45	55	65	75	85	95

8 a

h	0	1	2	3
C	18	33	48	63

b

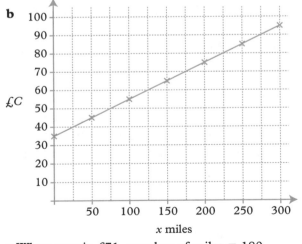

c When cost is £71, number of miles = 180.

b

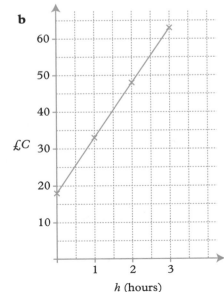

c When C = £55·50, Jeff worked $2\frac{1}{2}$ hours.

page 164 **Exercise 4Ⓔ**

1 a D **b** C **c** F **d** E **e** A **f** B **2** B **3** D **4 a** C **b** A **c** D **d** B

5 a i B **ii** A **b** from $7\frac{1}{2}$ secs to $18\frac{1}{2}$ secs **c** B **d** A **6** X → B Y → C Z → A

7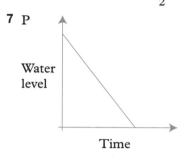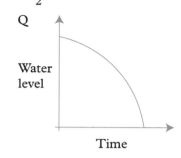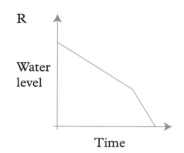

page 169 **Exercise 5Ⓒ**

1 a 7 **b** 5 **c** 15 **d** 6·5 **e** − 1 **f** 7 **g** 3 **2 a** 16 years **b** 170 cm **c** 50 kg
3 a 5 **b** 15 **4** 3·5 Sally does with 20p. **5** Multiple answers possible. **6** 17 kg

page 170 **Exercise 6Ⓒ**

1 a 1 **b** 6 **c** 4 **d** 4 **e** red **f** 1000 **2** 6 **3 a** 6 **b** 16 °C **c** 31
4 a 3 **b** 3 **5** Multiple answers possible.

6

Number of chicks	Tally	Frequency
0	||	2
1	||||	4
2	ⅢⅡ |	6
3	ⅢⅡ ||	7
4	ⅢⅡ ||||	9
5	|||	3
6	|	1

Modal number of chicks per nest = 4

page 172 **Exercise 7Ⓒ**

1 a 7 **b** 10 **c** 4 **d** 7·8 **e** 1·3 **f** 4 **2** Girls = 7 Boys = $6\frac{2}{3}$ (*or* 6·7 to 1 dp) **3** 44·2 kg
4 Multiple answers possible. **5 a** 10 **b** 13 **6 a** 5 **b** 35 **7** 5·9 grams **8** 70 kg **9** 161 cm

page 173 **Exercise 8Ⓒ**

1 10 − 2 = 8 **2 a** 26 **b** 49 **c** 41 **3** 7 °C **4** Multiple answers possible.
5 a Impossible **b** Possible **c** True **6 a** Impossible **b** Possible **c** Possible
7 a 17 **b** 10 to 19 **c** We do not actually know the highest or lowest scores obtained.

page 175 **Exercise 9Ⓒ**

1

	Mean	Median	Mode
a	6	5	4
b	9	7	7
c	6·5	8	9
d	3·5	3·5	4

e Median, as it is central to the data and not affected by the one very high value of 12.

2 a 2°C **b** 5°C **3** Mean = 17 Median = 3   Median best describes the set. **4 a** 11 **b** 4
5 2 or 45 **6** 4   **7** 6   **8 a** 1·60 m   **b** 1·634 m   **9 a** 7·2   **b** 5   **c** 6 **10** £2·10
11 Multiple answers possible. e.g. 4 4 5 7 10   **12 a** False **b** Possible **c** False **d** Possible
13 a mean = £47,920   Median = £22,500
  b The mean does not give a fair average, as 4 of the 5 employees earn less than half of this amount.
14 a Mean = 157·1 kg (1 dp)   Median = 91 kg
  b The farmer has used the mean weight. Not fair, as 25 of the 32 cattle are less than this weight.

page 178 **Exercise 10Ⓔ**

1 96·25 kg   **2** 51·9p   **3** 4·82 cm
4 a Mean = 3·025   Median = 3   Mode = 3   **b** Mean = 17·75   Median = 17   Mode = 17
5 a Mean = 6·6 (1 dp)   Median = 8   Mode = 3   **b** The mode   **6 a** 9   **b** 15

page 180 **Exercise 11Ⓔ**

1 a

Number of words	Frequency f	Midpoint x	fx
1–5	6	3	18
6–10	5	8	40
11–15	4	13	52
16–20	2	18	36
21–25	3	23	69
Totals	20	–	215

b 10·75 words

2 68·25   **3** 3·8 letters (1 dp)

page 181 **Exercise 12Ⓒ**

1 $\frac{5}{6}$ **2** $\frac{3}{4}$ **3** $\frac{1}{2}$ **4** $\frac{4}{5}$ **5** $\frac{3}{5}$ **6** $\frac{5}{6}$ **7** $\frac{4}{5}$ **8** $\frac{1}{3}$ **9** $\frac{1}{4}$ **10** $\frac{5}{12}$
11 $\frac{1}{2}$ **12** $\frac{2}{3}$ **13** $\frac{5}{6}$ **14** $\frac{1}{6}$ **15** $\frac{1}{8}$ **16** $\frac{1}{9}$ **17** $\frac{2}{9}$ **18** $\frac{7}{12}$ **19** $\frac{7}{72}$ **20** $\frac{3}{20}$
21 12 **22** 45 **23** 40 **24** 20 **25** 11 **26** 36 **27** 240° **28** 300° **29** 135° **30** 30°
31 252° **32** 150°

page 182 **Exercise 13Ⓒ**

1 a i $\frac{1}{6}$ **ii** $\frac{1}{3}$ **iii** $\frac{1}{8}$ **iv** $\frac{3}{8}$ **b i** $\frac{1}{6}$ **ii** $\frac{1}{3}$ **iii** $\frac{1}{8}$ **iv** $\frac{3}{8}$
  c i 20   **ii** 40   **iii** 15   **d i** 20   **ii** 40   **iii** 15   **iv** 45
2 a i $\frac{1}{4}$ **ii** $\frac{1}{6}$ **iii** $\frac{1}{3}$ **b i** $\frac{1}{4}$ **ii** $\frac{1}{6}$ **iii** $\frac{1}{4}$ **iv** $\frac{1}{3}$ **c i** 9   **ii** 6   **iii** 9   **iv** 12
3 a i $\frac{1}{8}$ **ii** $\frac{1}{6}$ **iii** $\frac{5}{12}$ **iv** $\frac{1}{12}$ **v** $\frac{1}{8}$ **vi** $\frac{1}{12}$ **b i** £45   **ii** £60   **iii** £150   **iv** £30
4 a £425   **b** £150   **c** £250   **d** £75
5 a £13,333 (nearest pound)   **b** £15,000   **c** £6,667 (nearest pound)   **d** £10,000   **e** £12,000
6 a i 8 mins   **ii** 34 mins   **iii** 10 mins   **b** 18°

page 184 **Exercise 14Ⓒ**

1 a i 45°   **ii** 200°   **iii** 110°   **iv** 5°   **b** Check pie chart has a title and that angles are
                    Spurs 45°, Chelsea 200° Man Utd 110°, York 5°.

2 a $\frac{3}{10}, \frac{2}{5}, \frac{1}{5}, \frac{1}{10}$ **b** Check pie chart has a title and that angles are sixth form 108°, employment 144°, FE colleges 72°, unemployed 36°.

3 $x = 60°$, $y = 210°$ **4** Check pie chart has a title and that angles are Oats 90°, Barley = 60°, Wheat = 45°, Rye = 165°.

5 a Number of boys choosing red = 20   **b** Number of boys choosing blue = 25
Number of girls choosing red = 12   Number of girls choosing blue = 15
∴ Tony is wrong.   ∴ Mel is right.

page 186 **Exercise 15 C**

1 a Sharon   **b** £11   **c** Half of a £ symbol

2 a 2   **b**

Make	Number of cars	
Ford	4	🚗 🚗
Renault	6	🚗 🚗 🚗
Toyota	6	🚗 🚗 🚗
Audi	3	🚗 🚙

3 Any suitable pictogram – check students' work.

page 187 **Exercise 16 C**

1 a 5   **b** 19   **c** 23   **d** 55   **e** $\frac{6}{23}$

2

Score	Tally	Frequency
1	⊥⊥⊤ I	6
2	I I I I	4
3	I I I I	4
4	I I I	3
5	⊥⊥⊤ I I I	8
6	⊥⊥⊤	5

3 a

No. of letters	Tally	Freq.
1	⊥⊥⊤ I	6
2	⊥⊥⊤ I I	7
3	⊥⊥⊤ I I I	8
4	⊥⊥⊤ ⊥⊥⊤ I I	12
5	⊥⊥⊤ I	6
6	⊥⊥⊤ I I I	8
7	I I I I	4
8	I I	2
9	I I	2
10	I	1
	Total	56

b 4

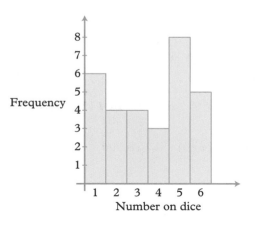

4 a

Score	Tally	Freq	Score	Tally	Freq
2	||	2	8	L̶H̶T̶ |	6
3	||||	4	9		0
4	L̶H̶T̶	5	10	L̶H̶T̶	5
5	L̶H̶T̶ ||	7	11	L̶H̶T̶ |	6
6	L̶H̶T̶ ||||	9	12	|||	3
7	L̶H̶T̶ L̶H̶T̶ |||	13		TOTAL	60

b

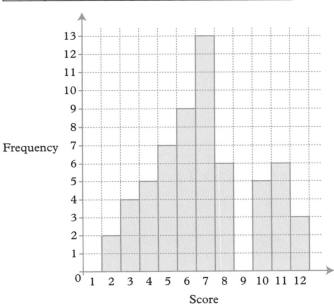

Frequency

Score

5 a £3,400

 b The shop's profits increase drastically in the run-up to Christmas. However, in January, sales drop-off as no more presents are being bought. Sales begin to improve as the year progresses, but a loss is still being made as late as April.

page 190 **Exercise 17Ⓔ**

1 a 5   **b** 24   **c** 35
2 a D   **b** A   **c** A   **d** D   **e** B

3

Weight (gm)	Tally	Frequency
$60 \leqslant w < 70$	||||	4
$70 \leqslant w < 80$	L̶H̶T̶ |	6
$80 \leqslant w < 90$	L̶H̶T̶	5
$90 \leqslant w < 100$	L̶H̶T̶ L̶H̶T̶	10
$100 \leqslant w < 110$	L̶H̶T̶ L̶H̶T̶ ||||	14
$110 \leqslant w < 120$	L̶H̶T̶ L̶H̶T̶ |	11
		50

Field A was treated with the new fertilizer.

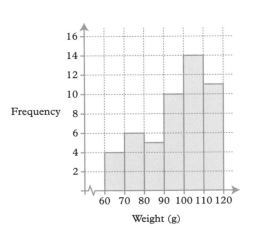

Frequency

Weight (g)

4 Should go to Isola, as it has 16 days with more than 40 cm of snow compared to only 3 days in Les Jets.

5 **a** Multiple answers possible. **b** Multiple answers possible.
 c Check conclusion against answer (b).

page 195 **Exercise 18Ⓔ**

1
1	5
2	3 7 8 9
3	2 5 6 8 9
4	0 1 2 5 6 7 8
5	1 2 3 4 9
6	5 6

2 **a**
2	0 4 5 8
3	1 7 9
4	0 4 6
5	2 5 8 9
6	1 5 7 8
7	3 5

b
2	2 8 9
3	0 5 8
4	1 4 6 7 7
5	3 4 9
6	7
7	2

3 **a** 50 kg **b** 15 **c** 50 kg

4
1	4 8
2	4 4 8
3	1 3 3 7 8
4	0 5 5 6 6 7 9
5	1 2 5 8
6	2 3 7

a 4·5 **b** 5·3

5 **a** 13 **b** 78 **c** The pulse rate of women is on average higher than the pulse rate of men.

page 197 **Test yourself**

1 **a** $\frac{3}{4}$ **b** £15 **c** Cooker & Iron

2 **a** 5 **b** 6 **c** The mode, as this is the size he sells more of.

3 **b** 30 **c** 3 **d** 30·2

4 **a** 11 am to 12 noon **b** 2 miles

c i

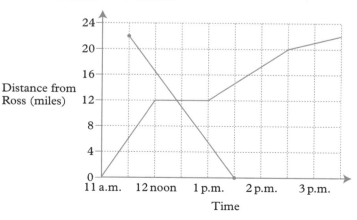

ii 12·25

5 **a** $\frac{1}{8}$ **b** 150° **c** 6

6 a

No. of people	Tally	Frequency
1					3
2	LHT ||	7
3					3
4	LHT	5
5				2
	TOTAL	20

b 2   **c** 7·4

7 a 9   **b** 42   **c** 35

8 a 0 | 5 7 8 8

1 | 0 0 0 0 2 5 5 5 6

2 | 0 0 0 4 5

3 | 3 5

where 1 | 5 = 15 minutes

b 0·95

9 a 178·5 seconds   **b** $150 < t \leqslant 180$. The median is halfway between the 20th and 21st observation – this lies in the second class interval.   **c** $\frac{3}{40}$

5 Number 2

page 201 **Exercise 1 C**

1 805   **2** 459   **3** 650   **4** 1333   **5** 2745   **6** 1248   **7** 4522
8 30 368   **9** 28 224   **10** 8568   **11** 46 800   **12** 66 281   **13** 57 602   **14** 89 516
15 97 525

page 202 **Exercise 2 C**

1 32   **2** 25   **3** 18   **4** 13   **5** 35   **6** 22 r 2   **7** 23 r 24
8 18 r 10   **9** 27 r 18   **10** 13 r 31   **11** 35 r 6   **12** 23 r 24   **13** 64 r 37   **14** 151 r 17
15 2961 r 15

page 202 **Exercise 3 C**

1 £47·04   **2** 46   **3** 7592   **4** 15 stamps, 20p change
5 8   **6** £85   **7** £14 million   **8** £80·64

page 204 **Exercise 4 C**

1 a a quarter, $\frac{1}{4}$ **b** a half, $\frac{1}{2}$ **c** four fifths, $\frac{4}{5}$ **d** three quarters, $\frac{3}{4}$

e one tenth, $\frac{1}{10}$ **f** one third, $\frac{1}{3}$ **g** two thirds, $\frac{2}{3}$ **h** three tenths, $\frac{3}{10}$

2 a A = $\frac{1}{4}$, B = $\frac{3}{4}$ **b** C = $\frac{4}{10}$, D = $\frac{9}{10}$ **c** E = $\frac{1}{3}$, F = $\frac{2}{3}$

3 Multiple answers possible – check students' work.

page 205 **Exercise 5** Ⓒ

1 2   **2** 2   **3** 2   **4** 6   **5** 5   **6** 4   **7** 6   **8** 8

9 a $\frac{1}{2} = \frac{2}{4} = \frac{4}{8} = \frac{8}{16}$ **b** $\frac{1}{2} = \frac{2}{4} = \frac{3}{6} = \frac{4}{8} = \frac{5}{10}$ **10** $\frac{3}{4}$ **11** $\frac{1}{4}$ **12** $\frac{4}{5}$ **13** $\frac{2}{5}$

14 $\frac{1}{4}$ **15** $\frac{2}{3}$ **16** $\frac{3}{5}$ **17** $\frac{1}{3}$ **18** $\frac{1}{3}$ **19** $\frac{8}{9}$ **20** $\frac{4}{5}$ **21** $\frac{1}{3}$ **22** $\frac{2}{9}$ **23** $\frac{2}{3}$

24 $\frac{1}{3}$ **25** $\frac{2}{5}$ **26** $\frac{3}{5}$ **27** $\frac{7}{8}$ **28** $\frac{3}{5}$ **29** $\frac{7}{9}$ **30** $\frac{2}{7}$ **31** $\frac{4}{5}$ **32** $\frac{2}{5}$ **33** $\frac{1}{4}$

34 $\frac{3}{4}$ **35** $\frac{1}{100}$ **36** $\frac{1}{2}$ **37** $\frac{1}{4}$ **38** $\frac{1}{6}$ **39** $\frac{3}{10}$ **40** $\frac{5}{12}$ **41** $\frac{7}{30}$ **42** $\frac{1}{10}$ **43** $\frac{1}{2}$

44 a $\frac{2}{4}$ **b** $\frac{2}{97}$ **45 a** $\frac{8}{12}, \frac{6}{12}, \frac{9}{12}$ **b** $\frac{1}{2}, \frac{2}{3}, \frac{3}{4}$ **46 a** $\frac{10}{12}, \frac{8}{12}, \frac{3}{12}$ **b** $\frac{1}{4}, \frac{2}{3}, \frac{5}{6}$

47 $\frac{1}{2}, \frac{3}{5}, \frac{7}{10}$ **48** $\frac{7}{12}, \frac{5}{6}, \frac{5}{4}$ **49** $\frac{3}{8}, \frac{1}{2}, \frac{3}{4}$ **50** $\frac{2}{15}, \frac{1}{3}, \frac{7}{15}$

page 208 **Exercise 6** Ⓔ

1 $3\frac{1}{2}$ **2** $1\frac{2}{3}$ **3** $2\frac{1}{3}$ **4** $1\frac{1}{4}$ **5** $2\frac{2}{3}$ **6** $1\frac{1}{3}$ **7** 3

8 $4\frac{1}{2}$ **9** $2\frac{1}{4}$ **10** 5   **11** $1\frac{2}{3}$ **12** $1\frac{3}{7}$ **13** $1\frac{5}{8}$ **14** $2\frac{1}{3}$

15 2   **16** 12   **17** $3\frac{1}{7}$ **18** $1\frac{2}{3}$ **19** $2\frac{2}{5}$ **20** $1\frac{1}{2}$ **21** $\frac{5}{4}$

22 $\frac{4}{3}$ **23** $\frac{9}{4}$ **24** $\frac{8}{3}$ **25** $\frac{15}{8}$ **26** $\frac{5}{3}$ **27** $\frac{22}{7}$ **28** $\frac{13}{6}$

29 $\frac{19}{4}$ **30** $\frac{15}{2}$ **31** $\frac{29}{8}$ **32** $\frac{22}{5}$ **33** $\frac{17}{5}$ **34** $\frac{33}{4}$ **35** $\frac{13}{10}$

page 208 **Exercise 7** Ⓒ

1

	Fraction of quantity required	Divide the quantity by
a	$\frac{1}{2}$	2
b	$\frac{1}{3}$	3
c	one-quarter	4
d	$\frac{1}{10}$	10
e	$\frac{1}{5}$	5

2 £6   **3** £10   **4** 5 litres
5 5 kg   **6** 24 cm   **7** £4
8 75 miles   **9** £10   **10** 70 kg
11 6 litres   **12** 3 kg   **13** £20
14 6 kg   **15** 70 eggs   **16** 110 hens
17 111 miles   **18** £12   **19** 40 pages
20 2, 3, 5, 6, 9, APPLE

page 210 **Exercise 8** Ⓒ

1 £46   **2** £48   **3** 9 kg   **4** £52   **5** 48 miles   **6** 60 cm   **7** £144
8 15 kg   **9** 35 hens   **10** 300 cm   **11** 48 pence   **12** 400 miles   **13** 28p   **14** £2500
15 27p   **16** £80   **17** 16 kg   **18** 603 km   **19** 54   **20** 180   **21** £4
22 72   **23** 12, 15, 16, 18, 30, 32, 40 BECKHAM

page 211 **Exercise 9Ⓔ**

1 a $\frac{6}{7}$ **b** $\frac{5}{9}$ **c** $\frac{3}{5}$ **d** 1   **e** $\frac{5}{8}$ **f** $\frac{5}{11}$ **g** $\frac{1}{3}$ **h** $\frac{4}{15}$

2 a $\frac{2}{8} + \frac{1}{8} = \frac{3}{8}$ **b** $\frac{5}{10} + \frac{4}{10} = \frac{9}{10}$ **c** $\frac{6}{15} + \frac{5}{15} = \frac{11}{15}$

     d $\frac{5}{10} + \frac{2}{10} = \frac{7}{10}$ **e** $\frac{6}{8} + \frac{1}{8} = \frac{7}{8}$ **f** $\frac{3}{6} + \frac{1}{6} = \frac{4}{6} = \frac{2}{3}$

3 a $\frac{5}{6}$ **b** $\frac{5}{12}$ **c** $\frac{7}{12}$ **d** $\frac{9}{20}$ **e** $1\frac{1}{15}$ **f** $\frac{13}{20}$ **g** $\frac{11}{14}$ **h** $3\frac{3}{4}$

4 a $\frac{1}{6}$ **b** $\frac{5}{12}$ **c** $\frac{3}{10}$ **d** $\frac{5}{12}$ **e** $\frac{1}{10}$ **f** $\frac{2}{15}$

5 $\frac{9}{10}$ m   **6** $\frac{3}{4}$ **7** $\frac{1}{4}$ **8** 5   **9** Length = $\frac{7}{8}$ inch, Width = $\frac{1}{2}$ inch

page 214 **Exercise 10Ⓔ**

1 a $\frac{2}{15}$ **b** $\frac{15}{28}$ **c** $\frac{8}{15}$ **d** $\frac{25}{42}$ **e** $\frac{10}{27}$ **f** $\frac{10}{33}$ **g** $\frac{21}{32}$ **h** $\frac{2}{3}$

2 a $\frac{2}{5}$ **b** $\frac{2}{7}$ **c** $\frac{1}{6}$ **d** $\frac{7}{9}$ **e** $1\frac{1}{5}$ **f** $\frac{5}{14}$ **g** $\frac{9}{13}$ **h** 1

3 a $\frac{3}{4}$ **b** $\frac{7}{8}$ **c** $\frac{3}{4}$ **d** $\frac{7}{10}$ **e** $1\frac{1}{2}$ **f** $1\frac{1}{5}$ **g** $3\frac{1}{2}$ **h** 6

4 a $1\frac{1}{2}$ **b** $\frac{9}{10}$ **c** $3\frac{1}{3}$ **d** $\frac{8}{9}$ **e** $1\frac{1}{9}$ **f** $\frac{20}{21}$ **g** $\frac{5}{8}$ **h** $9\frac{2}{3}$

5

×	$\frac{2}{3}$	$\frac{3}{4}$	$\frac{1}{5}$
$\frac{1}{2}$	$\frac{1}{3}$	$\frac{3}{8}$	$\frac{1}{10}$
$\frac{1}{4}$	$\frac{1}{6}$	$\frac{3}{16}$	$\frac{1}{20}$
$\frac{2}{5}$	$\frac{4}{15}$	$\frac{3}{10}$	$\frac{2}{25}$

6 £53·33   **7** 192 cm   **8** $\frac{1}{5}$ **9**

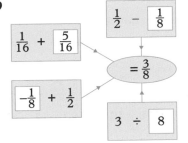

10 a $\frac{1}{2}$ **b** $\frac{1}{3}$ **c** $\frac{6}{7}$ **d** $\frac{3}{10}$ **e** $1\frac{3}{4}$ **f** $\frac{17}{24}$ **g** $2\frac{1}{5}$ **h** $\frac{7}{15}$ **11** $\frac{1}{36}$

page 216 **Exercise 11Ⓒ**

1 i $\frac{45}{100} = \frac{9}{20}$ **ii** $\frac{22}{100} = \frac{11}{50}$ **iii** $\frac{18}{100} = \frac{9}{50}$

2 60%   **3** 27%   **4** 33%   **5 a** 15%   **b** 5%   **c** 45%   **d** 60%   **e** 75%

6 a $\frac{30}{100} = \frac{3}{10}$ **b** $\frac{75}{100} = 75\%$ **c** $33\frac{1}{3}\%$ **d** $\frac{1}{100}$ **e** $\frac{80}{100}$ or $\frac{8}{10}$ or $\frac{4}{5}$ **f** 10%

7 A a $\frac{1}{2}$ **b** 50%   **B a** $\frac{4}{5}$ **b** 80%   **C a** $\frac{1}{4}$ **b** 25%   **D a** $\frac{1}{5}$ **b** 20%

     E a $\frac{3}{10}$ **b** 30%   **F a** $\frac{1}{4}$ **b** 25%   **G a** $\frac{1}{3}$ **b** $33\frac{1}{3}\%$ **H a** $\frac{2}{5}$ **b** 40%

8 a 25%   **b** 75%   **c** 60%   **d** 20%   **9 a** 75%   **b** 60%   **c** 75%   **d** $33\frac{1}{3}\%$ **e** 50%   **f** 25%

10 Multiple answers possible – check students' work.

page 218 **Exercise 12 C**

1 75% **2** 40% **3** $37\frac{1}{2}$% **4** 90% **5** 85% **6** 25% **7** 68% **8** 15% **9** 98%
10 7% **11** 25% **12** 32% **13** $67\frac{1}{2}$% **14** $12\frac{1}{2}$% **15** 23·5% **16** 64% **17** 90% **18** 40%
19 $22\frac{1}{2}$% **20** 34% **21** 93% **22 a** 44% **b** 65% **23** 21% **24 a** 50% **b** 40% **c** 10%
25 a 25 **b** 44% **c** 56% **26** Susan 70%, Jane 54%, Jackie 52%

page 219 **Exercise 13 C**

1 £12 **2** £8 **3** £10 **4** £3 **5** £2·40 **6** £24 **7** £45 **8** £72 **9** £244
10 £9·60 **11** $42 **12** $88 **13** 8 kg **14** 12 kg **15** 272 g **16** 45 m **17** 40 km **18** $710
19 4·94 kg **20** 60 g **21** €204 **22** £5 **23** 48 **24** 156 **25** £12 **26** £70

page 221 **Exercise 14 C**

1 £63 **2** £864 **3** £87·45 **4** £104 **5** £1960 **6** £792 **7** £132 **8** £45·75
9 £110·30 **10** £42 **11** £12·41 **12** £266·40 **13** £6·90 **14** £8160 **15** £12·88 **16** £79·20
17 £31·87 **18** £8·89 **19** £8·93 **20** £14·21

page 221 **Exercise 15 C**

1 £35·20 **2** £5724 **3** £171·50 **4** £88·35 **5** £58·50 **6** 24 **7** 59,400 **8** £20·72
9 3·348 kg **10** 13·054 kg **11** £2762·50 **12 a** 480 cm^2 **b** 384 cm^2

page 222 **Exercise 16 E**

1 £17·51 **2** £40·66 **3** £77·96 **4** £185·34

page 223 **Exercise 17 C**

1 £32 870 **2** £47 260 **3** £76 850 **4** £39 110 **5** £33 850 **6** £6390 **7** £5795 **8** £3080
9 £2770 **10** £6790 **11** £5740 **12** £7670 **13** £4900 **14** £5348 **15** £6470

page 224 **Exercise 18 C**

1 £2364 **2** £2099·40 **3** £3393 **4** £4530·60 **5** £4806 **6** £5589 **7** £27 555

page 225 **Exercise 19 E**

1 8% **2** 10% **3** 25% **4** 2% **5** 4% **6** 2·5% **7** 20% **8** 50% **9** 15% **10** 80%
11 25% **12** 20% **13** 12·5% **14** $33\frac{1}{3}$% **15** 80% **16** 5% **17** 6% **18** 20% **19** 5% **20** 2·5%

page 226 **Exercise 20 E**

1 36·4% **2** 19·0% **3** 19·4% **4** 22·0% **5** 12·2% **6** 9·4% **7** 14·0% **8** 17·4%
9 32·7% **10** 10·2% **11** 7·7% **12** 35·3% **13** 30·8% **14** 5·2% **15** 14·1% **16** 14·5%
17 19·1% **18** 3·6% **19** 31·1% **20** 6·5%

page 226 **Exercise 21 E**

1 12% **2** 29% **3** 30% **4** 0·25% **5** 15% **6** 61·1% **7** 15%

page 227 **Exercise 22Ⓔ**

1 a £2200   **b** £2420   **c** £2,662   **2 a** £5600   **b** £6272   **c** £7024·64   **3** £11 449
4 a £36 465·19   **b** £38 288·45   **5 a** £95 400   **b** £101 124   **c** £113,623 (nearest pound)
6 a £7582 (nearest pound)   **b** £397 (nearest pound)   **c** £583 200
7 £9212 to nearest £   **8** 8 years   **9** 11 years

page 229 **Exercise 23Ⓔ**

1 a 3:2   **b** 3:5   **c** 1:4   **d** 5:2   **e** 12:11   **f** 3:4   **g** 8:5
h 3:7   **i** 10:7   **j** 8:11   **k** 3:2:4   **l** 8:1:3   **m** 6:5:4   **n** 3:2:3
2 3:4   **3** 4:1   **4 a** 2:5   **b** 1:5   **c** 1:5   **d** 1:3   **e** 4:1   **f** 1:5   **g** 1:10   **h** 1:6   **i** 1:3
5 $\frac{3}{8}$ **6** $\frac{1}{5}$ **7** $\frac{3}{5}$ **8 a** $\frac{1}{4}$ **b** $\frac{1}{3}$ **c** $\frac{5}{12}$ **9** $\frac{2}{3}$

page 230 **Exercise 24Ⓔ**

1 £10:£20   **2** £45:£15   **3** Cat 330 g; Dog 550 g   **4** Sam $480; Chris $600
5 Steven 36 litres; Dave 90 litres.   **6** £10:£20:£30   **7** £70   **8** £50   **9** 3250
10 a 2:6 = 1:3   **b** 2:7   **c** 3:5

page 231 **Exercise 25Ⓔ**

1 8 girls   **2** 5 women   **3** 9 screws   **4** 30 g zinc and 40 g of tin   **5** 24   **6** 2·4 kg   **7** 6 eggs
8 18 cm   **9** 22·5 cm   **10** 12 cm   **11** 300 g   **12** 5:3   **13** £200   **14** 42p   **15** £175 000   **16** 0·25 m³

page 233 **Exercise 26Ⓔ**

1 200 m

3

	Map scale	Length on map	Actual length on land
a	1:10 000	10 cm	1 km
b	1:2000	10 cm	200 m
c	1:25 000	4 cm	1 km
d	1:10 000	6 cm	0·6 km

2 500 m
4 63 m
5 24 km   **6** 120 m
7 a 3·5 km   **b** 4·35 km   **c** 3·7 km

page 234 **Exercise 27Ⓔ**

1 1·5 m   **2** 1·25 m   **3** 28 cm   **4** 5·9 cm
5 a 60 cm   **b** 84 cm   **c** 56 cm   **d** 140 cm   **e** 100 cm   **6** 2·5 cm

page 235 **Exercise 28Ⓒ**

1 £24   **2** £1·08   **3** 315p   **4 a** £1·26   **b** £4·20   **5 a** £2·20   **b** £22   **6** £97·50   **7** 2750 g   **8** 1400
9 4·5 litres   **10** £3·45   **11** £1·61   **12** £3·99   **13** 125 seconds or 2 mins 5 secs   **14** $1\frac{1}{2}$ hours

page 236 **Exercise 29Ⓒ**

1 10   **2** 10   **3** 12 m²   **4** 9   **5** 100   **6** 160   **7** 450   **8 a** £267   **b** 11
9 a £127·50   **b** 2   **10 a** £2·24   **b** £4·20   **11 a** £1·60   **b** £3·60   **12** 35 litres   **13** 200 litres
14 70 gallons

page 237 **Test 1**

1 £3·50   **2** £4·95   **3** 24   **4** 20p, 10p, 10p   **5** $6\frac{1}{2}$ **6** $\frac{1}{100}$ **7** 50   **8** 75%   **9** 15
10 56p   **11** 87   **12** 6   **13** 120 m   **14** 200   **15** 16 m²   **16** 25   **17** $1\frac{1}{4}$ **18** £10   **19** 10
20 15   **21** 50 mm   **22** 16   **23** 8°C   **24** 4·5   **25** 105 or 75 (allow either)   **26** 20   **27** £2·40
28 82%   **29** £4000   **30** 8

page 238 **Test 2**

1 96   **2** 19   **3** 06:30   **4** £2·75   **5** £1·90   **6** 95°   **7** 5 018 001   **8** 15
9 £6   **10** 3·5p   **11** 55   **12** 800 g   **13** 74   **14** 280 miles   **15** 40   **16** 4
17 62   **18** 7·5   **19** 5   **20** 480   **21** 0·7   **22** 18   **23** 0·2   **24** 0·7
25 £84   **26** £2455   **27** 64   **28** £3·60   **29** 55 mph   **30** 28

page 239 **Test 3**

1 70   **2** 240   **3** 900   **4** 10 705   **5** Quarter to eleven   **6** 245
7 7   **8** £3·05   **9** 15   **10** 50p, 5p, 5p, 1p, *or* 20p, 20p, 20p, 1p   **11** 0·75
12 5   **13** Tuesday   **14** $1\frac{1}{2}$ kg   **15** £15 000·50   **16** 640 m   **17** 75%   **18** £30 000
19 4   **20** £1·10   **21** 23   **22** 9   **23** 91   **24** £6   **25** £1·40
26 £4·46   **27** £3·30   **28** £42   **29** 34   **30** 64

page 240 **Test 4**

1 £8·05   **2** 75   **3** 25   **4** 0·1 cm   **5** 24p   **6** 104   **7** 40 pence
8 £18   **9** Ten to six   **10** 270°   **11** North-east   **12** 92%   **13** £4·25   **14** 998
15 20 miles   **16** 200   **17** 22·5 cm   **18** 75 pence   **19** 10   **20** 16   **21** 20
22 £9·75   **23** 25 minutes   **24** 15 : 40   **25** 15   **26** 70p   **27** 200   **28** 35%
29 100 minutes   **30** £2500

page 241 **Test 5**

1 160   **2** 106   **3** 6011   **4** £2·10   **5** 92   **6** 12   **7** £1·01
8 1·55 m   **9** £30·11   **10** 4 cm   **11** 1500 m   **12** £25   **13** 7   **14** $2\frac{1}{2}$
15 64   **16** 12   **17** 12 litres   **18** 23%   **19** 31   **20** 3   **21** 64
22 84 square yards   **23** 100 000   **24** 27 hours   **25** 6   **26** £10   **27** 500   **28** 180
29 9 or 18 or 27 etc.   **30** Saturday

page 242 **Test 6**

1 60°   **2** 15   **3** 75%   **4** 8000   **5** £26   **6** 11   **7** Half past 7   **8** 20%
9 8·15 pm   **10** 0·11   **11** 27   **12** £2·50   **13** 1·8 litres (allow 1·795)   **14** 2104   **15** £40
16 150   **17** 36°   **18** 60   **19** 270°   **20** £9   **21** 50 mph   **22** 250   **23** 25
24 37   **25** 96p   **26** £20   **27** 12   **28** 50p, 20p, 5p, 1p   **29** 35°   **30** 20

page 243 **Test 7**

1 82°   **2** 66p   **3** 2107   **4** 1000   **5** 75%   **6** 0·23   **7** 15   **8** 8   **9** 24
10 0·9   **11** Trapezium   **12** 89 990   **13** False   **14** $\frac{5}{6}$ **15** 21   **16** 63 m²   **17** 6

18 1000   **19** 4 hours   **20** 9   **21** £8·70   **22** 5·5   **23** 2550 grams   **24** £40 000
25 120°   **26** 1100 or 1000 (Accept either)   **27** 8   **28** 5   **29** $\frac{1}{4}$ **30** 2·1

page 244 **Test 8**

1 61   **2** 39   **3** 10   **4** 0·82   **5** 120   **6** 55   **7** £3·80   **8** 63   **9** 154   **10** 1100
11 15   **12** 165   **13** 355   **14** 4·1   **15** 40   **16** 200   **17** £2·10   **18** 19   **19** 24   **20** 24

page 244 **Test 9**

1 130   **2** 77   **3** 1   **4** 2300   **5** 49   **6** £9·50   **7** 342   **8** 0·8   **9** 300   **10** 27
11 30   **12** 10 000   **13** 7   **14** 25   **15** 40   **16** 5·5   **17** 19·2   **18** 8   **19** 45   **20** 99

page 244 **Test 10**

1 84   **2** 35   **3** 36   **4** 100   **5** 210   **6** 5000   **7** 84   **8** 20   **9** 90   **10** 22
11 415   **12** 1   **13** 1000   **14** 227   **15** 1600   **16** 60   **17** 23   **18** 8·8   **19** 5·2   **20** 26

page 244 **Test 11**

1 220   **2** 20   **3** 56   **4** 3·85   **5** 200   **6** 199   **7** 121   **8** 315   **9** 500   **10** 60
11 400   **12** 6   **13** 1800   **14** 69   **15** 32   **16** 101   **17** 10·7   **18** 1   **19** 3600   **20** 120

page 245 **Exercise 30 Ⓒ**

1 18   **2** 23   **3** 42   **4** 3   **5** 225   **6** 36   **7** 8   **8** 57   **9** 4   **10** 1
11 18   **12** 111   **13** 18   **14** 6   **15** 713   **16** 63   **17** 3   **18** 5742   **19** 20   **20** 60
21 14   **22** 129   **23** 153   **24** 10   **25** 4   **26** 33   **27** 2   **28** 4   **29** 44   **30** 6
31 24   **32** 57   **33** 34   **34** 28   **35** 331   **36** 37   **37** 18   **38** 12   **39** 23   **40** 8

page 246 **Exercise 31 Ⓒ**

1 200   **2** 400   **3** 5000   **4** 7000   **5** 8000   **6** 5000   **7** 400   **8** 30   **9** 40
10 200   **11** 400   **12** 4000   **13** 700   **14** 700   **15** 7000   **16** 30   **17** 4000   **18** 20 000
19 9000   **20** 700   **21** 500   **22** 2000   **23** 7000   **24** 90   **25** 200   **26** 60   **27** 20
28 60   **29** 5000   **30** 2000   **31** 6000   **32** 1000
33 a 1493·2 → 1500 m³   23·41 → 23°C   2108 → 2000   5173 → 5000
     b 100, 58·23, 2008

page 247 **Exercise 32 Ⓔ**

1 2·35   **2** 0·814   **3** 26·2   **4** 35·6   **5** 113   **6** 211   **7** 0·825   **8** 0·0312
9 5·9   **10** 1·2   **11** 0·55   **12** 0·72   **13** 0·14   **14** 1·8   **15** 25   **16** 31
17 487   **18** 500   **19** 2·89   **20** 3·11   **21** 0·0715   **22** 3·04   **23** 2460   **24** 489 000
25 0·513   **26** 5·8   **27** 66   **28** 588   **29** 0·6   **30** 0·07   **31** 5·84   **32** 88
33 2500   **34** 52 700   **35** 0·006   **36** 7000

page 248 **Exercise 33 Ⓔ**

1 5·38   **2** 11·05   **3** 0·41   **4** 0·37   **5** 8·02   **6** 87·04   **7** 9·01   **8** 0·07   **9** 8·4
10 0·7   **11** 0·4   **12** 0·1   **13** 6·1   **14** 19·5   **15** 8·1   **16** 7·1   **17** 8·16   **18** 3·0

19 0·545　**20** 0·0056　**21** 0·71　**22** 6·83　**23** 0·8　　**24** 19·65　**25** 0·0714　**26** 60·1　**27** 7·3
28 5·42　　**29 a i** length 6 cm, width 3·2 cm　　**ii** length 5 cm, width 3 cm
　　　　　　　b i 19·2 cm²　　　　　　　　　　　**ii** 15 cm²

page 249 **Exercise 34**Ⓔ

1 0·57　　**2** 3·45　　**3** 431　　**4** 19·3　　**5** 0·22　　**6** 3942·7　　**7** 53　　**8** 18·4
9 0·059　　**10** 1·1　　**11** 6140　　**12** 127·89　　**13** 20·3　　**14** 47·6　　**15** 71·1　　**16** 0·16

page 250 **Exercise 35**Ⓔ

1 B　　**2** A　　**3** C　　**4** B　　**5** C　　**6** A　　**7** B　　**8** B　　**9** A　　**10** C　　**11** B
12 A　**13** A　**14** C　**15** C　**16** B　**17** C　**18** A　**19** B　**20** B　**21** C　**22** B
23 B　**24** A　**25** B　**26** B　**27** B　**28** C　**29** A　**30** C　**31** B　**32** C　**33** C

page 251 **Exercise 36**Ⓔ

1 £8000　**2** £6　**3** £440　**4** £5200　**5 a** 89·89　**b** 4·2　　**c** 358·4　　**d** 58·8　　**e** 0·3　　**f** 2·62
6 a 4·5　　　　**b** 462　　　　**c** 946·4　　　　**d** 77·8　　　**e** 0·2　　　**f** 21
7 B　　　　　**8** C　　　　　**9** B　　　　　**10** B　　　**11** C　　**12** B　　　**13** A　　　　**14** A
15 Area ≈ 30 × 40 = 1200 m²　　　　　　　**16** £3900 (accept £4000)
　　Cost ≈ 1200 × 7 = £8400
　　Estimate job at £8500

page 252 **Exercise 37**Ⓔ

1 a 1670　**b** 90·8　**c** 32·6　**d** 5·29 (2dp)　　**e** 44·7　　**2 a** smaller　**b** larger　**c** smaller
3 a OK　**b** OK　**c** OK　**d** Highly unlikely　**e** Impossible　**f** Highly unlikely

page 253 **Exercise 38**Ⓒ

1 42 kg　　**2** £120　　**3 a** False, unless $n = 1\frac{1}{2}$　　**b** True　**c** True　**d** False, unless $n = 1\frac{1}{2}$

e False, unless $a = b$　**f** False, unless $n = 0$ or $\dfrac{1 \pm \sqrt{5}}{2}$　　**4 a** 8 hectares　**b** 24 tonnes　**5** £345

6 $83 200　　**7** £1·80　　**8 a** 600　　**b** £204　　**9 a**　　　　**b**　　　　**c**

page 254 **Exercise 39**Ⓒ

1 a 15　　**b i** 16·7% (1 dp)　　**ii** 30%　　**2 a** Multiply by 3　　**b** 177 147　　**c** 1 594 323
3 a 36　　**b** 24　　**c** 240　　**4 a** 38,62　　**b** 64,125　　**c** 81,64　　**d** $\dfrac{6}{7}, \dfrac{7}{8}$　　**5** 120°
6 (1, 2) (2, 1·1) (3, 1·1) (4, 2) (4, 3) (3, 4) (4, 5) (4, 6) (3, 6·9) (2, 6·9) (1, 6)　　**7** 2　　**8** 6; 14p
9 100　　**10 a** 90　**b** 1　**c** 23　**d** 44　**e** 77　**f** 111

page 256 **Exercise 40**Ⓒ

1 410 calories　　　　**2 a** 64　**b** 1　**c** 100　**d** 3000　**e** 32　**f** 81
3 a 13, 14　**b** 7, 8, 9　**c** 10, 11, 12, 13　**4** 10 cm²　**5 a** 273　**b** 7457　**c** 84·5　**d** 305
6 a 30　**b** 32　**c** 5　**7 a** £162　**b** 200 litres　**c** €280　**8** 5 hours 34 minutes　**9** 50 m　**10** 60

page 257 **Exercise 41 C**

1 a 80g **b** 5·2 calories **c** 416 calories **2 a** 7 **b** 50 **c** 1 **d** 5 **3 a** 12 **b** 8 **c** 48

4 Put 2 coins on the scales. If they don't balance, the heavier one is the
fake. If they do balance, the 3rd coin is the fake.

5 200 litres **6** £21 600 **7** 5·4 km

8

×	6	3	4	7
5	30	15	20	35
9	$\sqrt{4}$	27	36	63
2	12	6	8	14
8	48	24	32	56

9 16 **10** A solution is:

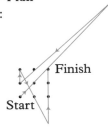

Finish

Start

page 259 **Exercise 42 C**

1 a 5 m **b** 50 m **c** 6 km **2** 9 **3** 51·4°
4 a 40 acres **b** 15 acres **c** 10% **d** 22·5% **5** £3285 **6** £246 **7** 78
8 £5·85 **9 a** $99 + \frac{9}{9}$ **b** $6 + \frac{6}{6}$ **c** $55 + 5$ **d** $55 + 5 + \frac{5}{5}$ **e** $77 \div 77$ **f** $88 \div 8$

10 a

2	3
+ 5	4
7	7

b

1	7
+ 4	6
6	3

c

5	8	2
+ 1	3	6
7	1	8

d

4	7	4
+ 3	5	0
8	2	4

e

8	6
− 3	4
5	2

f

8	8	2
− 6	5	0
2	3	2

page 261 **Test yourself**

1 a $\frac{2}{8}, \frac{6}{24}$ **b** 0·25 **2 a i** $\frac{4}{8} = \frac{1}{2}$ **ii** any answer with 2 squares shaded

b i 40 **ii** 150 000 **iii** 6·55 **iv** $\frac{3}{8}$ **3 a** 74% **b i** $\frac{5}{12}$ **ii** $\frac{1}{16}$ **4** $\frac{1}{2} + \frac{1}{3} = \frac{3}{6} + \frac{2}{6} = \frac{5}{6}$

5 £5 for Ken and £15 for Susan **6** £42 + £21 = £63 **7 a** 295 **b** 11 676 **c** 60

8 60% of £40 = 6 × £4 = £24
$\frac{2}{5}$ of £55 = 2 × £11 = £22
Therefore 60% of £40 is larger.

9 £10 948 **10 a** 9000 **b i** 60 × £10 = £600 **ii** Smaller, as both the number of books
and the cost have been underestimated.

11 Approximate calculation = $\frac{30}{4 \times 0·5} = \frac{30}{2} = 15$. Therefore Gemma is correct.

12 $\frac{800 \times 20}{100} = 160$ **13** £9720 **14 a** 300 m **b** 24 cm **15** £14·72

16 Alan £320, Brendon £256, Chloe £192

17 a $\frac{3}{4}$ **b** £15·54 **c i** 45 mph **ii** 5·3 amps **18** 72·7% (1 dp)

19 200g flour; 150g almonds; 225g sugar; 150g butter; 10 pears **20** £9272·88

21 a $\frac{1}{6}, \frac{3}{8}, \frac{1}{2}, \frac{2}{3}, \frac{3}{4}$ **b** $\frac{3}{4}, 65\%, \frac{2}{3}, 0·72, \frac{3}{4}$ **22 a** 6 hours **b** £255·84

6 Algebra 2

page 267 **Exercise 1Ⓔ**

1 12 cm
2 15 cm × 5 cm
3 a 26 cm, 13 cm   **b** 16 cm, 8 cm   **c** 32 cm, 16 cm   **d** 9 cm, 4·5 cm   **e** 6·5 cm, 3·25 cm
4 a 6 cm, 5 cm   **b** 12 cm, 11 cm   **c** 20 cm, 19 cm   **d** 6·5 cm, 5·5 cm   **e** 8·7 cm, 7·7 cm
5 9·2 cm, 8·2 cm (1 dp)

page 269 **Exercise 2Ⓔ**

1 a 9·5   **b** 7·6   **2 a** 3·4   **b** 4·6   **c** 6·7   **3 a** 1·7   **b** 3·0   **c** 6·1
4 a 5·1   **b** 3·8   **c** 4   **d** 3·3   **5** $x^3 = 526$ $x = 8·1$ **6** 3·58 cm   **7 a** 3·2   **b** 2·7   **c** 3·6
8 a 3 cm²   **b** $x(x + 3) - 6$

$$\text{Therefore } x(x + 3) - 6 = 20$$
$$\text{so} \quad x(x + 3) = 26$$
$$x = 3·8$$

page 271 **Exercise 3Ⓔ**

1 $x^2 + 4x + 3$ **2** $x^2 + 5x + 6$ **3** $y^2 + 9y + 20$ **4** $x^2 + 7x + 12$
5 $x^2 + 7x + 10$ **6** $x^2 + 10x + 16$ **7** $a^2 + 12a + 35$ **8** $z^2 + 11z + 18$
9 $x^2 + 6x + 9$ **10** $k^2 + 22k + 121$ **11** $2x^2 + 7x + 3$ **12** $3x^2 + 10x + 8$
13 $2y^2 + 5y + 3$ **14** $49y^2 + 14y + 1$ **15** $9x^2 + 12x + 4$ **16** $x^2 + 9x + 20$
17 $2x^2 + 6x + 4$ **18** $6x^2 + 15x + 9$ **19** $24y^2 + 28y + 8$ **20** $6x^2 + 14x + 4$

page 271 **Exercise 4Ⓔ**

1 $x^2 + 8x + 16$ **2** $x^2 + 4x + 4$ **3** $x^2 + 2x + 1$ **4** $4x^2 + 4x + 1$
5 $y^2 + 10y + 25$ **6** $9y^2 + 6y + 1$ **7** $3x^2 + 12x + 12$ **8** $x^2 + 6x + 9$
9 $9x^2 + 12x + 4$ **10** $2x^2 + 4x + 2$ **11** $2x^2 + 6x + 5$ **12** $2x^2 + 10x + 13$
13 $5x^2 + 8x + 5$ **14** $2y^2 + 14y + 25$ **15** $5x^2 + 16x + 13$ **16** $2x^2 + 8x + 10$

page 272 **Exercise 5Ⓔ**

1 $2(3x + 2y)$ **2** $3(3x + 4y)$ **3** $2(5a + 2b)$ **4** $4(x + 3y)$
5 $5(2a + 3b)$ **6** $6(3x - 4y)$ **7** $4(2u - 7v)$ **8** $5(3s + 5t)$
9 $8(3m + 5n)$ **10** $9(3c - 8d)$ **11** $4(5a + 2b)$ **12** $6(5x - 4y)$
13 $3(9c - 11d)$ **14** $7(5u + 7v)$ **15** $4(3s - 8t)$ **16** $8(5x - 2t)$
17 $12(2x + 7y)$ **18** $4(3x + 2y + 4z)$ **19** $3(4a - 2b + 3c)$ **20** $5(2x - 4y + 5z)$
21 $4(5a - 3b - 7c)$ **22** $8(6m + n - 3x)$ **23** $7(6x + 7y - 3z)$ **24** $3(2x^2 + 5y^2)$
25 $5(4x^2 - 3y^2)$ **26** $7(a^2 + 4b^2)$ **27** $9(3a + 7b - 4c)$ **28** $6(2x^2 + 4xy + 3y^2)$
29 $8(8p - 9q - 5r)$ **30** $12(3x - 5y + 8z)$ **31** $3x(3 + 2y - x)$ **32** $x(x - 5)$
33 $x(2x - 3)$ **34** $x(7x + 1)$ **35** $y(y + 4)$ **36** $2x(x + 4)$
37 $4y(y - 1)$ **38** $p(p - 2)$ **39** $2a(3a + 1)$ **40** $a(2b - 1)$
41 $x(3y + 2)$ **42** $3t(1 + 3t)$ **43** $4(1 - 2x^2)$ **44** $5x(1 - 2x^2)$
45 $\pi r(4r + h)$ **46** $\pi r(r + 2)$ **47 a** $x - 1$ **b** $x + 2$ **c** $2(x + 5)$

page 273 **Exercise 6Ⓔ**

1 $x = e - b$ **2** $x = m + t$ **3** $x = a + b + f$ **4** $x = A + B - h$ **5** $x = y$

6 $x = b - a$ **7** $x = m - k$ **8** $x = w + y - v$ **9** $x = \dfrac{b}{a}$ **10** $x = \dfrac{m}{h}$

11 $x = \dfrac{(a + b)}{m}$ **12** $x = \dfrac{(c - d)}{k}$ **13** $x = \dfrac{(e + n)}{v}$ **14** $x = \dfrac{(y + z)}{3}$ **15** $x = \dfrac{r}{p}$

16 $x = \dfrac{(h - m)}{m}$ **17** $x = \dfrac{(a - t)}{a}$ **18** $x = \dfrac{(k + e)}{m}$ **19** $x = \dfrac{(m + h)}{u}$ **20** $x = \dfrac{(t - q)}{e}$

21 $x = \dfrac{(v^2 + u^2)}{k}$ **22** $x = \dfrac{(s^2 - t^2)}{g}$ **23** $x = \dfrac{(m^2 - k)}{a}$ **24** $x = \dfrac{(m + v)}{m}$ **25** $x = \dfrac{(c - a)}{b}$

26 $x = \dfrac{(y - t)}{s}$ **27** $x = \dfrac{(z - y)}{c}$ **28** $x = \dfrac{a}{h}$ **29** $x = \dfrac{2b}{m}$ **30** $x = \dfrac{(cd - ab)}{k}$

31 $x = \dfrac{c}{a} + b$ **32** $x = \dfrac{e}{c} + d$ **33** $x = \dfrac{n^2}{m} - m$ **34** $x = \dfrac{t}{k} + a$ **35** $x = \dfrac{k}{n} + h$

36 $x = \dfrac{n}{m} - b$ **37** $x = 2a$ **38** $x = \dfrac{d}{c} - a$ **39** $x = \dfrac{e}{m} - b$

page 274 **Exercise 7Ⓔ**

1 $x = tm$ **2** $x = en$ **3** $x = pa$ **4** $x = tam$ **5** $x = abc$

6 $x = ey^2$ **7** $x = a(b + c)$ **8** $x = t(c - d)$ **9** $x = m(s + t)$ **10** $x = k(h + i)$

11 $x = \dfrac{ab}{c}$ **12** $x = \dfrac{mz}{y}$ **13** $x = \dfrac{hc}{d}$ **14** $x = \dfrac{em}{n}$ **15** $x = \dfrac{hb}{e}$

16 $x = c(a + b)$ **17** $x = m(h + k)$ **18** $x = \dfrac{mu}{y}$ **19** $x = t(h - k)$ **20** $x = (a + b)(z + t)$

21 $x = \dfrac{e}{t}$ **22** $x = \dfrac{e}{a}$ **23** $x = \dfrac{h}{m}$ **24** $x = \dfrac{cb}{a}$ **25** $x = \dfrac{ud}{c}$

26 $x = \dfrac{m}{t^2}$ **27** $x = \dfrac{h}{\sin 20°}$ **28** $x = \dfrac{e}{\cos 40°}$ **29** $x = \dfrac{m}{\tan 46°}$ **30** $x = \dfrac{b^2 c^2}{a^2}$

page 275 **Exercise 8Ⓔ**

1 $x = \pm\sqrt{\dfrac{h}{c}}$ **2** $x = \pm\sqrt{\dfrac{f}{b}}$ **3** $x = \pm\sqrt{\dfrac{m}{t}}$ **4** $x = \pm\sqrt{\dfrac{a + b}{y}}$ **5** $x = \pm\sqrt{\dfrac{t + a}{m}}$

6 $x = \pm\sqrt{a + b}$ **7** $x = \pm\sqrt{t - c}$ **8** $x = \pm\sqrt{z - y}$ **9** $x = \pm\sqrt{a^2 + b^2}$ **10** $x = \pm\sqrt{m^2 - t^2}$

11 $x = \pm\sqrt{a^2 - n^2}$ **12** $x = \pm\sqrt{\dfrac{c}{a}}$ **13** $x = \pm\sqrt{\dfrac{n}{h}}$ **14** $x = \pm\sqrt{\dfrac{z + k}{c}}$ **15** $x = \pm\sqrt{\dfrac{c - b}{a}}$

16 $x = \pm\sqrt{\dfrac{e + h}{d}}$ **17** $x = \pm\sqrt{\dfrac{m + n}{g}}$ **18** $x = \pm\sqrt{\dfrac{z - y}{m}}$ **19** $x = \pm\sqrt{\dfrac{f - a}{m}}$ **20** $x = \pm\sqrt{b^2 - a^2}$

21 $x = a - y$ **22** $x = h - m$ **23** $x = z - q$ **24** $x = b - v$ **25** $x = k - m$

26 $x = \dfrac{h - d}{c}$ **27** $x = \dfrac{y - c}{m}$ **28** $x = \dfrac{k - h}{e}$ **29** $x = \dfrac{a^2 - d}{b}$ **30** $x = \dfrac{m^2 - n^2}{t}$

31 $x = \dfrac{v^2 - w}{a}$ **32** $x = y - y^2$ **33** $x = \dfrac{k - m}{t^2}$ **34** $x = \dfrac{b - e}{c}$ **35** $x = \dfrac{h - z}{g}$

36 $x = \dfrac{c - a - b}{d}$ **37** $x = \dfrac{v^2 - y^2}{k}$ **38** $x = \dfrac{d - h}{f}$ **39** $x = b - \dfrac{c}{a}$ **40** $x = m - \dfrac{n}{h}$

page 275 **Exercise 9Ⓔ**

1 a $a = \dfrac{v - u}{t}$ **b** 2 **2 a** $k = \dfrac{py}{m}$ **b** $y = \dfrac{mk}{p}$ **3 a** $n = pR + d$ **b** $n = 1255$

page 276 **Exercise 10ⓔ**

1 $x = \dfrac{h + d}{a}$ **2** $y = \dfrac{m - k}{z}$ **3** $y = \dfrac{f}{d} - e$ **4** $k = \dfrac{d}{m} - a$ **5** $m = \dfrac{c - a}{b}$

6 $e = \pm\sqrt{\dfrac{b}{a}}$ **7** $t = \pm\sqrt{\dfrac{z}{y}}$ **8** $x = \pm\sqrt{e + c}$ **9** $y = \dfrac{b + n}{m}$ **10** $z = \dfrac{b}{a} - a$

11 $x = \dfrac{a}{d}$ **12** $k = mt$ **13** $u = mn$ **14** $x = \dfrac{y}{d}$ **15** $m = \dfrac{a}{t}$

16 $g = \dfrac{d}{n}$ **17** $t = k(a + b)$ **18** $e = \dfrac{v}{y}$ **19** $y = \dfrac{m}{c}$ **20** $a = mb$

21 $m = \dfrac{b}{g} - a$ **22** $g = \dfrac{x^2}{h} - h$ **23** $t = y - z$ **24** $e = \pm\sqrt{\dfrac{c}{m}}$ **25** $x = \dfrac{t}{a} - y$

26 $v = \dfrac{y^2 + t^2}{u}$ **27** $k = \pm\sqrt{c - t}$ **28** $w = k - m$ **29** $n = \dfrac{b - c}{a}$ **30** $y = \dfrac{c}{m} - a$

31 $x = pq - ab$ **32** $k = \dfrac{a^2 - t}{b}$ **33** $z = \dfrac{w}{v^2}$ **34** $u = t - c$ **35** $c = \dfrac{t}{x}$

36 $w = \dfrac{k}{m} - n$ **37** $m = \dfrac{v - t}{x}$ **38** $y = \dfrac{c}{a} - b$ **39** $c = a - \dfrac{e}{m}$ **40** $a = \dfrac{ce}{b}$

41 $p = \dfrac{a}{q}$ **42** $n = \pm\sqrt{\dfrac{a}{e}}$

page 277 **Exercise 11ⓔ**

1 a $<$ **b** $>$ **c** $>$ **d** $>$ **e** $>$ **f** $>$
2 a $x > 2$ **b** $x \leqslant 5$ **c** $x < 100$ **d** $-2 \leqslant x \leqslant 2$ **e** $x \geqslant -6$ **f** $3 < x \leqslant 8$

3

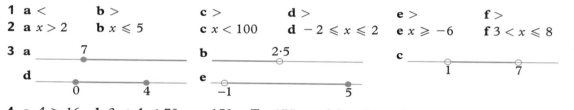

4 a $A \geqslant 16$ **b** $3 < A \leqslant 70$ **c** $150 < T < 175$ **d** $h \geqslant 1{\cdot}75$ **5 a** True   **b** True   **c** True

page 278 **Exercise 12ⓔ**

1 $x > 13$ **2** $x < -1$ **3** $x < 12$ **4** $x \leqslant 2\tfrac{1}{2}$ **5** $x > 3$ **6** $x \geqslant 8$ **7** $x < \tfrac{1}{4}$

8 $x \geqslant -3$ **9** $x < -8$ **10** $x < 4$ **11** $x > -9$ **12** $x < 8$ **13** $x > 3$ **14** $x \geqslant 1$

15 $x < 1$ **16** $x > 2\tfrac{1}{3}$ **17** $x < -3$ **18** $x > 7\tfrac{1}{2}$ **19** $x > 0$ **20** $x < 0$ **21** $x > 5$

22 $5 \leqslant x \leqslant 9$ **23** $-1 < x < 4$ **24** $5\tfrac{1}{2} \leqslant x \leqslant 6$

25 $1\tfrac{1}{3} < x < 8$ **26** $-8 < x < 2$ **27** $\tfrac{1}{4} < x < 1\tfrac{1}{5}$

page 279 **Exercise 13ⓔ**

1 $x > 8$ **2** 1, 2, 3, 4, 5, 6   **3** 7, 11, 13, 17, 19   **4** 4, 9, 16, 25, 36, 49   **5** $-4, -3, -2, -1$
6 2, 3, 4, 5, 6, 7, 8, 9,10, 11, 12   **7** 2, 3, 5, 7, 11   **8** 2, 4, 6, 8, 10, 12, 14, 16, 18
9 1, 2, 3, 4 **10** 5 **11 a** 16   **b** -16 **c** 20   **d** -5 **12** $>$ **13** $\tfrac{1}{2}$ (other answers are possible)
14 19   **15** 17   **16** $x > 3\tfrac{2}{3}$ **17** 7   **18** 5   **19** 6   **20** 3, 4, 5

page 281 **Exercise 14 C**

1 $A: y = 7$, $B: y = 3$, $C: y = 1$

2 $P: x = 3$, $Q: x = 5$, $R: x = -3$

3
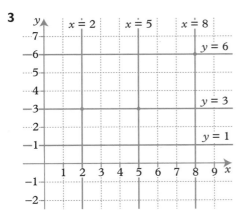

(b) $(2, 3)$
(c) $(5, 1)$
(d) $(8, 6)$

4 **a** $x = 1$ **b** $y = 7$ **c** $x = 2$ **d** $x = 7$ **e** $x = 3$ **f** $y = 3$ **g** $y = 5$ **h** $y = 0$

page 283 **Exercise 15 C**

1

x	0	1	2	3	4	5	6
y	3	5	7	9	11	13	15
coordinates	(0, 3)	(1, 5)	(2, 7)	(3, 9)	(4, 11)	(5, 13)	(6, 15)

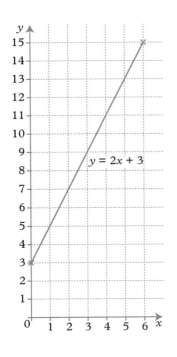

2

x	0	1	2	3	4	5	6	7
y	3	4	5	6	7	8	9	10
coordinates	(0,3)	(1,4)	(2,5)	(3,6)	(4,7)	(5,8)	(6,9)	(7,10)

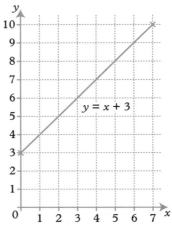

3

x	0	1	2	3	4	5
y	0	2	4	6	8	10
coordinates	(0,0)	(1,2)	(2,4)	(3,6)	(4,8)	(5,10)

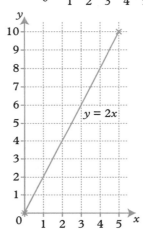

4

x	0	1	2	3	4	5
y	4	6	8	10	12	14

5

x	0	1	2	3	4	5
y	−1	1	3	5	7	9

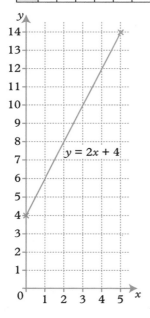

6

x	0	1	2	3
y	1	4	7	10

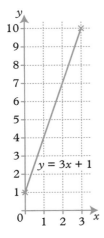

$y = 3x + 1$

7

x	0	1	2	3	4	5
y	2	4	6	8	10	12

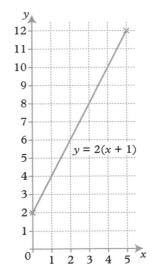

$y = 2(x + 1)$

8

x	0	1	2	3	4
y	6	9	12	15	18

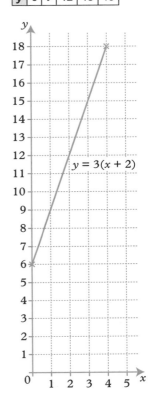

$y = 3(x + 2)$

9

x	0	1	2	3	4	5	6	7
y	0	$\frac{1}{2}$	1	$1\frac{1}{2}$	2	$2\frac{1}{2}$	3	$3\frac{1}{2}$

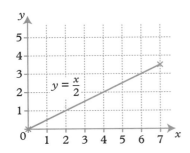

$y = \dfrac{x}{2}$

10

x	0	1	2	3	4	5
y	5	6	7	8	9	10

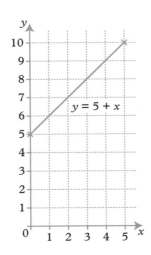

$y = 5 + x$

11

x	0	1	2	3	4	5	6
y	6	5	4	3	2	1	0

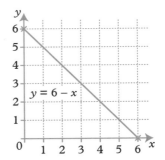

$y = 6 - x$

12

x	0	1	2	3	4	5
y	−3	−1	1	3	5	7

$y = 2x - 3$

page 284 **Exercise 16 C**

1

x	−1	0	1	2	3	4
y	−1	1	3	5	7	9

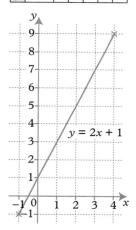

$y = 2x + 1$

2

x	−2	−1	0	1	2	3
y	−4	−1	2	5	8	11

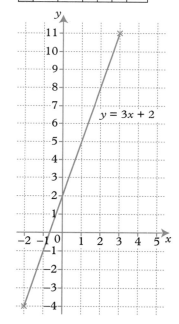

$y = 3x + 2$

3

x	−2	−1	0	1	2	3	4	5
y	−5	−4	−3	−2	−1	0	1	2

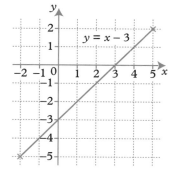

$y = x - 3$

4

x	−3	−2	−1	0	1	2	3
y	3	4	5	6	7	8	9

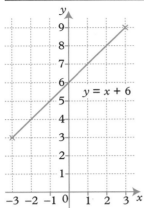

$y = x + 6$

5

x	−3	−2	−1	0	1	2	3
y	−7	−5	−3	−1	1	3	5

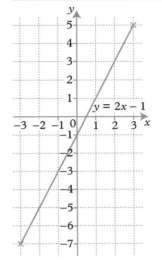

$y = 2x − 1$

6

x	−2	−1	0	1	2
y	−7	−3	1	5	9

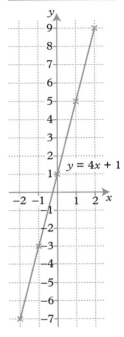

$y = 4x + 1$

7

x	0	1	2	3	4	5
y	5	4	3	2	1	0

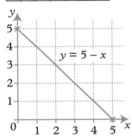

$y = 5 − x$

8

x	0	1	2	3	4	5	6
y	10	8	6	4	2	0	−2

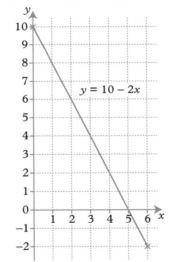

$y = 10 − 2x$

9

x	0	1	2	3
y	4	2	0	−2

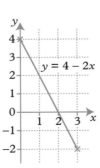

$y = 4 − 2x$

10

x	0	1	2	3	4	5	6
y	12	10	8	6	4	2	0

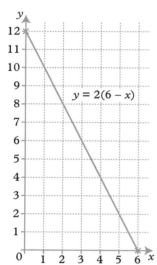

$y = 2(6 - x)$

11

x	-2	-1	0	1	2	3	4
y	-2	0	2	4	6	8	10

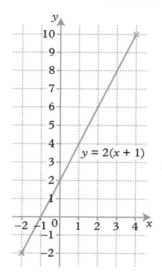

$y = 2(x + 1)$

12

x	-2	-1	0	1	2
y	-9	-6	-3	0	3

$y = 3(x - 1)$

13

x	0	1	2	3	4
y	1	3	5	7	9

c $(3\frac{1}{2}, 8)$

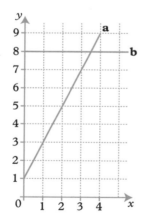

14

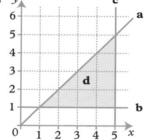

b (4,4)

15

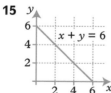

$x + y = 6$

16

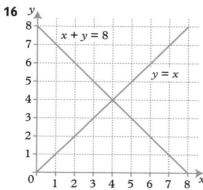

$x + y = 8$

$y = x$

page 286 **Exercise 17**Ⓔ

1 AB $\frac{1}{5}$ BC $\frac{5}{2}$ AC $-\frac{4}{3}$ **2** PQ $\frac{4}{5}$ PR $-\frac{1}{6}$ QR -5 **3 a** 3 **b** $\frac{3}{2}$ **c** $\frac{1}{2}$ **d** $-\frac{5}{2}$

4 1 **5 a** $\dfrac{n-4}{3}$ **b** 4

page 288 **Exercise 18**Ⓔ

1 a $y = 2x + 1$ **b** $y = \frac{1}{2}x$ **c** $y = 4x - 2$ **d** $y = -2x + 5$ **e** $y = 7x$
  f $y = -x + 1$ **g** $y = 2x + 3$ **h** $y = x - 4$ **i** $y = 3x - 10$
2 $y = 3x - 7$ and $y = 3x$ **3 a** $y = 4x - 3$ **b** $y = -3x + 5$ **c** $y = \frac{1}{3}x - 2$
4 Any line of the form $y = 5x + c$

page 289 **Exercise 19**Ⓔ

1 a 1 **b** 3

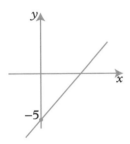

2 a 1 **b** -2

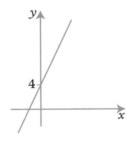

3 a 2 **b** 1

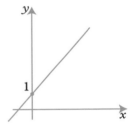

4 a 2 **b** -5

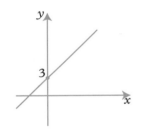

5 a 3 **b** 4

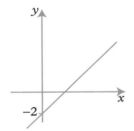

6 a $\frac{1}{2}$ **b** 6

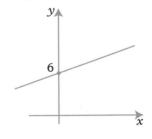

7 a 3 **b** -2

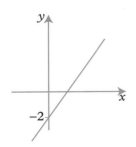

8 a 2 **b** 0

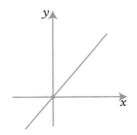

9 a $\frac{1}{4}$ **b** -4

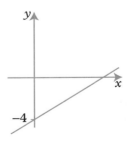

10 a −1   **b** 3

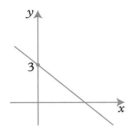

11 a −2   **b** 6

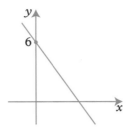

12 a −1   **b** 2

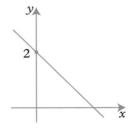

13 a −2   **b** 3

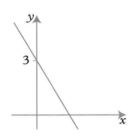

14 a −3   **b** − 4

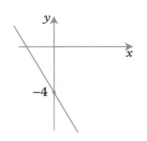

15 a $\frac{1}{2}$ **b** 3

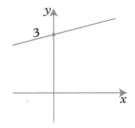

16 a $-\frac{1}{3}$ **c** 3

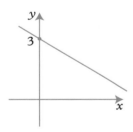

17 a 4   **c** −5

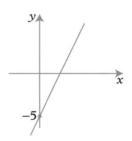

18 a $\frac{3}{2}$ **b** −4

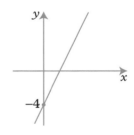

19 a 10   **b** 0

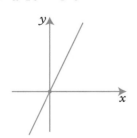

20 a 0   **b** 4

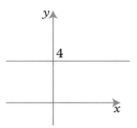

21 A: $y = 3x − 4$
   B: $y = x + 2$

22 C: $y = \frac{2}{3}x − 2$
   D: $y = −2x + 4$

page 291 **Exercise 20Ⓔ**

1 a $x = 3, y = 7$ **b** $x = 1, y = 3$ **c** $x = 11, y = -1$ **2** $x = 2, y = 4$ **3** $x = 2, y = 3$
4 $x = 3, y = 1$ **5** $x = 1, y = 5$ **6** $a = 5, b = 3$
7 a $x = 4, y = 0$ **b** $x = 1, y = 6$ **c** $x = -2, y = -3$ **d** $x = 8, y = -1$ **e** $x = -0.6, y = 1.2$

page 293 **Exercise 21Ⓔ**

1 $x = 2, y = 1$ **2** $x = 4, y = 2$ **3** $x = 3, y = 1$ **4** $x = -2, y = 1$ **5** $x = 3, y = 2$
6 $x = 5, y = -2$ **7** $x = 2, y = 1$ **8** $x = 5, y = 3$ **9** $x = 3, y = -1$ **10** $a = 2, b = -3$
11 $a = 5, b = \frac{1}{4}$ **12** $a = 1, b = 3$ **13** $m = \frac{1}{2}, n = 4$ **14** $w = 2, x = 3$ **15** $x = 6, y = 3$
16 $x = \frac{1}{2}, z = -3$ **17** $m = 2, n = 1$ **18** $c = \frac{39}{23}, d = \frac{-58}{23}$

page 294 **Exercise 22Ⓔ**

1 $x = 2, y = 4$ **2** $x = 1, y = 4$ **3** $x = 2, y = 5$ **4** $x = 3, y = 7$ **5** $x = 5, y = 2$
6 $a = 3, b = 1$ **7** $x = 1, y = 3$ **8** $x = 1, y = 3$ **9** $x = -2, y = 3$ **10** $x = 4, y = 1$
11 $x = 1, y = 5$ **12** $x = 0, y = 2$ **13** $x = \frac{5}{7}, y = 4\frac{3}{7}$ **14** $x = 1, y = 2$ **15** $x = 2, y = 3$
16 $x = 4, y = -1$ **17** $x = 3, y = 1$ **18** $x = 1, y = 2$ **19** $x = 2, y = 1$ **20** $x = -2, y = 1$

page 295 **Exercise 23Ⓔ**

1 $9\frac{1}{2}$ and $5\frac{1}{2}$ **2** 6 and 3 (also $5\frac{2}{5}$ and $2\frac{2}{5}$ works)   **3** 5 and 2   **4** 8 and 5
5 ? $= 10\frac{1}{2}$, $\star = 7\frac{1}{2}$ **6** $a = 2, c = 7$ **7** $m = 4, c = -3$

page 296 **Exercise 24Ⓔ**

1 $y = x^2 + 4$

x	−3	−2	−1	0	1	2	3
x²	9	4	1	0	1	4	9
4	4	4	4	4	4	4	4
y	13	8	5	0	5	8	13

2 $y = x^2 - 7$

x	−3	−2	−1	0	1	2	3
x²	9	4	1	0	1	4	9
−7	−7	−7	−7	−7	−7	−7	−7
y	2	−3	−6	−7	−6	−3	2

3 $y = x^2 + 3x$

x	−4	−3	−2	−1	0	1	2
x²	16	9	4	1	0	1	4
3x	−12	−9	−6	−3	0	3	6
y	4	0	−2	−2	0	4	10

4 $y = x^2 + 5x$

x	−4	−3	−2	−1	0	1	2
x²	16	9	4	1	0	1	4
5x	−20	−15	−10	−5	0	5	10
y	−4	−6	−6	−4	0	6	14

5 $y = x^2 - 2x$

x	−3	−2	−1	0	1	2	3
x²	9	4	1	0	1	4	9
−2x	6	4	2	0	−2	−4	−6
y	15	8	3	0	−1	0	3

6 $y = x^2 - 4x$

x	−2	−1	0	1	2	3	4
x²	4	1	0	1	4	9	16
−4x	8	4	0	−4	−8	−12	−16
y	12	5	0	−3	−4	−3	0

7 $y = x^2 + 2x + 3$

x	−3	−2	−1	0	1	2	3
x²	9	4	1	0	1	4	9
−2x	−6	−4	−2	0	2	4	6
3	3	3	3	3	3	3	3
y	6	3	2	3	6	11	18

8 $y = x^2 + 3x − 2$

x	−4	−3	−2	−1	0	1	2
x²	16	9	4	1	0	1	4
3x	−12	−9	−6	−3	0	3	6
−2	−2	−2	−2	−2	−2	−2	−2
y	2	−2	−4	−4	−2	2	8

9 $y = x^2 + 4x − 5$

x	−3	−2	−1	0	1	2	3
x²	9	4	1	0	1	4	9
+ 4x	−12	−8	−4	0	4	8	12
−5	−5	−5	−5	−5	−5	−5	−5
y	−8	−9	−8	−5	0	7	16

10 $y = x^2 − 2x + 6$

x	−3	−2	−1	0	1	2	3
x²	9	4	1	0	1	4	9
−2x	6	4	2	0	−2	−4	−6
6	6	6	6	6	6	6	6
y	21	14	9	6	5	6	9

11

x	−3	−2	−1	0	1	2	3
y	−3	−4	−3	0	5	12	21

12

x	−4	−3	−2	−1	0	1	2
y	40	27	16	7	0	−5	−8

13

x	−3	−2	−1	0	1	2	3
y	17	12	9	8	9	12	17

14

x	−4	−3	−2	−1	0	1	2
y	5	1	−1	−1	1	5	11

15

x	−3	−2	−1	0	1	2	3
y	27	17	9	3	−1	−3	−3

16

x	−2	−1	0	1	2	3	4
y	5	−1	−5	−7	−7	−5	−1

17

x	−5	−4	−3	−2	−1	0	1
y	22	13	6	1	−2	−3	−2

18

x	−3	−2	−1	0	1	2	3
y	19	9	3	1	3	9	19

page 298 **Exercise 25Ⓔ**

1

x	−3	−2	−1	0	1	2	3
y	11	6	3	2	3	6	11

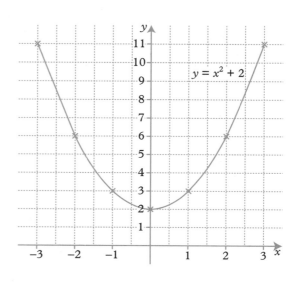

$y = x^2 + 2$

2

x	−3	−2	−1	0	1	2	3
y	14	9	6	5	6	9	14

3

x	−3	−2	−1	0	1	2	3
y	5	0	−3	−4	−3	0	5

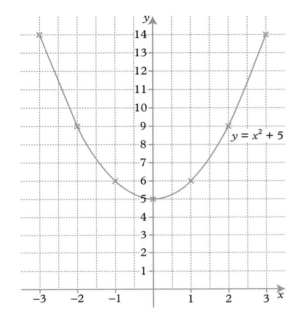

$y = x^2 + 5$

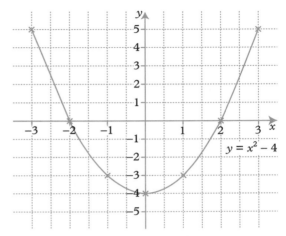

$y = x^2 - 4$

4

x	−3	−2	−1	0	1	2	3
y	1	−4	−7	−8	−7	−4	1

5

x	−4	−3	−2	−1	0	1	2
y	8	3	0	−1	0	3	8

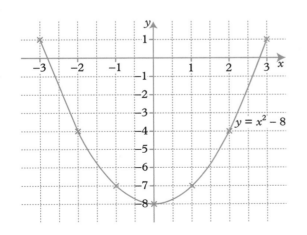

$y = x^2 - 8$

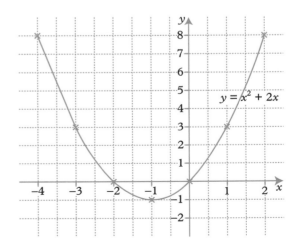

$y = x^2 + 2x$

6

x	−5	−4	−3	−2	−1	0	1
y	5	0	−3	−4	−3	0	5

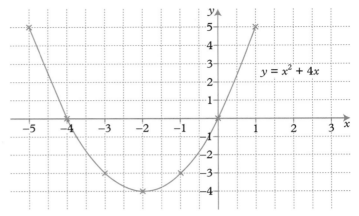

$y = x^2 + 4x$

7

x	−2	−1	0	1	2	3	4
y	−5	−4	−1	4	11	20	31

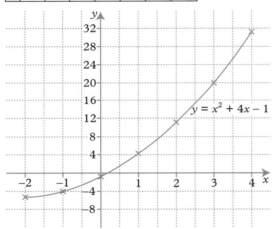

$y = x^2 + 4x - 1$

8

x	−4	−3	−2	−1	0	1	2
y	3	−2	−5	−6	−5	−2	3

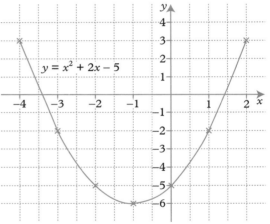

$y = x^2 + 2x - 5$

9

x	−4	−3	−2	−1	0	1	2
y	5	1	−1	−1	1	5	11

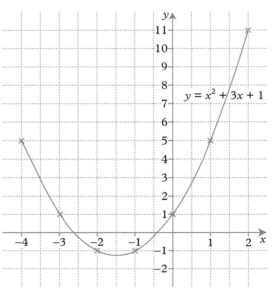

$y = x^2 + 3x + 1$

Exercise 26Ⓔ

1 a $x = -1.6$ or 3.6 **b** $x = -0.4$ or 2.4 **c** $x = -1$ or 3
2 a $x = -0.4$ or 2.4 **b** $x = 0$ or 2 **3 a** $x = 0$ or 3 **b** $x = -0.8$ or 3.8
4 a $x = -1.6$ or 3.6 **b** $x = -0.4$ or 2.4
5 a $x = 0.5$ or 6.5 **b** The curve $y = x^2 - 7x$ does not meet the line $y = -14$.

6 a b

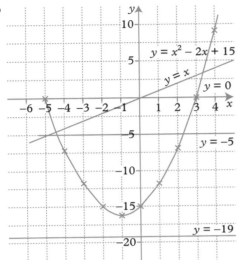

c i -4.4, 2.4 **ii** $-5, 3$
iii No solutions   **iv** $-4.4, 3.4$

7 a b

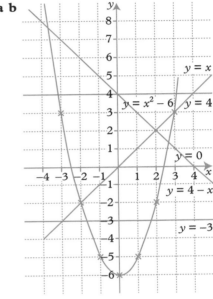

c i $x = \pm 3.2$ **ii** $x = \pm 2.4$
iii $x = \pm 1.7$ **iv** $x = -2$ or 3
v $x = -3.7$ or 2.7

Test yourself

1 $y(3 + y)$ **2 a** $n < 4$ **b** $1, 2, 3$ **3 a** 5 **b** $x = \dfrac{y - c}{m}$ **4 a** $-1, 0, 1, 2, 3$ **b** $x \geqslant 5$

5 $x = 3, y = -2$ **6** $x = 2\frac{1}{2}, y = 4$ **7 a i** p^7 **ii** $4t^3$ **b** $x > -5$ **c** $w = 2s - y$

8 a 2 **b** $y = 2x + 3$ **9** $y = 2x + 6$

10 a $x = 8$ **b** $y = \frac{1}{2}x + c$ where $c \neq 1$ **c** $x = \dfrac{y - c}{a}$ **11** $x = 4.6$ **12** $x = 3.7$

13 a i $5(2x + 1)$ **ii** $x(x + 7)$ **b** $a^5 b^3$ **c** $a = \dfrac{v - u}{t}$ **d i** $x = 48$ **ii** $x = \dfrac{13}{2}$

14 a Volume $= x \times x \times (x + 1)$
Therefore $x^2(x + 1) = 230$
$\therefore x^3 + x^2 = 230$ **b** $x = 5.8$

15 a

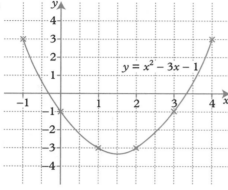

x	−1	0	1	2	3	4
y	3	−1	−3	−3	−1	3

b

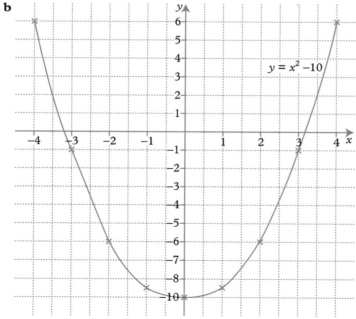

$y = x^2 - 3x - 1$

c $x = -0{\cdot}3$ or $3{\cdot}3$

16 a

x	− 4	− 3	− 2	− 1	0	1	2	3	4
y	6	− 1	− 6	− 9	− 10	− 9	− 6	− 1	6

b

$y = x^2 - 10$

c $x = \pm 3{\cdot}2$

7 Shape and space 2

page 307 **Exercise 1 C**

1
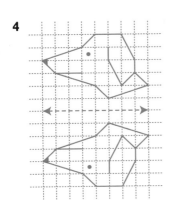

2
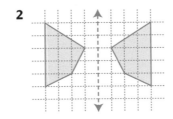

3
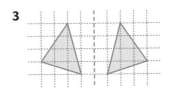

4
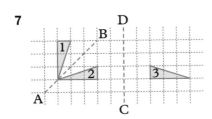

5
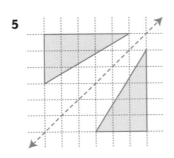

6
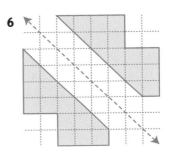

7
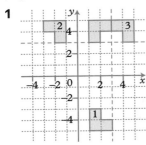

page 307 **Exercise 2 C**

1
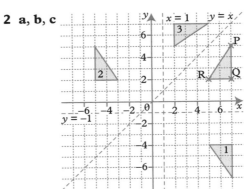

2 a, b, c

d i $(7, -7)$
ii $(-5, 5)$
iii $(5, 7)$

3 a, b, c

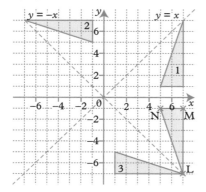

d i $(7, 7)$ **ii** $(-7, 7)$ **iii** $(7, -7)$

4 a–f

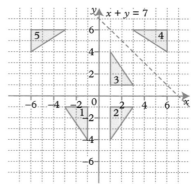

g $(-3, 6), (-6, 6), (-6, 4)$

5 a–f

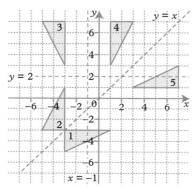

g $(3, 1), (7, 1), (7, 3)$

6 a $y = 0$
  b $x = 1$
  c $y = 1$
  d $y = -x$

page 309 **Exercise 3 C**

1

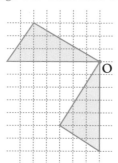

2

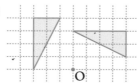

3

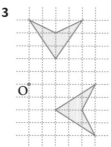

4

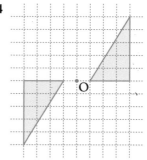

5

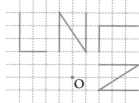

6

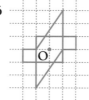

7 Shape 1: Centre C, 90°, clockwise
Shape 2: Centre B, 180°, clockwise
Shape 3: Centre A, 90°, anticlockwise
Shape 4: Centre E, 90°, clockwise
Shape 5: Centre F, 180°, clockwise

page 310 **Exercise 4 C**

1

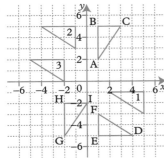

2 a–d

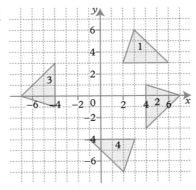

e $(-1, -4)$,
$(3, -4)$,
$(2, -7)$

3 a–d

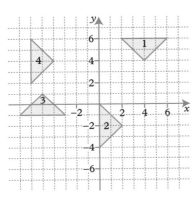

e $(-6, 2), (-4, 4), (-6, 6)$

page 311 **Exercise 5 C**

1

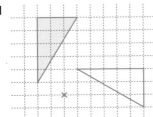

2

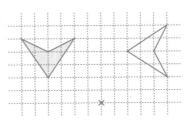

3

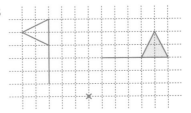

4 a $(0, 0)$ **b** $(1, 2)$ **c** $(0, 0)$ **d** $(-1, 1)$

5 a

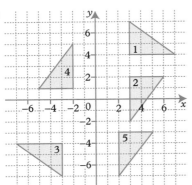

b i $(3, 3)$
ii $(0, 0)$
iii $(2, 0)$
iv $(-1, 1)$

6 a

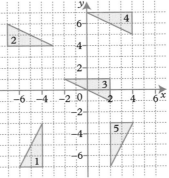

b i $(0, 0)$
ii $(-5, 0)$
iii $(-7, 4)$
iv $(-1, -5)$

page 313 **Exercise 6 C**

1 a Yes   **b** No   **c** Yes   **d** No   **2** $x = 75\,\text{mm}$ **3 a** $y = 66\,\text{mm}$ **b** $z = 18\,\text{mm}$

4

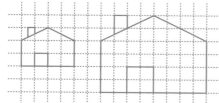

5

6 a unchanged
   b not congruent
7 OA′ = 2 × OA, OB′ = 2 × OB

8

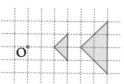

9 a

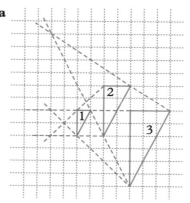

b 1·5

page 317 **Exercise 7 C**

1

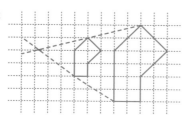

2

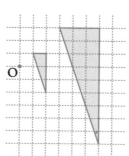

3

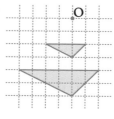

4

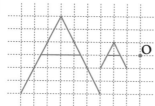

5

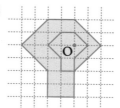

6

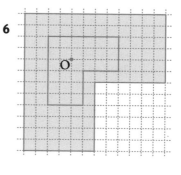

7 a–d

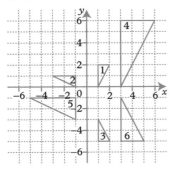

e $(3, 0)$, $(-5, -1)$, $(3, -1)$
f They are the same.

8 a–d

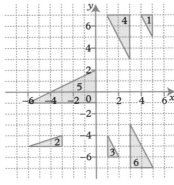

e $(3, 3)$, $(-6, -1)$, $(3, -3)$

9 a–d

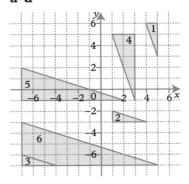

e $(3, -1)$, $(2, -1)$, $(5, -7)$

page 319 **Exercise 8(E)**

1

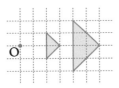

2

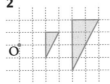

3

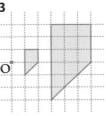

4

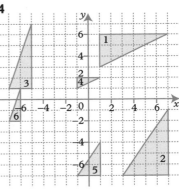

page 320 **Exercise 9(E)**

1 a $\begin{pmatrix} 4 \\ 6 \end{pmatrix}$ **b** $\begin{pmatrix} 6 \\ 4 \end{pmatrix}$ **c** $\begin{pmatrix} 0 \\ 3 \end{pmatrix}$ **d** $\begin{pmatrix} 6 \\ 0 \end{pmatrix}$ **e** $\begin{pmatrix} 5 \\ -2 \end{pmatrix}$ **f** $\begin{pmatrix} 1 \\ 2 \end{pmatrix}$ **g** $\begin{pmatrix} -2 \\ 5 \end{pmatrix}$

h $\begin{pmatrix} 2 \\ -2 \end{pmatrix}$ **i** $\begin{pmatrix} -4 \\ -3 \end{pmatrix}$ **j** $\begin{pmatrix} 2 \\ -6 \end{pmatrix}$ **k** $\begin{pmatrix} 1 \\ -8 \end{pmatrix}$ **l** $\begin{pmatrix} -6 \\ -1 \end{pmatrix}$ **m** $\begin{pmatrix} 0 \\ -4 \end{pmatrix}$ **n** $\begin{pmatrix} 6 \\ 1 \end{pmatrix}$

page 320 **Exercise 10(E)**

1 a, b

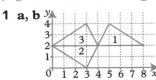

c Reflection in the line $x = 4$
3 a Reflection in the
   line $y = 0$ (x-axis),
   followed by a
   translation $\begin{pmatrix} -7 \\ 0 \end{pmatrix}$.
   b Yes

2 a, b

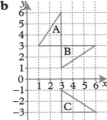

c Rotation 90° clockwise
   about $(0, 0)$

4

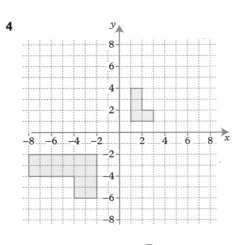

Enlargement with centre $(0, 0)$ and scale factor 2, followed by reflection in the line $y = -x$.

page 322 **Exercise 11 C**

1 a Radius = 6 cm, diameter = 12 cm   **b** Radius = 9 cm, diameter = 18 cm
c Radius = 5 cm, diameter = 10 cm   **d** Radius = 2 m, diameter = 4 m
2 Radius = 1·4 cm, diameter = 2·8 cm   **3** Radius = 1·1 cm, diameter = 2·2 cm
4 45°   **5 a** chord   **b** tangent   **c** ninety degrees   **d** segment   **e** arc   **f** sector

page 323 **Exercise 12 E**

1 34·6 cm   **2** 25·1 cm   **3** 37·7 cm   **4** 15·7 cm   **5** 31·4 cm   **6** 40·8 cm   **7** 6·28 cm
8 13·8 cm   **9** 28·3 cm   **10** 53·4 cm   **11** 44·6 m   **12** 72·3 m   **13** 1·9 cm   **14** 8·48 m
15 400 m   **16** 90·4 cm   **17** 823 m   **18** 212   **19 A** = 10π cm   **B** = 13π cm   **20** 642·74 cm

page 326 **Exercise 13 E**

1 95·0 cm²   **2** 78·5 cm²   **3** 28·3 m²   **4** 38·5 m²   **5** 113 cm²   **6** 201 cm²
7 19·6 m²   **8** 380 cm²   **9** 29·5 cm²   **10** 125 m²   **11** 21·46 cm²   **12** 4580 g
13 30 if not staggered   **a** 508 cm²   **b** 508 cm²   **14** 40·8 m²   **15** 13·73 cm²   **16** 42·92 cm²

page 328 **Exercise 14 E**

1 23·1 cm   **2** 38·6 cm   **3** 20·6 m   **4** 8·23 cm   **5** 28·6 cm   **6** 39·4 m
7 17·9 cm   **8** 28·1 m   **9** 24·8 cm   **10** 46·3 cm   **11** 28·8 cm

page 330 **Exercise 15 E**

1 48 cm³   **2** 30 cm³   **3** 60 cm³   **4** 120 cm³   **5** 110 cm³   **6** 27 cm³   **7** 27 cm³
8 4 cm³   **9** 8 cm³   **10** 9 cm³   **11** 6 cm³   **12** 12 cm³   **13** 9 or 10 cm³
14 1 hour 40 minutes   **15** 27 cm³   **16 a** 3 cm   **b** 4 cm   **c** 2 cm   **17** 333
18 1 000 000   **19** No, the garage would be solid brick inside.

page 332 **Exercise 16 E**

1 150 cm³   **2** 60 m³   **3** 480 cm³   **4** 300 cm³   **5** 28 m³   **6** 280 cm³
7 140 cm³   **8** 57 cm³   **9** 1700 m³   **10** 2400 cm³   **11 a** 200 m²   **b** 2400 m³
12 145 cm³   **13** 200 cm³   **14** 108 m³

page 334 **Exercise 17Ⓔ**

1 62·8 cm³　　**2** 113 cm³　　**3** 198 cm³　　**4** 763 cm³　　**5** 157 cm³　　**6** 385 cm³
7 770 cm³　　**8** 176 m³　　**9** 113 litres　　**10 a** 141 cm³　　**b** 25·1 cm³　　**11** 770 cm³
12 a 2·25 cm³　　**b** 0·451 cm³　　**13** 125　　**14** 139 000 cm³

page 336 **Exercise 18Ⓔ**

1 a 126 cm²　**b** 132 cm²　**c** 94·2 cm²　**d** 75·4 cm²　**2 a** 151 cm²　　**b** 408 cm²　　**c** 81·7 cm²
3 237 cm²　　**4 a** 6　　**b** 96 cm²　　**5 a** 40 cm²　　**b** 52 cm²　　**c** 82 cm²
6 a 6　　**b** 12　　**c** 8　　**d** 1　　**7** 18 cm²

page 339 **Exercise 19Ⓒ**

Agnes 035°, Belinda 070°, Carlo 155°, Derek 220°, Ernie 290°, Louise 340°,
Ann 040°, Bill 065°, Chris 130°, Duncan 160°, Eric 230°, Fred 330°

page 339 **Exercise 20Ⓒ**

1 a 147°　　**b** 122°　　**c** 090°　　　**2 a** 285°　　**b** 225°　　**c** 154°
3 a 061°　　**b** 327°　　**4 a** 302°　　**b** 344°　　**c** 046°

page 340 **Exercise 21Ⓒ**

1 10 km　　**2** 5 km　　**3** 11·5 km　　**4** 14·1 km　　**5 a** 12·5 km　**b** 032°

page 341 **Exercise 22Ⓒ**

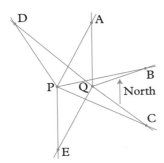

page 342 **Exercise 23Ⓔ**

1　　　　　　**2 a**　　　　　　**b** 9　　**3 a** Equilateral triangle　**4 a, b**
　　　　　　　　　　　　　　　　　　　b square

page 343 **Exercise 24(E)**

1 $a = 80°$, $b = 140°$, $c = 170°$, $d = 100°$, $e = 100°$, $f = 110°$, $g = 120°$
2 a $36°$ **b** Angle ECB $= 108° - 36° = 72°$
                Angle ABC $= 108°$
                So angle ABC + angle ECB $= 180°$
                Therefore EC is parallel to AB.
3 a $270°$ **b** $100°$ **c** $a = 119°$, $b = 25°$

page 345 **Exercise 25(E)**

1 $360°$ **2** $72°$ **3** $a = 80°$, $b = 70°$, $c = 65°$, $d = 86°$, $e = 59°$ **4 a** $36°$ **b** $144°$
5 a $40°$ **b** $20°$ **c** $6°$ **6** $p = 101°$, $q = 79°$, $x = 70°$, $m = 70°$, $n = 130°$ **7 a** $30°$ **b** $150°$
8 24 **9** 20 **10** 9

page 347 **Exercise 26(E)**

1 Circle, radius 3 cm centred on X   **2** Perpendicular bisector of PQ
3 Angle bisector of angle BAC
4 a   **b**   **c**

5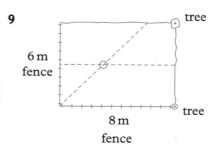
D   C

A   B

6 Intersection of circle radius 40 km centred on
L and the circle radius 20 km centred on C.

7 b i Perpendicular bisector of LN
    ii Angle bisector of angle LNM
    iii Circle radius 4 cm centred on M
8 Line perpendicular to AB through P

9

page 349 **Exercise 27(E)**

1 a 7.21 cm   **b** 5 cm   **c** 5.39 cm   **d** 12.21 cm   **e** 13 cm   **f** 5.20 cm   **g** 6.26 cm   **h** 12.73 cm
2 11.31 cm   **3 a** 8.32 cm   **b** 12.32 cm   **4** 6.32 m   **5** 114.13 m

page 351 **Exercise 28**Ⓔ

1 a 5·74 cm **b** 4·47 cm **c** 5·66 cm **d** 8·49 cm **e** 4 cm **f** 7·5 cm **g** 5 cm **h** 4·65 cm
2 3·46 m **3** 7·07 cm

page 351 **Exercise 29**Ⓔ

1 10 cm **2** 8·94 cm **3** 5·66 cm **4** 5·66 cm **5** 4·24 cm **6** 9·90 cm **7** 4·58 cm
8 5·20 cm **9** 9·85 cm **10** 9·80 cm **11** 4·27 m **12** 40·31 km **13** 5·70 cm

page 352 **Exercise 30**Ⓔ

1 a 5 units **b** 40 square units **2** 9·49 cm **3** 32·56 cm **4** 5·39 units
5 PQ = 5·83 units, QR = 6 units, PR = 5·83 The triangle is isosceles. **6** 6·63 units
7 5·57 units **8** 8·7 units **9** 5·66 units **10** 6·63 units **11** 4·69 units **12 a** 5 cm **b** 7·81 cm

page 355 **Exercise 31**Ⓔ

1 16 cm^2 **2** 27 cm^2 **3** 4·5 cm^2 **4** 15 cm^2 **5** 480 cm^3 **6** 540 cm^3 **7** 160 cm^3 **8** 4500 cm^3

page 356 **Exercise 32**Ⓔ

1 a area **b** length
2 a length **b** area **c** area **d** length **e** area **f** length **g** area **h** volume
  i volume **j** area **k** length **l** area **m** area **n** volume **o** area **p** length

3

expression	length	area	volume	none of these
$7r$	✓			
$3r^2$		✓		
$\pi r^2 h$			✓	
abc^2				✓
$\pi(a^2 + c^2)$		✓		

page 358 **Exercise 33**Ⓔ

1 A (2, 4, 0)   B (0, 4, 3)   C (2, 4, 3)   D (2, 0, 3)
2 a B (3, 0, 0)   C (3, 4, 0)   Q (3, 0, 2)   R (3, 4, 2)
  b i (0, 2, 0)   **ii** (0, 4, 1)   **iii** (1·5, 4, 0)
  c i (1·5, 2, 0)   **ii** (1·5, 2, 2)   **iii** (1·5, 4, 1)
  d i (1·5, 2, 1)

3 a C (2, 2, 0),   R (2, 2, 3),   B (2, −2, 0),   P (0, −2, 3),   Q (2, −2, 3)
  b i (2, −2, 1·5) **ii** (1, −2, 3)

page 359 **Test yourself**

1 a i $4\,\text{cm}^2$ **ii** $10\,\text{cm}$ **b** $28\,\text{cm}^3$ **2** $150\,\text{cm}^3$ **3 a** $9.4\,\text{cm}$ **b** For example, $2\,\text{cm}$, $3\,\text{cm}$ and $6\,\text{cm}$.
4 a $31.4\,\text{cm}$ **b** $3.18\,\text{cm}$ **5 a** $254°$ **b** $728\,\text{m}$

6 a $050°$, $12\,\text{km}$ **b**
7 $170\,\text{g}$

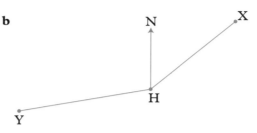

8 a

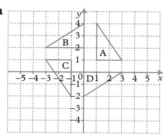

b Rotation $180°$ about $(0, 1)$

9 a Reflection in the line $y = 0$ (x-axis)
  b Rotation $90°$ anticlockwise about $(0, 0)$
  c Enlargement scale factor 3 with centre $(5, 4)$

10 a, b

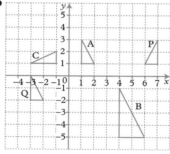

c 2
d Rotation $90°$ anticlockwise about $(0, 0)$

11 $116.9\,\text{cm}$ **12 a** $13.8\,\text{m}$ **b** $15.2\,\text{m}^2$ **13 a** $201\,\text{cm}^2$ **b** $12.6\,\text{cm}$ **14** $61.7\,\text{cm}$
15 Perpendicular bisector of PQ

16

17 $45°$ **18 a** 12   **b** $13\,\text{cm}$
19 a $41\,\text{m}^2$ **b i** $5\,\text{m}$ **ii** $30\,\text{m}$
20 $\begin{aligned} RQ^2 + QS^2 &= SR^2 \\ 12^2 + QS^2 &= 13^2 \\ QS^2 &= 169 - 144 \\ QS &= \sqrt{25} \\ QS &= 5 \end{aligned}$ $\begin{aligned} PS^2 + QS^2 &= PQ^2 \\ 4^2 + 5^2 &= PQ^2 \\ 16 + 25 &= PQ^2 \\ \sqrt{41} &= PQ \end{aligned}$

8 Data handling 2

page 368 **Exercise 1 Ⓔ**

1 Many possible answers.

2 a

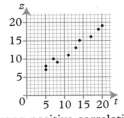

Strong positive correlation

b

No correlation

c

Weak negative correlation

3 a No correlation **b** Strong negative correlation **c** No correlation **d** Weak positive correlation

4 a

b 9

5 a

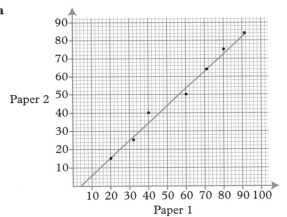

b 44

6 a

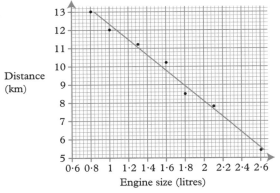

b 6·8 km

7 a

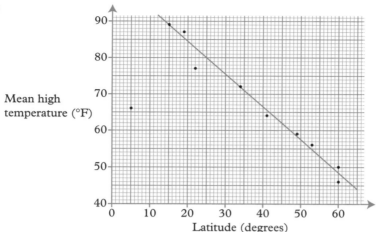

b Bogota, high altitude
c 75°F

8 a Positive correlation   **b** Positive correlation   **c** No correlation
d Negative correlation   **e** Positive correlation

page 371 **Exercise 2** ⓒ

1

	Shaded	Unshaded
Squares	2	4
Triangles	2	2

2 a

	Football	Hockey	Swimming
Boy	5	2	1
Girl	2	3	2

b 8
c 25%

3 a

	Girls	Boys	Total
Can cycle	95	120	215
Cannot cycle	179	82	261
Total	274	202	476

b 34·7%
c 59·4%

4 a

	Men	Women	Total
Accident	75	88	163
No accident	507	820	1327
Total	582	908	1490

b 12·9%
c 9·7%
d Men are more likely to have accidents.

page 379 **Exercise 3** ⓒ

1 No choices, such as beach, winter, activity.
2 Too vague: Do you like the new head teacher? yes/no
3 Choices should include less than 2 hours and more than 6 hours.
   They also shouldn't overlap. No option for 4–5 hours.
4 There should be a choice for between £1 and £2·50.
5 Leading question. Which is the most important subject at school? Maths,
   English, Science, other.
6 Not clear what 'often' means. How often do you or your parents hire
   DVDs from a shop? Never, 1–2 times a month, 3–4 times a month, 5 or
   more times a month.

7 Leading question. Do you get too much homework? yes/no
8 Name is not needed. There should be a choice of age ranges to choose from. The choices for question 1 are not specific enough. Question 2 should have some options including Other. Question 3 is leading. Question 4 is too vague. A better question would be 'How often do you watch nature programs?' with some numeric choices.
9 What type of TV programe do you most watch? Comedy, sport, news, soaps, films, factual, other.
10 Many possible answers.

page 381 **Exercise 4Ⓔ**

1 Not satisfactory. She will not get the opinions of students that no longer have school dinners.
2 Not satisfactory. People travelling abroad will have different opinions to those that have never been abroad.
3 Not satisfactory. 17-year-old drivers will not be on the electoral roll.
4 She will not get the views of people at work or school.

page 383 **Exercise 5Ⓔ**

1 a Females under 30 = 8 = 57·6°
   Males under 30 = 9 = 64·8°
   Females 30 or over = 10 = 72°
   Males 30 and over = 23 = 165·6°
   b Most employees are males 30 and over.   **c i** 240   **ii** 270   **iii** 300   **iv** 690

2 a

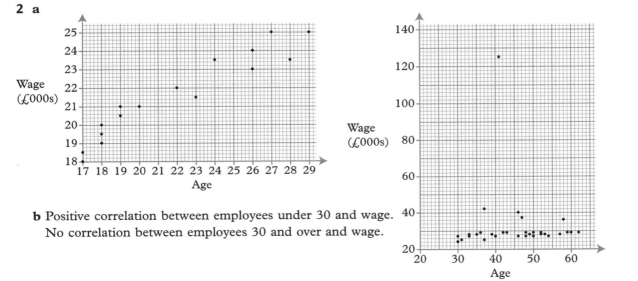

   b Positive correlation between employees under 30 and wage.
   No correlation between employees 30 and over and wage.

3 a

Wage, £	Mid-value, x	Frequency, f	fx
15 000–20 000	17 500	4	70 000
20 000–25 000	22 500	12	270 000
25 000–30 000	27 500	29	797 500
30 000–35 000	32 500	0	0
35 000–40 000	37 500	2	75 000
40 000–45 000	42 500	2	85 000
125 000–130 000	127 500	1	127 500
Total		50	1 425 000

b £28 500   **c** £28 430

d The mean from the raw data is more reliable. The value from the table is an estimate.

4

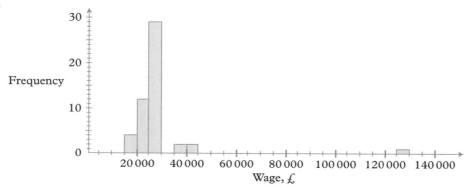

5 a £27 500   **b** £107 000   **c** median   **d** The mean is distorted by one high value.

page 386 **Test yourself**

1 a

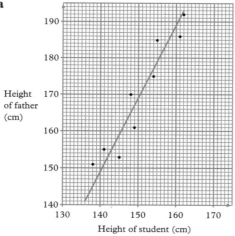

2 a, c

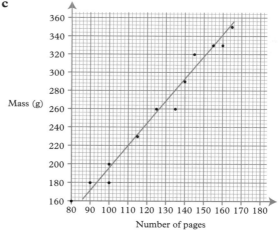

b Positive correlation
d 171 cm

b Strong positive correlation
d i 134 pages   **ii** 220 g

3 a Leading question
   b Which do you think are more important, tennis courts or squash courts?

4 Response boxes should be, for example, less than £1, between £1 and £5, over £5. The question should also include a period of time, for example, 'How much do you normally spend in the canteen in a week?'

5 a 1 Too personal   2 Leading question

  b Do you think shops should open on a Sunday? yes/no

6 a Not enough choices

  b His survey doesn't include a wide range of ages or locations.

7

	France	Germany	Spain	Total
Female	2	23	9	34
Male	15	2	9	26
Total	17	25	18	60

8

	French	German	Spanish	Total
Boys	23			
Girls				
Total				

9

	Red	Green	Yellow	Total
Male	3	2	2	7
Female	2	3	1	6
Total	5	5	3	13

9   Number 3

page 391 Exercise 1 ⓒ

1 a 0.25   **b** 0.4   **c** 0.375   **d** 0.41$\dot{6}$ **e** 0.1$\dot{6}$ **f** 0.$\dot{2}$85 71$\dot{4}$

2 a $\frac{1}{5}$ **b** $\frac{9}{20}$ **c** $\frac{9}{25}$ **d** $\frac{1}{8}$ **e** $1\frac{1}{20}$ **f** $\frac{7}{1000}$

3 a 25%   **b** 10%   **c** 72%   **d** 7.5%   **e** 2%   **f** 33.$\dot{3}$%

4 a 45 000   **b** 396 000   **c** $\frac{1}{30}$

5

Fraction	Decimal	Percentage
$\frac{1}{4}$	0.25	25%
$\frac{1}{5}$	0.2	20%
$\frac{4}{5}$	0.8	80%
$\frac{1}{100}$	0.01	1%
$\frac{3}{10}$	0.3	30%
$\frac{1}{3}$	0.3	33.3%

6 a $\frac{1}{5}$ **b** $\frac{4}{10}$ **c** $\frac{11}{33}$ **d** $\frac{7}{12}$

7 A   **8** $\frac{1}{3}$ off is cheaper

9 $\frac{1}{6}$ of £5000   **10** 25%   **11** 20%

page 393 Exercise 2 ⓒ

1 a 110 miles   **b** 340 m   **c** 315 km   **d** 60 cm

2 a 14 mph   **b** 45.5 km/h   **c** 25 m/s   **d** 10.5 cm/min   **3** 120 miles

4 500 m   **5** 300 m   **6** 360 m   **7** 2000 km   **8** 12 km   **9** 61 km/h

10 7.5 m/s   **11** 83 km/h   **12** 0.5 cm/s   **13** 3 hours   **14** 6 hours   **15** 250 km

16 a 18 hours   **b** 7 hours   **c** 5 seconds   **d** 30 minutes   **17** 88 mph   **18** $4\frac{1}{2}$ km

page 394 **Exercise 3**(E)

1 a $1 \cdot 5 \, \text{g/cm}^3$ **b** $9 \, \text{g/cm}^3$ **c** $5 \cdot 6 \, \text{g/cm}^3$ **2** $8 \, \text{kg}$ **3** $36 \, \text{g}$ **4** £90 **5** £15 000
6 £700 **7** £329 000 **8** £30 000 **9 a** 924 people/square mile **b** 43 400

page 396 **Exercise 4**(E)

1 4×10^3 **2** 5×10^2 **3** 7×10^4 **4** 6×10 **5** $2 \cdot 4 \times 10^3$
6 $3 \cdot 8 \times 10^2$ **7** $4 \cdot 6 \times 10^4$ **8** $4 \cdot 6 \times 10$ **9** 9×10^5 **10** $2 \cdot 56 \times 10^3$
11 7×10^{-3} **12** 4×10^{-4} **13** $3 \cdot 5 \times 10^{-3}$ **14** $4 \cdot 21 \times 10^{-1}$ **15** $5 \cdot 5 \times 10^{-5}$
16 1×10^{-2} **17** $5 \cdot 64 \times 10^5$ **18** $1 \cdot 9 \times 10^7$ **19** $1 \cdot 1 \times 10^9$ **20** $1 \cdot 67 \times 10^{-24} \, \text{g}$
21 $5 \cdot 1 \times 10^8 \, \text{km}^2$ **22** $2 \cdot 5 \times 10^{-10} \, \text{cm}$ **23** $6 \cdot 023 \times 10^{23}$ **24** $3 \times 10^{10} \, \text{cm/s}$ **25** £$3 \cdot 6 \times 10^6$

page 397 **Exercise 5**(E)

1 $1 \cdot 5 \times 10^7$ **2** 3×10^8 **3** $2 \cdot 8 \times 10^{-2}$ **4** 7×10^{-9} **5** 2×10^6
6 4×10^{-6} **7** 9×10^{-2} **8** $6 \cdot 6 \times 10^{-8}$ **9** $3 \cdot 5 \times 10^{-7}$ **10** 1×10^{-16}
11 8×10^9 **12** $7 \cdot 4 \times 10^{-7}$ **13** $4 \cdot 9 \times 10^{11}$ **14** $4 \cdot 4 \times 10^{12}$ **15** $1 \cdot 5 \times 10^3$
16 2×10^{17} **17** $1 \cdot 68 \times 10^{13}$ **18** $4 \cdot 25 \times 10^{11}$ **19** $9 \cdot 9 \times 10^7$ **20** $6 \cdot 25 \times 10^{-16}$
21 $2 \cdot 88 \times 10^{12}$ **22** $6 \cdot 82 \times 10^{-7}$ **23** $1 \cdot 2 \times 10^{10}$ days **24** $1 \cdot 3 \times 10^{-4}$
25 $a = 5 \cdot 12 \times 10^4$, $b = 4 \cdot 78 \times 10^5$, $c = 4 \cdot 9 \times 10^4$ **26** 13
27 a $4 \cdot 2 \times 10^4$ **b** 8×10^7 **c** 2×10 **28 a** 1×10^3 **b** 4×10^7
29 50 minutes **30** 600 **31 a** $9 \cdot 5 \times 10^{12} \, \text{km}$ **b** $1 \cdot 44 \times 10^8 \, \text{km}$

page 399 **Exercise 6** (C)

1 400 cm **2** 2400 cm **3** 63 cm **4** 0·25 cm **5** 7 cm **6** 20 mm **7** 1200 m
8 700 cm **9** 500 m **10** 8·15 m **11** 0·65 km **12** 2·5 cm **13** 5000 g **14** 4200 g
15 6400 g **16** 3000 g **17** 800 g **18** 0·4 kg **19** 2000 kg **20** 0·25 kg **21** 500 kg
22 620 kg **23** 7000 g **24** 1·5 kg **25** 8000 ml **26** 2000 ml **27** 1 litre **28** 4500 ml
29 320 cm **30** 5·5 cm **31** 1400 g **32** 110 mm
33 a kilometres **b** millilitres **c** grams **d** millimetres **e** tonnes **f** square metres

page 400 **Exercise 7** (C)

1 24 **2** 16 **3** 30 **4** 4480 **5** 48 **6** 17 600 **7** 36 **8** 70 **9** 4 **10** 880
11 3 **12** 2 **13** 3520 **14** 80 **15** 140 **16** 12 **17** 48 **18** 22 400 **19** 5280 **20** 2

page 401 **Exercise 8**(E)

1 25 cm **2** 50 litres **3** 6 pounds **4** 8 pounds **5** 5 miles **6** 1 kg
7 20 miles **8** 10 cm **9** 4 kg **10** 5 litres **11** 25 miles **12** 20 gallons
13 6 pounds **14** 8 km **15** 80 gallons **16** 2 ounces **17** 300 g **18** 1 gallon
19 40 pounds **20** 10 inches **21** 30 cm **22** 5000 miles **23** 50 mph **24** Super market B
25 Bake in a 30 cm dish, 500 g flour, 120 g sugar, 8 large apples (about 10 cm in diameter.)
26 10 g **27** 7 m **28** 500 ml **29** 100 miles **30** 3 mm
31 a 2·75 kg **b** 15 litres **c** 3 pounds **32** $\frac{1}{8}$

page 404 **Exercise 9**(E)

1 195·5 cm **2** 36·5 kg **3** 95·55 m **4** 3·25 kg **5** 28·65 s
6 a 1·5 °C, 2·5 °C **b** 2·25 g, 2·35 g **c** 63·5 m, 64·5 m **d** 13·55 s, 13·65 s **7** B
8 C **9 a** No **b** 1 cm

10 Yes, the largest the card could be is 11·25 cm, the smallest the envelope could be is 11·5 cm
11 16·5 kg, 17·5 kg   **12** 255·5 km, 256·5 km   **13** 2·35 m, 2·45 m
14 0·335 g, 0·345 g   **15** 2·035 m/s, 2·045 m/s   **16** 7·5 cm, 8·5 cm
17 81·35 °C, 81·45 °C   **18** 0·25 kg, 0·35 kg   **19** 4·795 cm, 4·805 cm
20 0·065 cm, 0·075 cm   **21** 0·65 tonne, 0·75 tonne   **22** 614·5 seconds, 615·5 seconds
23 7·125 m, 7·135 m   **24** 51 500, 52 500

page 406 **Exercise 10 C**

1 4p per can   **2** 11p per ruler   **3** 9p per card   **4** 10p per can   **5** 4 p per paper
6 18p per box   **7** 6p per kg   **8** 13p per kg   **9** £1·75 per T-shirt   **10** 36p per dozen
11 2·5p per orange   **12** £7·25   **13** 15p per 100 ml   **14** 5p per kg   **15** 16p per m
16 89p per kg   **17** 14p per m   **18** 2p per apple   **19** £3·60   **20** 15p per tin

page 407 **Exercise 11 C**

1

5	×	12	→	60
×		÷		
20	+	24	→	44
↓		↓		
100	×	$\frac{1}{2}$	→	50

2

7	×	6	→	42
÷		÷		
14	−	3	→	11
↓		↓		
$\frac{1}{2}$	×	2	→	1

3

19	×	2	→	38
−		÷		
12	×	4	→	48
↓		↓		
7	−	$\frac{1}{2}$	→	$6\frac{1}{2}$

4

17	×	10	→	170
−		÷		
9	÷	100	→	0·09
↓		↓		
8	−	0·1	→	7·9

5

0·3	×	20	→	6
+		−		
11	÷	11	→	1
↓		↓		
11·3	−	9	→	2·3

6

$\frac{1}{2}$	×	50	→	25
−		÷		
0·1	+	$\frac{1}{2}$	→	0·6
↓		↓		
0·4	×	100	→	40

7

7	×	0·1	→	0·7
÷		×		
4	÷	0·2	→	20
↓		↓		
1·75	+	0·02	→	1·77

8

1·4	+	8	→	9·4
−		×		
0·1	×	0·1	→	0·01
↓		↓		
1·3		0·8	→	2·1

9

100	×	0·3	→	30
−		×		
2·5	÷	10	→	0·25
↓		↓		
97·5	+	3	→	100·5

10

3	÷	2	→	1·5
÷		÷		
8	÷	16	→	$\frac{1}{2}$
↓		↓		
$\frac{3}{8}$	+	$\frac{1}{8}$	→	$\frac{1}{2}$

11

$\frac{1}{4}$	−	$\frac{1}{16}$	→	$\frac{3}{16}$
×		×		
$\frac{1}{2}$	÷	4	→	$\frac{1}{8}$
↓		↓		
$\frac{1}{8}$	+	$\frac{1}{4}$	→	$\frac{3}{8}$

12

0·5	−	0·01	→	0·49
+		×		
3·5	×	10	→	35
↓		↓		
4	÷	0·1	→	40

page 408 **Exercise 12 🄒**

1 a
```
    5 7 3 2
  + 2 6 9 6
  ─────────
    8 4 2 8
```
b
```
      8 3 5
    − 2 6 2
    ───────
      5 7 3
```
c $245 \div 7 = 35$

2 6p **3** 15 litres **4 a** 36 **b** £11 111 **5 a** £1600 **b** 8% **6** 70 mm
7 24 **8** One **9 a** 0·3 **b** 0·003 **c** 0·115 **d** 70 000 **10** 1·29

page 409 **Exercise 13 🄒**

1 20 **2** 200 g **3** Six **4** 400 **5** 3854 **6 a** 323 g **b** 23 **c** 67p **d** 29
7 20 cm^2 **8** 22 222 **9** 234 − 65 **10 a** £93·20 **b** £8·60

page 410 **Exercise 14 🄒**

1 a
```
      4 3 2
      2 3 1
    + 7 3 5
    ───────
    1 3 9 8
```
b
```
      4 3 6
      5 2 1
    + 1 1 5
    ───────
    1 0 7 2
```
2 1221
3 a $x = 1$ **b** $x = 15$
4 £18

5

	Passed	Failed	Total
Boys	350	245	595
Girls	416	180	596
Total	766	425	1191

6 a 18, 19 **b** 24, 25, 26, 27 **c** 11, 37 **d** 3, 16
7 8 **8** 120 **9** 125° **10** 57%

Test yourself

1

Fraction	Decimal	Percentage
$\frac{1}{2}$	0·5	50%
$\frac{3}{4}$	0·75	75%
$\frac{1}{4}$	0·25	25%
$\frac{9}{100}$	0·09	9%

2 a g or kg **b** cm or mm **c** tonne
3 a square metres **b** grams **c** kilometres **d** litres
4 a 92 000 000 **b** $3·05 \times 10^{-1}$ mm **5** $1·2 \times 10^8$

6 a $\frac{1}{4}$ **b** 0·75 **c** 75% **d i** 9 litres **ii** 16 litres
7 10 pounds (or 11 pounds) **8** 50 mph
9 98 km/h **10 a** 30 cm/s **b** 90 m

11 a $3·8 \times 10^{-11}$ **b** $2·4528 \times 10^7$ km **12** 24·5 kg, 25·5 kg
13 a 114·5 cm, 115·5 cm **b** The space could be 1·15 m wide, but **14 a** 7 gallons **b** 48 km
the wardrobe could be 1·115 m wide.

10 Probability

page 416 **Exercise 1 🄒**

1–12 Many possible answers.

page 417 **Exercise 2 🄒**

1 $\frac{1}{6}$ **a, b** Many possible answers. **2** $\frac{1}{4}$ **3** Many possible answers. **4** $\frac{6}{36} = \frac{1}{6}$

5 There is a 1 in 10 chance of spinning a zero on the spinner. Both Steve and Mike reported relative frequencies of 2 in 10. Mike did a greater number of trials and so his results should be closer to the expected relative frequency of 0·1. It isn't. Therefore, Mike recorded his results incorrectly.

page 419 **Exercise 3 C**

1 a $\frac{1}{3}$ **b** $\frac{1}{3}$ **2 a** $\frac{1}{4}$ **b** $\frac{1}{4}$ **c** $\frac{1}{4}$ **3 a** $\frac{2}{3}$ **b** $\frac{1}{3}$ **4 a** $\frac{1}{6}$ **b** $\frac{1}{6}$ **c** 0

5 a $\frac{1}{8}$ **b** $\frac{1}{8}$ **c** $\frac{1}{2}$ **6 a** $\frac{3}{4}$ **b** $\frac{1}{2}$ **c** $\frac{1}{4}$ **d** $\frac{1}{3}$ **e** $\frac{1}{8}$ **f** $\frac{1}{2}$ **g** $\frac{11}{12}$ **h** $\frac{3}{5}$

7 a True   **b** False   **c** False   **8 a** $\frac{1}{13}$ **b** $\frac{1}{52}$ **c** $\frac{1}{4}$ **9 a** $\frac{1}{9}$ **b** $\frac{1}{3}$ **c** $\frac{4}{9}$ **d** $\frac{2}{9}$

10 a $\frac{5}{11}$ **b** $\frac{2}{11}$ **c** $\frac{4}{11}$ **11 a** $\frac{4}{17}$ **b** $\frac{3}{17}$ **c** $\frac{11}{17}$ **12 a** $\frac{4}{17}$ **b** $\frac{8}{17}$ **c** $\frac{5}{17}$

13 a $\frac{2}{9}$ **b** $\frac{2}{9}$ **c** $\frac{1}{9}$ **d** 0   **e** $\frac{5}{9}$ **14 a** $\frac{1}{13}$ **b** $\frac{2}{13}$ **c** $\frac{1}{52}$ **d** $\frac{5}{13}$

page 423 **Exercise 4 C**

1 a $\frac{1}{7}$ **b** $\frac{4}{7}$ **c** $\frac{6}{7}$ **2 a** $\frac{1}{6}$ **b** $\frac{1}{3}$ **c** $\frac{1}{6}$ **3 a** $\frac{1}{5}$ **b** $\frac{1}{5}$ **c** $\frac{2}{5}$ **4 a** $\frac{1}{10}$ **b** $\frac{3}{10}$ **c** $\frac{3}{10}$

5 a $\frac{3}{13}$ **b** $\frac{5}{13}$ **c** $\frac{8}{13}$ **6 a i** $\frac{5}{13}$ **ii** $\frac{6}{13}$ **b i** $\frac{5}{12}$ **ii** $\frac{1}{12}$ **7 a** $\frac{1}{12}$ **b** $\frac{1}{40}$ **c** $\frac{1}{4}$ **8** $\frac{9}{20}$

9 Megan   **10** 3 white, 6 black   **11** $\frac{1}{7}$ **12 a** 1   **b** He may have just not picked the blue ball.

13 a i $\frac{1}{4}$ **ii** $\frac{1}{4}$ **iii** $\frac{1}{4}$ **b** $\frac{1}{4}$ **c** $\frac{6}{27}$ **14** 15

page 426 **Exercise 5 C**

1 a 150   **b** 50   **2** 25   **3** 50 people   **4** 40   **5 a** $\frac{3}{8}$ **b** 25

6 a $\frac{1}{2}$ **b** $\frac{1}{4}$ **7 a** 15   **b** 105   **8 a** 20   **b** 5   **c** 50   **d** 40

page 428 **Exercise 6 C**

1 a

10p	20p	50p
H	H	H
H	H	T
H	T	H
T	H	H
T	T	H
T	H	T
H	T	T
T	T	T

b $\frac{1}{8}$ **2** 16 different outcomes   **3 a, b** 4   **c** $\frac{1}{9}$

4 a (1, 1), (1, 2), (1, 3), (2, 1), (2, 2), (2, 3),   **b** 3   **c** $\frac{1}{4}$
(3, 1), (3, 2), (3, 3), (4, 1), (4, 2), (4, 3)

5 $(w, x), (w, y), (w, z), (x, y), (x, z), (y, z)$

6 a

Dice

	1	2	3	4	5	6
1	1,1	1,2	1,3	1,4	1,5	1,6
2	2,1	2,2	2,3	2,4	2,5	2,6
3	3,1	3,2	3,3	3,4	3,5	3,6
4	4,1	4,2	4,3	4,4	4,5	4,6
5	5,1	5,2	5,3	5,4	5,5	5,6

Spinner

b $\frac{1}{2}$

7 a (A, C, F), (B, C, F), (A, D, F), (B, D, F), (A, E, F), (B, E, F), **b** $\frac{1}{12}$
   (A, C, G), (B, C, G), (A, D, G), (B, D, G), (A, E, G), (B, E, G)

8 a 15 **b** 676   **9 a** 24 different ways **b** 12   **10** Yes   **11** No   **12** 4 wins

page 433 **Exercise 8 C**

1 $\frac{4}{5}$ **2 a** $\frac{1}{13}$ **b** $\frac{12}{13}$ **c** $\frac{3}{13}$ **d** $\frac{10}{13}$ **3** $\frac{35}{36}$ **4** 0·76   **5** 0·494

6 a $\frac{1}{4}$ **b** $\frac{3}{4}$ **c** $\frac{1}{4}$ **d** $\frac{3}{4}$ **e** 0   **f** 1   **7 a** $\frac{1}{17}$ **b** $\frac{16}{17}$ **c** $\frac{4}{17}$

8 a 0·3   **b** 0·9   **9 a i** 0·24   **ii** 0·89   **b** 575

page 435 **Test yourself**

1 a 1 **b** $\frac{1}{2}$ **c** Near 0 **d** $\frac{5}{6}$ **2 a** A Impossible, B Unlikely, C Even, D Likely, E Certain **b** Likely

3 a $\frac{1}{2}$ **b** Randomness   **4 a** Red **b** $\frac{1}{2}$ **c** Probability on first is $\frac{1}{5}$, on second it is $\frac{1}{6}$.

5 a $\frac{1}{520}$ **b** $\frac{7}{13}$ **6 a** 0·3 **b** 68   **7 a** $\frac{23}{40}$ **b** $\frac{2}{5}$ **c** It could have been an unusually bad day.

8 a GBC, GCB, BGC, BCG, CBG, CGB   **b i** $\frac{7}{20}$ **ii** 0·6

9 a 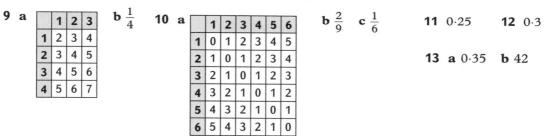 **b** $\frac{1}{4}$ **10 a**   **b** $\frac{2}{9}$ **c** $\frac{1}{6}$ **11** 0·25   **12** 0·3

13 a 0·35 **b** 42

11   Using and applying mathematics

page 440   **11.1.1 Opposite corners**

The difference in an $n \times n$ square is $(3n - 3)^2$.

page 441   **11.1.2 Hiring a car**

Hav-a-car is cheapest up to 950 miles. Snowdon is cheapest between
950 and 1300 miles. Gibson is cheapest over 1300 miles.

page 441   **11.1.3 Half-time score**

12 different scores were possible at half-time.
The general rule is $(a + 1)(b + 1)$ for a final score $a - b$.

page 442 **11.1.4 Squares inside squares**

Area is always $a^2 + b^2$.

page 443 **11.1.5 Maximum box**

The corner squares should be $\frac{1}{6}$th the size of the square sheet.

page 444 **11.1.6 Diagonals**

889 squares

page 444 **11.1.7 Painting cubes**

27 cubes: 8 have 3 red faces, 12 have 2 red faces, 6 have 1 red face, 1 has 0 red faces.

n^3 cubes: 8 have 3 red faces, $12(n-2)$ have 2 red faces, $6(n-2)^2$ have 1 red face, $(n-2)^3$ have 0 red faces.

page 445 **11.2.1 Crossnumbers**

A

5	2	5		3	5		2
2		2	4	1	7		1
1	7	2		0		6	4
8		1	7	0		8	
	1	0	1		5	4	9
8	7		5	5	5		
7		4		3	0	4	6
7	2	9		5		6	6

B

5	8	4		3	5		3
8		3	4	2	7		6
3	9	9		8		3	4
6		1	8	3		1	
	7	6	4		9	5	0
8	7		5	6	7		
5		9		6	8	7	5
8	2	4		7		3	6

C

1	1	5		3	9		1
8		1	4	4	0		2
6	5	0		7		6	5
0		5	7	0		4	
	2	1	9		6	6	6
1	4		3	4	1		
2		4		9	0	5	0
7	0	5		0		8	1

D

1	7	5		2	4		0
2		9	9	6	0		0
9	9	9		9		3	7
5		9	0	5		8	
	1	6	0		8	0	7
3	6		5	7	2		
0		2		2	3	4	5
5	0	1		9		2	2

page 447   **11.2.2 Crossnumbers without clues**

Part A

2	4	7	1	5		1	2	7
5	2		3	3	8	4	7	2
	1	1	1		2		6	1
6	0	9		9	1	8		8
5		8	5	9	2	0	7	0
3	8	2	4		5	7	1	
	9	6	0	3		2	4	9
4	1	4	7	2	5		3	0
2	6		3	8	7	5	6	6

Part B

4	2	1	4		9	1	3	1
1		5	7	3	2	0		3
	2	6	3		1	6	2	4
7	1		4	2	7		1	9
2	3	4				3	8	5
7	6		2	1	5		6	3
5	2	1	6		8	7	2	
4		8	3	6	4	2		2
2	0	0	7		1	5	0	6

Part C

7	3	1	5		3	8	9	7	5
7	0		6	4	2	6		2	8
5	6	4	3	5		7	6	1	8
4		9	1	7		7	6	4	3
3	7	5	5		5	0	4		5
	2	6		5	8	5	3	5	
3	8	1	9		5	6	0	7	3
7	4		8	2	7			4	7
3	6	5	4		7	7	4	3	5
5	8	2	7	7	8		9	5	1

Part D

	6	2	3	1		5	6	7	8		2
1	7	4		4	5	0		7	6	3	0
2	8	7		7	8	9		2	1	0	4
1	9		7	0		6	2		2	3	
		1	7	0		8	4	6	7	8	9
3	7	1	4			5	1	7	8	9	
1	8		2	4	5	6					
	9	9	2	1		8	7	6	4	5	2
6	6	4	5	6	7	8		2	3	5	6
1	7		9	8	4	3	8		1	4	7
2		3	2	4		6	7	8		8	5
	4	7	1	1		9	0	1	2		9

page 450   **11.2.4 Calculator words**

1 SELL	**2** SLOSH	**3** BELL	**4** GOES	**5** HOLE
6 HELLO	**7** GEESE	**8** LEGIBLE	**9** OBSESS	**10** OBOE
11 SHE LIES	**12** BIG LOG	**13** LOOSE	**14** BIGGISH	**15** EGGSHELL
16 IGLOO	**17** GLOSS	**18** BILGE	**19** LEGLESS	**20** SHOES
21 SIEGE	**22** HE DID	**23** OBLIGE	**24** LIBEL	**25** BESIEGE
26 HE IS SO BIG	**27** BOO HOO	**28** BOOZE	**29** EEL	**30** GOOSE

31 GOOD BIE (GOOD BYE)   **32** HE SELLS

12 Revision

12.1 Revision exercises

Exercise 1

1 £25·60, £6·70, 4, £55·30   **2 a** 30, 37   **b** 12, 10   **c** 7, 10   **d** 8, 4   **e** 26, 33
3 £6·50   **4** £172   **5 a** 1810 seconds   **b** 72·4 seconds   **6** 0·8 cm
7 a £13   **b** £148   **c** £170   **8 a** 5·9   **b** 6   **c** 7
9 a −11   **b** 23   **c** −10   **d** −20   **e** 6   **f** −14   **10 a** 3   **b** 5   **c** −6   **d** −7
11 a $x = 9$ **b** $x = 11$ **c** $x = 3$ **d** $x = 7$ **12** Net **c** or **a**

13 a   **b** 1, 4, 9, 16   **c** Square numbers   **d** 49

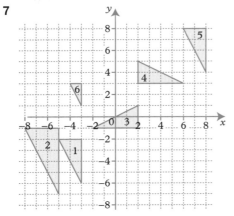

Exercise 2

1 a $x = 7$ **b** $x = \frac{1}{4}$ **c** $x = \frac{4}{5}$ **2 a** 7·21 cm   **b** 9·22 cm   **c** 7·33 cm

3 a $\frac{3}{8}$ **b** $\frac{5}{8}$ **4 a** $\frac{2}{11}$ **b** $\frac{5}{11}$ **c** $\frac{9}{11}$ **5 a** 2·088   **b** 3·043

6 a Reflection in line $y = 0$ (x-axis)   **b** Reflection in line $x = -1$ **c** Reflection in line $y = x$
   d Rotation 90° clockwise about $(0, 0)$ **e** Reflection in line $y = -1$ **f** Rotation 180° about $(0, -1)$

7

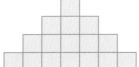

a Enlargement with centre $(1, -4)$,   **b** Rotation 90° clockwise   **c** Reflection in line $y = -x$
   scale factor 1·5   about centre $(0, -4)$

d Translation $\begin{pmatrix} 11 \\ 10 \end{pmatrix}$ **e** Enlargement with centre   **f** Rotation 90° anticlockwise
                                                        $(-3, 8)$, scale factor $\frac{1}{2}$ about centre $\left(\frac{1}{2}, 6\frac{1}{2}\right)$
g Enlargement with centre $(-2, 5)$, scale factor 3

Exercise 3

1 a £10·40   **b** 29 minutes   **2 a** 2 cm   **b** 8 m   **3 i** 9   **ii** 50   **iii** $(7 \times 11) - 6 = 72 - 1$

4 a, b

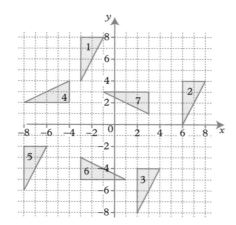

c △2: $(6, 0)$, $(6, 4)$, $(8, 4)$ △3: $(2, -8)$, $(2, -4)$, $(4, -4)$ △4: $(-4, -2)$, $(-8, 2)$, $(-4, 4)$
△6: $(-3, -3)$, $(-3, -5)$, $(1, -5)$ △7: $(3, 1)$, $(3, 3)$, $(-1, 3)$
5 $17 \cdot 7$ cm   **6 a** $197 \cdot 9$ cm^3 **b** $1 \cdot 4$ cm^3 **c** 145

page 455   **Exercise 4**

1 a £66 000   **b** 198   **c** £330   **d** £65 340   **2 a** 4   **b** 19   **3 a** 99p   **b** 760 miles
4 A: Cross-channel swimmer   B: Car ferry from Calais   C: Hovercraft from Dover
    D: Train from Dover   E: Marker buoy outside harbour   F: Car ferry from Dover
5 a 560 kg   **b** 57 kg   **6 a** A: 28 274   B: £79·15   C: £85·42   **b** November

page 456   **Exercise 5**

1 a $\frac{1}{8}$ **b** $\frac{3}{4}$ **c** 2   **d** $\frac{1}{2}$ **e** $\frac{1}{64}$ **2 a** 0·05 m/s   **b** 1·6 seconds   **c** 172·8 km

3 a 14   **b** 18   **c** 28   **4** A: $y = 6$ B: $y = \frac{1}{2}x - 3$ C: $y = 10 - x$ D: $y = 3x$

5 a $s = rt + 3t$ **b** $r = \dfrac{S - 3t}{t}$ **6** 12 cm^2 **7** 9·9 cm

page 457   **Exercise 6**

1 20·7 litres   **2 a** 8   **b** 140   **c** 29   **d** 42   **3** 25

4 a ⬚⬚⬚   **b** $6, 10, 14, 18, 22, 26$ **c i** 42 **ii** 62   **d** $n = 4x + 2$

5 a $t + t + t = 3t$ **b** $a^2 \times a^2 = a^4$ **c** Correct   **6 a** 5·45   **b** 5   **c** 5
7 $a = 45°$, $b = 67 \cdot 5°$ **8 a** $z = x - 5y$ **b** $k = \dfrac{11 - 3m}{m}$ **c** $z = \dfrac{C}{T}$

page 458 **Exercise 7**

1 a i Consett **ii** Durham **iii** Consett **b i** 55 km **ii** 40 km

c i 80 km/h **ii** 55 km/h **iii** 70 km/h **iv** 80 km/h **d** $1\frac{3}{4}$ hours

2 One ton is 17 kg heavier than one tonne.

3 3 hours **4 a** 0·198 **b** 0·0160 **c** 64·9 **d** 0·0585 **5** $2·3 \times 10^9$

6 a $(600 \times 50) \div 50 = 600$ **b** $(10 + 1000) \times 9 = 9090$ **c** $\sqrt{\frac{90}{10}} = 3$

d $3 \times \sqrt{(5^2 + 20^2} = 3 \times \sqrt{425} \approx 60$ **7 a** 40° **b** 100° **8** $\frac{x}{x+5}$

9 A: $4y = 3x - 16$ B: $2y = x - 8$ C: $2y + x = 8$ D: $4y + 3x = 16$

page 459 **Exercise 8**

1 1*lb* jar is better value. **2 a** $\frac{5}{3}$ **b** 20 cm **3 a** $6x + 15 < 200$ **b** 29 **4** 28 cm²

5 SB = 100 km **6 a** 0·5601 **b** 3·215 **c** 0·6954 **d** 0·4743 **7 a** 84 **b** 19·2 **8** 54

page 461 **12·2 Multiple choice tests**

page 461 **Test 1**

1 C	**2** A	**3** A	**4** C	**5** C	**6** A	**7** B	**8** D	**9** B	**10** B
11 C	**12** A	**13** D	**14** C	**15** C	**16** B	**17** A	**18** C	**19** B	**20** D
21 A	**22** B	**23** C	**24** B	**25** B					

page 463 **Test 2**

1 B	**2** C	**3** B	**4** A	**5** D	**6** C	**7** A	**8** D	**9** B	**10** C
11 A	**12** D	**13** A	**14** C	**15** C	**16** C	**17** B	**18** B	**19** B	**20** B
21 C	**22** D	**23** B	**24** A	**25** C					

Index